Memphis Medicine

A History *of* Science and Service

Memphis Medicine

A History of Science and Service

PATRICIA LaPOINTE McFARLAND, M.A.
MARY ELLEN PITTS, Ph.D.

LEGACY PUBLISHING COMPANY
Birmingham, Alabama

MEMPHIS MEDICINE

A HISTORY *of* SCIENCE AND SERVICE

Published By
LEGACY PUBLISHING COMPANY
100 Oxmoor Road, Suite 110, Birmingham, Alabama 35209 • (205) 941-4623

JOHN COMPTON, *Publisher*

Published with the cooperation and assistance of the
THE MEMPHIS MEDICAL SOCIETY
MICHAEL CATES, CAE, *Executive Vice President*
JAMES K. ENSOR, JR., M.D., *2010 President*

THE MEMPHIS MEDICAL SOCIETY AD HOC COMMITTEE

THOMAS C. GETTELFINGER, M.D., *Chair*

Members: R. Franklin Adams, M.D., Reed C. Baskin, M.D.,
George Cowan, M.D., Dan A. Dunaway, M.D., Nicholas Gotten, Jr., M.D.,
W. Webster Riggs, Jr., M.D.

Authors
PATRICIA LaPOINTE McFARLAND, M.A.
MARY ELLEN PITTS, Ph.D.

Contributor
PAULENE KELLER

Medical History Writers
PAULENE KELLER, MARY ELLEN PITTS, Ph.D.

Design By
ROBIN McDONALD

Editor
ALISON M. GLASCOCK

Photo Editor
RICHARD RAICHELSON, Ph.D.

Index
JIM SUMMERVILLE

COPYRIGHT © 2011 BY LEGACY PUBLISHING COMPANY
All Rights Reserved. Published 2011.

Printed in the United States of America by
TAYLOR PUBLISHING COMPANY, DALLAS, TEXAS

FIRST EDITION

ISBN 978-0-9668380-9-1

Library of Congress Control Number: 2010941077

MEMPHIS MEDICINE
A HISTORY *of* SCIENCE AND SERVICE

TABLE OF CONTENTS

	ACKNOWLEDGMENTS	vi
	FOREWORD	viii
Chapter 1	MEMPHIS IN THE AGE OF JACKSON: 1819-1839	2
Chapter 2	A BOOMING ECONOMY: 1840-1860	12
Chapter 3	A CIVIL WAR HOSPITAL CENTER: 1861-1865	22
Chapter 4	A DECADE OF DISEASE: 1870-1879	34
Chapter 5	THE GILDED AGE: 1880-1899	46
Chapter 6	THE PROGRESSIVE ERA: 1900-1915	60
Chapter 7	WORLD WAR I AND THE INFLUENZA PANDEMIC: 1917-1919	78
Chapter 8	1920s PROSPERITY TO 1930s GREAT DEPRESSION: 1920-1939	90
Chapter 9	THE WORLD AT WAR AND THE BATTLE AGAINST INFECTION: 1940-1949	110
Chapter 10	MEMPHIS: A CITY OF THE 1950s: 1950-1959	128
Chapter 11	ST. JUDE HOSPITAL, MEDICARE, AND SOCIAL CHANGE: 1960-1969	148
Chapter 12	TRANSPLANTS, NEW IMAGING, AND CHRONIC DISEASES: 1970-1979	164
Chapter 13	SURGERY CENTERS, MANAGED CARE, AND MEDICAL BUSINESS: 1980-1989	178
Chapter 14	BIOTECHNOLOGY, MINIMAL INTERVENTION, AND INSURANCE: 1990-1999	196
Chapter 15	ICONIC CHANGE, TRANSLATIONAL MEDICINE, AND ECONOMIC IMPACTS: 2000-2010	208
	EPILOGUE	228
	MEMPHIS MEDICAL HISTORIES	230
	IN MEMORIAM	305
	APPENDICES	311
	BIBLIOGRAPHY	325
	INDEX	330
	INDEX TO MEMPHIS MEDICAL HISTORIES	341
	PICTURE CREDITS	342
	PATRONS	344

Page i photos: Early color postcards of (left to right) Baptist Memorial Hospital, the Memphis U.S. Marine Hospital, and Gartly-Ramsay Hospital. Page ii-iii photo: Methodist University Hospital is the flagship facility of Methodist Le Bonheur Healthcare.

MEMPHIS MEDICINE
A HISTORY *of* SCIENCE AND SERVICE

ACKNOWLEDGMENTS

Special Thanks

The Memphis Medical Society wishes to express our sincere appreciation to the following physicians and individuals who helped make the publication of this book possible. We wish to convey special gratitude to Thomas C. Gettelfinger, M.D. His tireless efforts and relentless pursuit of factual information from both primary and secondary sources helped bring this book to life. His unwavering dedication guided the project from its inception to finish. The board and members of the Society are forever in debt to him for all his hard work and dedication.

Thanks also go to the other members of the Ad Hoc Committee which guided the progress of the book: R. Franklin Adams, M.D.; Reed C. Baskin, M.D.; George Cowan, M.D.; Dan A. Dunaway, M.D.; W. Webster Riggs, Jr., M.D.; and especially Nicholas Gotten, Jr., M.D.

Special thanks go to Patricia LaPointe McFarland and to Richard Nollan, University of Tennessee Health Science Center Library.

Part One: Chapters 1-8

We all owe a great debt to the late Dr. Simon Rulin Bruesch, professor at University of Tennessee Health Science Center (UTHSC) and medical historian. His research into the medical history of Tennessee is the foundation on which any subsequent history of Memphis medicine rests. He generously shared this knowledge in 1983 and 1984 during the writing of *From Saddlebags to Science: A Century of Health Care in Memphis, 1830-1930.*

The authors thank the Doctors Ad Hoc Committee—Tom Gettelfinger, chair, Nicholas Gotten, Webster Riggs, George Cowan, Dan Dunaway, Frank Adams, and Reed Baskin—for their encouragement and assistance with the scientific details of medicine. Thanks also to Michael Cates and Victor Carrozza of the Medical Society for generous assistance in locating information on the history of the Society and its affiliated organizations.

As Curator of Collections in the Memphis and Shelby County Room of the Public Library, I spent much of my career assisting others researching local history. I thank the History Department staff, especially Joan Cannon and Wayne Dowdy, for their assistance and helpful suggestions.

Coauthor Mary Ellen Pitts shared professional expertise as well as large doses of humor to late night writing. Richard Nollan, library colleague at UTHSC, facilitated research in the University of Tennessee Library's special collections. Our thanks, also, to Lori Estes of Memphis Eye and Cataract Associates for her competent handling of the many clerical tasks that facilitated our work. Photo editor Richard Raichelson contributed his knowledge of the written and visual history of Memphis.

My introduction to medical history was a gift from my late husband, Dr. Gordon LaPointe. My brother, Dr. David Meadows, who shares this interest, stands in the long line of UTHSC graduates who have used their medical skills to serve others. Lastly, great appreciation is due my husband Jack McFarland, A.I.A., for his support and patience throughout the writing process. His lifetime in architectural practice gives him a unique perspective on the built history of Memphis and the contributions of the arts and sciences to this city.

—*Patricia LaPointe McFarland, M.A.*

Part Two: Chapters 9-15

Part two of this book owes much to the patience of the mentors who worked with me on the seven chapters from the 1940s to the present. As a cultural historian of science and a student of literature and science, I began with broad concepts, later connecting them to Memphis. With the

The Regional Medical Center at Memphis, known as The MED, is the state's oldest hospital, opening in 1830.

patience of Henry Higgins coaching Eliza Doolittle, the six physicians who became the "Saturday Group" helped narrow the focus to Memphis and fill in the names of physicians who shaped Memphis medicine. To Dr. Reed Baskin, thanks for an incredibly broad knowledge of Memphis medicine, its problems, and its successes; to Dr. Webster Riggs, Jr., thanks for in-depth knowledge of radiology and its practitioners, plus the rich legacy of Memphis families in medicine; to Dr. Frank Adams, thanks for the many names of major practitioners, their roles in shaping Memphis medicine, and much about the legacy of Mid-South medical families; to Dr. George Cowan, thanks for gentle but important technical suggestions, knowledge of developments in recent surgery, and insight into other specialties in Memphis and elsewhere; to Dr. Nick Gotten, thanks for a historian's love of details, knowledge of specific contributions of many physicians, and tales from the rich history of medical families in Memphis; and to Dr. Tom Gettelfinger, thanks for intense devotion to the project, many careful, helpful readings of the text, and endless hours spent making this history as good as it could be. To the many physicians interviewed, thanks for sharing your time and knowledge.

Victor Carrozza at the Medical Society provided in-depth research and exquisite lists, some of which became appendices. Mike Cates of the Medical Society supported us through difficult moments. Lori Estes of Memphis Eye and Cataract Associates devoted time and intricate knowledge of Microsoft Word to all of us. Dick Raichelson offered useful suggestions. Patricia LaPointe McFarland was a helpful, knowledgeable coauthor, generous of her time and painstaking with historical details. Pamela and Nick Gotten graciously opened their home to the "Saturday Group," and Pamela could somehow concoct a superb berry pie on short notice. Paulene Keller generously shared materials from some delightful interviews.

—*Mary Ellen Pitts, Ph.D.*

MEMPHIS MEDICINE

A HISTORY *of* SCIENCE AND SERVICE

FOREWORD

The history of medicine in Memphis is a great story. It's been told before, but never quite this way, in pictorial format with a heavy reliance on photographs.

There have been three notable predecessors. The standard, published in 1971, is the *History of Medicine in Memphis*, edited by Drs. Marcus Stewart and William T. Black, with particular contributions by Memphis's best-known doctor-historian, Simon Rulin Bruesch.

In 1984, *From Saddlebags to Science: A Century of Health Care in Memphis, 1830-1930*, by Patricia M. LaPointe, was published in conjunction with the opening of the Health Sciences Museum Foundation exhibit on Memphis medicine at the Pink Palace Museum. Published in 1986, *The University of Tennessee, Memphis, 75th Anniversary—Medical Accomplishments*, by Dr. James Edward Hamner, concentrated on the University of Tennessee.

Now, The Memphis Medical Society offers this new overview of the last 200 years in *Memphis Medicine, A History of Science and Service*, published in a year of great change in American medical care, with historic legislation on the future of that care. There are two writers: the first half is written by Patricia LaPointe McFarland, the author of *From Saddlebags to Science*, and the second half, from 1940 forward, is written by Dr. Mary Ellen Pitts. The photographs, equally important, were selected by Dr. Richard Raichelson.

Their work was assisted by an ad hoc committee of the Memphis Medical Society—Dr. Tom Gettelfinger, chair, Drs. Nick Gotten, Webster Riggs, George Cowan, Frank Adams, Dan Dunaway, and Reed Baskin—acting as mentors. Dr. Pat Wall and Richard Nollan of the University of Tennessee Health Science Center and Michael Cates and Victor Carrozza of The Memphis Medical Society also assisted. There were numbers of others called upon who are noted, at least in part, in the authors' acknowledgments on the preceeding pages. Drs. Dan Dunaway and James E. Bailey should also be credited with keeping alive interest in Memphis medical history through an annual lecture presented by the History Committee of the Society.

This is part of a family of similar books on medicine and law, including one on the Memphis legal community, published by John Compton of Legacy Publishing Company, Birmingham, Alabama. The format of the book is a history of medicine in Memphis followed by profiles submitted by practices, clinics, and hospitals.

In these pages you'll get a sense of the unhealthful conditions faced by the early settlers of Memphis, which was falsely advertised as a healthy place, of the colorful steamboat era, of the Civil War and Memphis's first national recognition for its medical care, of the devastation of the 1878 yellow fever epidemic that resulted in the loss of the city's charter, and of the city's subsequent recovery. Here are the stories of the development of Tennessee's largest medical education complex, of the great achievements in private medical practice, particularly in orthopedics at the Campbell Clinic and in neurosurgery at Semmes-Murphey, of Memphis doctors' contributions in World War I and World War II, of the big three hospitals—Baptist Memorial, Methodist, and Saint Francis and their growth into regional systems—accounts of the death of Elvis, and of the tragedy of Martin Luther King.

The 1950s brought epidemics of polio, the age of antibiotics, and an ear operation recognized around the world. The 1960s saw the ability to treat chronic disease more effectively, the opening, in 1962, of St. Jude Children's Research Hospital, the world's leading childhood disease research hospital, and landmark Medicare and Medicaid legislation.

In the 1970s, there were organ transplants, legislation encouraging managed care, and, in a bold move, Saint Joseph East, later called Saint Francis, opened in the eastern part of the city.

Entertainer Danny Thomas, founder of St. Jude Children's Research Hospital, unveils the statue of St. Jude Thaddeus, February 4, 1962.

The 1980s found AIDS, more managed care, outpatient surgery facilities, and the accelerated migration to the suburbs. In the 1990s came TennCare, a Nobel Prize in Medicine, the Resource Based Relative Value Scale, a radical change in physician payment, and major hospital consolidation. The 2000s saw the closing and then implosion of Baptist Memorial Hospital, once the largest private hospital in the country by number of admissions, new heart and gastrointestinal disorder centers, a liver transplant program the equal of any in the country, and the increasing burden of paying for care.

These, and more, are the stories you'll read and see in *Memphis Medicine, A History of Science and Service*. But this book cannot be comprehensive. It is an overview and no doubt omits important developments and individuals, and for that we apologize.

It is our hope that this look at the past will give an appreciation of who we are in Memphis medicine, where we have come from and why, and, with that, an understanding of how better to shape the future.

Yes, the history of medicine in Memphis is a great story.
—*Tom Gettelfinger, Chairman
for the Ad Hoc Committee*

CHAPTER ONE

MEMPHIS IN THE AGE OF JACKSON

1819–1839

Situated on the Fourth Chickasaw Bluff between St. Louis and New Orleans, location has defined the economic, political, and cultural history of Memphis. The center of a large agricultural area, the Memphis economy was built on cotton, hardwood lumber, wholesale groceries, and small manufacturing. Location has also been significant in shaping the city's health history and development as a medical center.

Founded as a real estate venture in 1819 by General Andrew Jackson, Judge John Overton, and General James Winchester, publicity for the new town proclaimed Memphis "on the high road for all the commerce of the vast and fertile valley through which the river flows." When James Winchester chose the name Memphis for his new town, linking it with the ancient city on the Nile, he could not have envisioned that it would become a great medical center in the twentieth century, as was its Egyptian namesake.

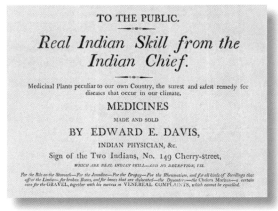

By the time Memphis was founded, existing patterns of medical care developed during the colonial period accompanied the thousands of settlers who moved into West Tennessee following the 1818 Chickasaw Cession. Negotiated by Andrew Jackson and Isaac Shelby (for whom Shelby County is named), the Chickasaw Cession transferred ownership of 6,848 million acres between the Tennessee and Mississippi rivers to the national government. Settlers living in thinly populated areas were dependent upon home remedies, including native botanical remedies and medical books written for the layman. One of the most successful of such books came from the pen of Knoxville, Tennessee, physician Dr. John C. Gunn. Published in 1830, *Gunn's Domestic Medicine, or Poor Man's Friend*, was a classic of the home medicine genre and was often carried by doctors in the South and Southwest. Gunn recommended calomel (a chlorine salt of mercury), opium, ipecac, bleeding, purging, and other "heroic" treatments, but he also recommended herbal medicines and stressed the importance of maintaining good health. Dr. Gunn's book went through some 70 printings between 1830 and 1920.

An important earlier treatment book, *Domestic Medicine; or, the Family Physician*, was the work of Scottish physician William Buchan (1729-1805). *Primitive Physic* ('physic' meaning medicine) by English Evangelist John Wesley, the founder of Methodism, first published anonymously in 1747 and later under his name in 1760, described various diseases, their symptoms, and treatment. Having spent time

ABOVE: *General Andrew Jackson on horseback in full battle dress with the Tennessee forces on the Hickory Grounds, Alabama, 1814.* OPPOSITE PAGE: *Portion of an advertisement for Dr. Edward E. Davis, Indian physician, around 1800, place unknown. He maintained that his remedies were composed of simples, i.e., plant products, and did not contain any mercury. In the body of the ad, Davis referred to William Buchan's* Family Physician.

DR. GUNN'S *DOMESTIC MEDICINE*

On August 30, 1830, Knoxville physician John C. Gunn (c. 1795-1863) filed a copyright with the United States District Court in East Tennessee for *Domestic Medicine, or Poor Man's Friend*, destined to become one of the most widely read books on home medical care published in this country. Keenly aware that frontier and rural families rarely had access to a doctor, Gunn penned a work of great breadth covering not only medical treatments but also essays on morality, marital compatibility, and the effect of emotions on the individual's health. It was even translated into German for the benefit of great numbers of immigrants coming into this country in the nineteenth century.

Dedicated to President Andrew Jackson, Gunn stated: "This book points out, in plain language, free from Doctor's terms, the diseases of men, women, and children, and the latest and most approved means used in their cure, and is expressly written for the benefit of families in the Western and Southern States [and] also contains, descriptions of Medicinal roots and herbs . . . and how they are to be used in the cure of Diseases." Emphasizing a holistic approach, Gunn stressed the responsibility of each individual to maintain good health.

Gunn was well educated and widely read in the medical literature of the time. He referred to the writings of noted Philadelphia doctors Benjamin Rush, Nathaniel Chapman, and Benjamin Barton, author of the first American book on botany. Gunn had studied in New York, and his book includes selections from the writings of their prominent physicians, Samuel Latham Mitchill, Valentine Mott, and David Hosack. Gunn began practice in Virginia, later moving to Knoxville in 1827.

Domestic Medicine describes treatments used in the early nineteenth century by doctors of regular, or allopathic, medicine and includes a diagram for bleeding and instructions for purging and catheterization. While Gunn freely prescribed "heroic" remedies such as calomel, a chloride of mercury, opium, and ipecac, a Brazilian root preparation that induced vomiting, he argued for the use of native medicinal plants, especially those found in Tennessee and the Southeast, as superior to imported drugs. This large compendium of treatments and medicines provided a valuable reference, especially for rural doctors whose patients were in widely scattered locations.

Much of Gunn's medical advice focused on women and included information on pregnancy, childbirth, and female diseases. He provided detailed instructions on the difficulties and complications that often accompanied childbirth in order to assist midwives who were attendants at most deliveries. Writing about childhood diseases such as measles, whooping cough, and common fevers, Gunn noted that most would pass with reasonable care; however, he noted that infantile diarrhea was the most dangerous of childhood illnesses, and one that caused much infant mortality. He was also a strong advocate of vaccination for smallpox. No history of nineteenth century American medicine would be complete without a reference to Gunn's *Domestic Medicine*.

Above left: *Dr. John C. Gunn, probably 1830s.* left: *The title page to Dr. Gunn's second edition of* Domestic Medicine, *1834.*

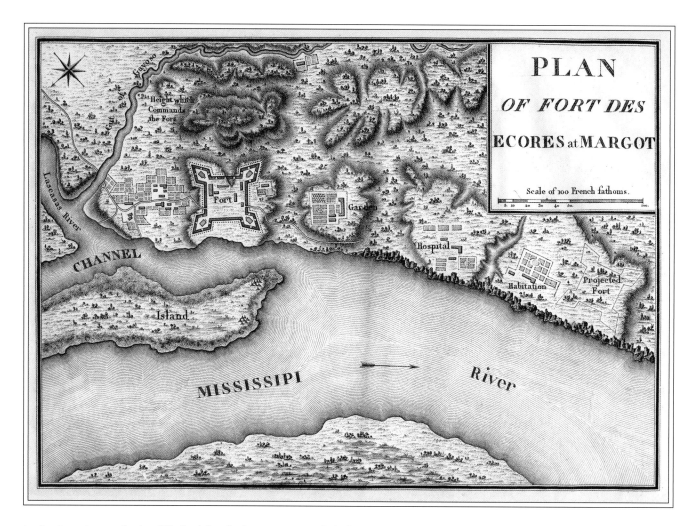

in the American colonies, Wesley's book also recommended basic medicines which should accompany immigrants to the colonies. The long alliance between Methodism and health care can be seen in Memphis today.

Early Medical Schools and Regular Medicine

By the last quarter of the eighteenth century, medical schools had been established in New England, New York, and Pennsylvania. Columbia University (1768), the University of Pennsylvania (1776), Harvard (1783), Dartmouth (1798), and Yale College Medical Department (1810) were the elite schools, and some of the better-trained faculty in these colleges had studied medicine in the prestigious schools of London, Edinburgh, or Leiden. They taught the regular system of medicine, later known as "heroic" and called by critics "bleed, blister and purge."

Medical training in the early nineteenth century began with an apprenticeship to an established doctor followed by a year or two in a medical school. Even in the best schools, courses were of short duration. Treatments included bleeding, done with a lancet to open a vein, or with leeches

Frenchman Georges H. Victor Collot sketched Fort San Fernando de las Barrancas in 1796. It was built by the Spanish close to where the Pyramid is now located, under an agreement with the Chickasaw. The Spanish burned the fort in 1797, but soon afterward, the United States constructed Fort Adams on this site. Note the bluffs which line the Mississippi River, along where the Fort, outer settlements, and hospital stood.

applied to the skin, emetics that induced vomiting, and harsh laxatives that depleted the body's fluid and electrolyte balance. Since internal surgery was unknown, "blistering" was used to cause burns, inducing infections, which produced pus that was thought to be beneficial. Laudanum, an alcohol preparation of ten percent opium and one percent morphine, was prescribed as a medicine. Despite the best of intentions, much eighteenth and nineteenth century medicine often did more harm than good, creating justifiable skepticism regarding traditional medical practice.

If colonists were at first reluctant to accept Indian remedies, their experience with native botanical remedies gave them credibility as the post-Revolutionary War population moved westward away from established medical practice. Most Indian medicines employed barks, herbs, roots, oils, and other natural materials and utilized the healing properties of spring waters for internal and external use. Indians had long known about the benefits of hot springs in treating arthritic and lumbar ailments and their skill in the use of herbs is affirmed in many sources. Writing about the Cherokees, Henry Thompson Malone states, "In medical practice some religious emphasis was attached to allegedly miraculous powers of certain medicinal drugs . . . the gensing root, for example, was highly regarded and its preparation prescribed by local conjurors." With the advent of cheap newspapers in the nineteenth century, some doctors advertised their knowledge of Indian techniques and botanical remedies. Indian motifs were often used to promote health products and patent medicines even into the early twentieth century.

Botanic Medicine—a New Form of Treatment

Advocating a milder form of treatment based primarily on herbal remedies, Samuel Thomson (1769-1843), a native of New Hampshire, became a leader of the botanic school of medicine. In 1822, he wrote *New Guide to Health: or Family Botanic Physician*. The treatment regimen of botanic doctors, generally referred to as Thomsonian, emphasized herbal healing, a movement away from the regular, and generally more harmful, medical treatments of bloodletting, purging, and the use of toxic chemical remedies such as calomel. The widespread appeal of the Thomsonian treatment was based on its simplicity and relation to Indian herbal practice. Botanic medicine and offshoots such as hydropathic, which utilized water as the chief form of treatment, and eclectic, which borrowed from all systems, garnered a large following in the nineteenth century.

New Medical Schools

With the great numbers of people moving west, medical schools opened to serve students in western areas. Transylvania University, a Kentucky school, which moved to Lexington in 1789, added a medical division in 1818, making it the first medical school west of the Appalachians. Daniel Drake, an outstanding pioneer physician and author of books on medicine and medical geography, founded the Medical College of Ohio at Cincinnati in 1819. Most medical schools were proprietary, as only a few eastern schools were associated with a university. Some proprietary schools of this period were largely diploma mills, requiring no recognized medical apprenticeship or stated period of class attendance to gain a diploma.

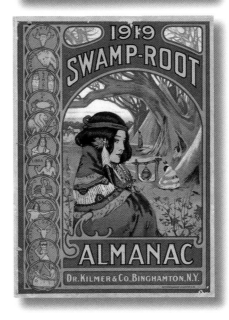

Above left: *Red Jacket Stomach Bitters used around 1864 to cure dyspepsia, as well as to prevent fever and ague, most likely from malaria. The label claims that Red Jacket, an Indian chief, gave the remedy to a Dr. Chapin. Indian herbal remedies were trusted by the public, a point used by patent medicine vendors for sales.* **Left:** *Kilmer's Swamp-Root Almanac from 1919 supports the contention that advertisements which pictured Native Americans and medicine was still a selling point in the twentieth century. The back cover shows additional symbols of Indian life—buffalo, teepees, and a male in full headdress shooting an arrow. Swamp root was a popular patent medicine used for kidney, liver, and bladder complaints.*

laws even rescinded them, making it easier to claim the practice of medicine. In 1830, Nashville physician Dr. Felix Robertson and other area doctors organized the Tennessee Medical Society as a means of establishing guidelines for the training and qualification of physicians. At the first meeting in Nashville on May 3, 1830, a constitution and code of ethics were adopted. No legislation required membership in order to practice, negating the Society's regulatory authority. Licensure as a requirement to practice medicine in this state was not enacted until 1889. The Tennessee Medical Society did provide an organization for trained doctors that distinguished them from unqualified doctors and "quacks," (a shortened form of the Dutch "quacksalver," the word for mercury).

Memphis's First Physicians

The first physicians to settle in Memphis were men who had received the best training available in the early 1800s and who, in their practices, demonstrated a high level of commitment to the well-being of the community. Arriving about 1821, George Franklin Graham (1794-1827), the son of a family prominent in the government of North Carolina, was the first trained physician to practice medicine in the newly founded town of Memphis. He had received a B.A. degree in 1815 from the University of North Carolina, and his M.D. degree from the University of Pennsylvania, considered the preeminent medical school of the period.

Dr. Wyatt Christian (1798-1846) came to Memphis in 1825 from his home in Charles City, Virginia. Like Graham, Christian was a graduate of the University of Pennsylvania, receiving his M.D. degree in 1821. An active and public-spirited citizen, Christian was instrumental in founding the Memphis Hospital, where he was an attending physician. He was a charter member of the Tennessee Medical Society. A colleague wrote of him, "Dr. Wyatt Christian [was] an excellent physician, ever usefully employed in relieving the ailments of suffering humanity."

Rueben Davis (1813-1890), born in Winchester, Tennessee, came to Memphis in 1829

ABOVE: *Dr. Samuel Thomson (1769-1843) as he appeared around the time of his book,* New Guide to Health: or Family Botanic Physician *(1822)* RIGHT: *Samuel Thomson's favorite herb was Lobelia which he recommended as an emetic. It was also widely used among the Indians and early settlers. Other uses included pulmonary diseases, such as asthma and whooping cough. Lewis and Clark used the herb to treat venereal disease.*

Few if any agencies existed to regulate medical training and licensing of graduates. During the Jacksonian Era (1828-1836), with its populist and egalitarian spirit, some states with medical licensure

ANDREW JACKSON: A MEDICAL HISTORY

Andrew Jackson, Tennessean, president (1829-1837), and namesake of an era, had an extensive medical history exemplifying diseases and treatment of the times. Following an 1806 duel, Jackson carried a bullet in his chest for the rest of his life. The bullet, which shattered two ribs, lodged near Jackson's heart and could not be removed. In an 1813 duel, Jackson was shot twice, and one bullet lodged against the bone of his left shoulder, causing a festering osteomyelitis and great pain for many years. His health was damaged from years of hard military service, numerous injuries, and debilitating intestinal ailments. Jackson was also thought to have tuberculosis, and indeed may have had a chronic case much of his life.

In November 1813, Jackson's Tennessee militia mounted an attack against the Red Stick Creeks in retaliation for their unspeakably brutal slaughter of some 300 men, women, and children at Fort Mims, Alabama. A Creek infant found with its dead mother on the battlefield was brought to Jackson, who took the boy, named Lyncoya, and raised him as his adopted son. The youth died of tuberculosis, a disease common among Native Americans, at the age of 17.

The triumph of Jackson's 1828 presidential victory was crushed by the death of his beloved wife Rachel on December 22, 1828. Although Jackson tried his best to shield her from the vicious slander heaped upon her by his political enemies, the emotional toll affected her health. Returning to the Hermitage, their home near Nashville, from a trip to that city where she had accidentally come upon some of the printed assaults on her character, Rachel was in great emotional distress. She suffered an apparent heart attack on December 8. Doctors came immediately and bled Rachel several times and the family physician, Dr. Samuel Hogg, remained in attendance at the Hermitage. Rachel rallied slightly on Sunday, December 22, but that evening she suffered a second attack. When Dr. Hogg reached her side, there was no pulse. Jackson begged him, "bleed her—bleed her again," but two incisions produced only a few drops of blood. Rachel had been released from her physical and mental suffering. She was buried in the garden of the Hermitage.

President Andrew Jackson, one of the proprietors of Memphis, at the Hermitage, April 15, 1845. He died the next month, on June 8, at the age of 78.

In 1836, Jackson's beloved niece Emily Donelson died from tuberculosis at her Poplar Grove home near the Hermitage. She and her husband, Andrew Jackson Donelson, the president's nephew, were Jackson's Washington family, and Emily had served as his White House hostess. Her death caused him terrible grief, made worse because Emily died before her husband reached Nashville. Donelson, who was the president's secretary, had been detained in Washington on official business.

During his second term of office, Jackson suddenly ordered his doctor to remove the bullet in his shoulder, and the procedure was carried out without anesthesia. Drainage relieved the pain somewhat. During his later years, Jackson suffered from chronic dysentery, as well as ills associated with aging. He died in 1845 at the age of 78, having sustained injuries and multiple episodes of disease that would have killed most other men. Raised on the Tennessee frontier, Jackson—a fighter for his country and committed to democracy for the common man, with his will, courage, and strength of purpose—was an American icon.

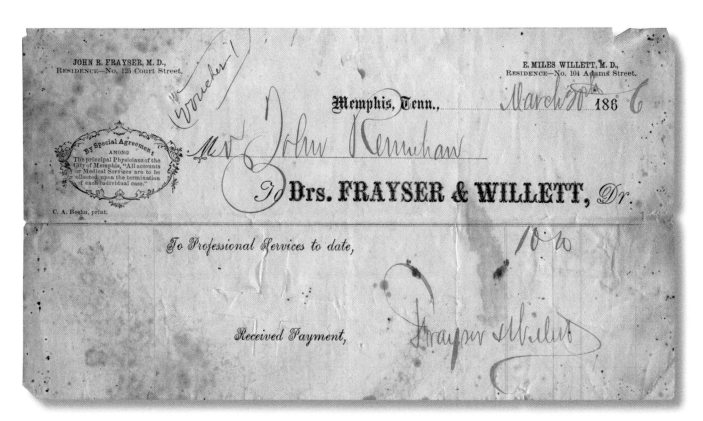

Above: *A medical bill from the offices of Memphis physicians John R. Frayser and E. Miles Willett, 1866. Dr. Frayser arrived in Memphis during the 1830s. He and Dr. Lewis Shanks were partners in the 1840s. Dr. Willett was a member of the Confederate Medical Corps and one of the founders of St. Joseph Hospital. Both died in 1888.* Right: *Dr. Solon Borland arrived in Memphis during the 1830s. This photograph is from 1848 to 1853, during the period he was a senator from Arkansas. He later returned to Memphis and joined the staff of the Memphis Medical College.*

on the recommendation of his brother, a lawyer in Somerville, to begin medical practice. Dismayed by what he found, Davis later wrote: "the locality and unsanitary condition of the town promised that disease and death would hold high carnival there."

One of a large family of doctors, Mark Brown Sappington (1812-1852) arrived in Memphis about 1834. His grandfather, also Dr. Mark Brown Sappington, located in Nashville about 1785, where he became the first physician to practice in Davidson County. The younger Dr. Mark Brown Sappington had studied medicine with his father and uncles in Nashville prior to coming to Memphis. A member of the first Board of Health, he was a strong advocate of vaccination for smallpox. Both Drs. Sappington and Christian served on the Board of Aldermen.

Other doctors who located in Memphis in the 1830s included William W. Tucker, John R. Frayser, Jeptha Fowlkes, Lewis Shanks, and Solon Borland. Dr. Shanks was a leader in the profession, a member of the Memphis Medical College faculty, and the first local physician to specialize in obstetrics and the diseases of women and children. His death in 1861 ended an active career in medicine and public service. When Dr. John R. Frayser died in 1888 at age 73, he was lauded as being hard working, upright, generous to the poor, and a physician "who served in every epidemic which afflicted Memphis in the last 50 years." The community of Frayser is named for the doctor, whose summer home was located in northeast Shelby County.

Two physicians who settled in Shelby County in 1831 were brothers Samuel and William Bond, who located in the Bartlett area. The railroad stop there was first called Bond's Station for this large land-owning family. Samuel Bond received his education at Cumberland College in Nashville (later the University of Nashville) and then attended a year of medical classes at Transylvania University in Kentucky. The Drs. Bond engaged in large-scale cotton farming as well as the practice of medicine. A collection of Bond family papers at the Memphis Public Library includes Dr. Samuel Bond's daybook with detailed entries on patients (slave and white) and charges for treatment and medicines.

Memphis's First Hospital

Because there was no facility to care for those who were sick when they arrived in Memphis or for the boatmen who regularly traveled the river, in 1829 local officials requested the Tennessee legislature to appropriate money to build a hospital. The legislature set aside $3,330 to build the first hospital in the state, which was located near Front Street. Patients were admitted in 1830. It was soon apparent that the Memphis Hospital was not able to care for the large number of transients who arrived in Memphis suffering from various diseases or for those injured in steamboat accidents on the river. Most hospitals were limited to the care of transients or indigents. Local residents were cared for in their homes.

In 1832, the Tennessee legislature recognized that "the great and increasing number of lunatics in this state have made it necessary to the well-being of society, as well as for the comfort and security of these unfortunate beings that a suitable hospital be created in this state." The first hospital for the mentally ill was built near Nashville in 1832. Fifteen years later, the state funded construction of Central State Hospital at Nashville (1847), followed by East Tennessee State Hospital in Knoxville (1873) and Western State Hospital at Bolivar (1887).

In addition to fevers such as malaria and typhoid and the contagious diseases smallpox and measles, cholera was another deadly disease that afflicted Memphis residents. Cholera traveled across the Atlantic with Irish immigrants destined for Canada. Moving down the Great Lakes and

ABOVE: *The Memphis Hospital, built in 1841, was torn down to make room for Forrest Park. General Nathan Bedford Forrest's statue stands about where the hospital was located. The two-story hospital had a separate female ward and four other wards located behind the main building. A one-story doctor's office building was near the southwest corner of Manassas and Union. In the photograph, from a 1907 postcard, the Memphis Hospital Medical College is seen in the background.*
OPPOSITE PAGE: *Seth Wheatley was one of the directors of the first hospital in Memphis. He was also president of the Farmers and Merchants Bank, a lawyer, and mayor of Memphis from 1831 to 1832.*

finally into the Mississippi River, it first appeared in Memphis in the fall of 1832. A water-borne disease, cholera was largely confined to small areas adjacent to the river. Following repeated outbreaks, strongly worded editorials in the *Memphis Inquirer* urged officials to address the sanitary needs of the town. In August 1838, the mayor and aldermen created a Board of Health composed of several of the town's practicing physicians: Wyatt Christian, Lewis Shanks, Mark Sappington, John R. Frayser, and Washington J. DeWitt. The board was authorized to regularly report on all causes of disease and all deaths; however, without funding and regulatory authority, it soon became inactive.

The Tennessee legislature petitioned Congress in 1836 for federal aid to expand Memphis Hospital and include a medical training facility. When the request was denied, the state approved the sale of the first Memphis Hospital for $4,225 and appropriated an additional $5,000 for a new hospital. An amended hospital charter provided for seven appointed directors. The board, composed of Alexander H. Bowman; Dr. Wyatt Christian; Petro W. Lucas; Robertson Topp; Lewis Trezevant; Seth Wheatley; and Marcus B. Winchester, purchased ten acres east of the town limit, a choice of far-reaching significance. That site, now Forrest Park, would be key to the establishment of the present medical center. Although a new hospital was authorized in 1836, the Memphis Hospital was not completed until 1841. A brick building on the north side of Union Avenue, it provided the primary clinical facility for medical students until replaced in 1898 by the new City Hospital on Madison Avenue.

Efforts to address matters of public health and improve sanitation in the 1830s were hampered by the lack of funding and enforcement authority. As the decade closed, Memphis had survived the financial panic of 1837 and bested Randolph in the race to become the principal river port between St. Louis and New Orleans. During the next two decades, Memphis experienced a period of rapid growth fueled by cotton and railroads. Medical education would be firmly established in the Bluff City during the boom decades.

CHAPTER TWO

A BOOMING ECONOMY

1840–1860

THE 1830S DOMINANCE OF RIVER flatboats as the principal means of transportation on the Mississippi River gave way in the 1840s to steamboat transportation. In Memphis, a more stable local government encouraged business growth. The opening of the new Memphis Hospital in 1841 on the site that is now Forrest Park was a first step in the later development of the medical center. The government authorized construction of a navy yard at the mouth of the Wolf River which added importance to Memphis as a major port on the Mississippi. Completed in 1846, the navy yard outfitted only one ship and was closed in 1854.

The rapid increase in the number of steamboats plying the Mississippi and its main tributaries moved enormous quantities of goods and people upriver from New Orleans and down the Ohio River from Cincinnati and Louisville to points south. By 1850, more than one thousand steamboats annually traveled the great river artery. With the increase in traffic came an increase in steamboat disasters—boiler explosions, fires, and hulls ripped by unseen snags in the river. Passengers and crew members injured in accidents which occurred near Memphis were brought to the city for treatment. The injuries, especially burns, frequently proved fatal.

The Explosion of the Steamboat *Pennsylvania*

In 1858, when the *Pennsylvania* exploded 60 miles south of Memphis, 40 of the wounded were brought to the city. Lacking adequate hospital facilities, the injured were placed in a large public hall where local doctors provided the best care available. One of the injured was Henry Clemens, brother of Samuel Langhorne Clemens, better known as Mark Twain. In *Life on the Mississippi*, Twain wrote about the kindness of the Memphis doctors, especially Dr. Thomas Peyton, who cared for his brother. Henry had been badly burned and his suffering was intense. In an effort to relieve Henry's suffering, a young medical student administered morphine measured out on the blade of a knife. In the nineteenth century there was no standard for the strength of the active ingredient in a drug or for the dosage. Though easing Henry's pain, the overdose of morphine hastened his death.

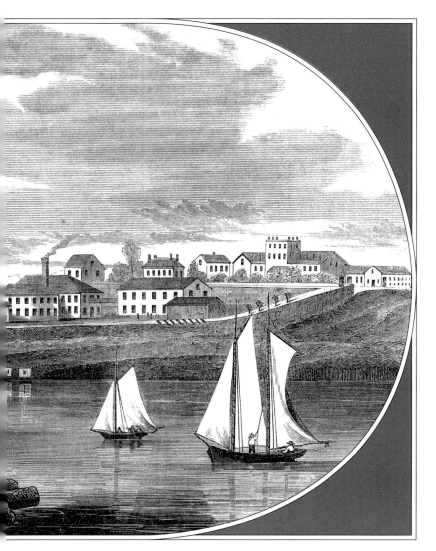

View of the Navy Yard from the Arkansas shore, 1850s. Lieutenant Matthew Fontaine Maury, a distinguished naval officer and relative of Dr. Richard B. Maury, was the force behind the construction of the facility at Memphis. Only one boat was built, the Allegheny. *The elongated building stretching to the left is a rope factory (from* Gleason's Pictorial Drawing-Room Companion, *1850s).*

Memphis—Railroads and Immigration

Expanding commerce and the national interest in building railroads was a catalyst for bringing large numbers of Irish and German immigrants to Memphis. The 1840 population of 1,799 had, in ten years, grown to 8,841. Several railroads out of Memphis were begun in the 1850s, none so important as the Memphis and Charleston Railroad. Its completion in 1857 provided a direct link between the lower Mississippi and a major Atlantic seaport. Civic celebrations in both cities marked completion of this important rail link. Steamboats continued to be the major transport for cargo and, by the 1850s, more than one million cotton bales were annually sent to New Orleans for shipment to the cotton mills in England.

As commercial activity expanded and large brick buildings appeared on the downtown skyline, some 30 additional doctors had established their practice in Memphis by the end of 1849. Two who made valuable contributions to the community were Drs. William J. Tuck and Howell R. Robards, both of whom had received their medical training at the Philadelphia College of Medicine. Tuck, an accurate observer of disease, was active in public health, serving as secretary of the Board of Health in 1855. His articles on various fevers and the use of quinine in the treatment of malaria appeared in Southern medical journals. After the navy yard closed in 1854, Tuck became an official weather observer for the Memphis station.

The Mexican War and Tennessee Volunteers

Howell Robards, who came to Memphis from Columbia, Tennessee, about 1845, enlisted as a volunteer surgeon at the beginning of the Mexican War (1846-1848). He wrote an account of his medical experience as surgeon to the Eagle Guards, a volunteer cavalry unit from West Tennessee. In "*Diseases of the Army of Occupation in the Summer of 1846,*" Robards describes his treatment of the fevers and other ills which afflicted the men on

MARK TWAIN AND NINETEENTH CENTURY AMERICAN MEDICINE

Mark Twain, American humorist and author of *The Adventures of Tom Sawyer* and *Huckleberry Finn*, was a keen observer of small-town American life. It may, however, come as a surprise to learn how much he tells his readers about nineteenth-century medicine. Born Samuel Langhorne Clemens on November 30, 1835, the early medical experiences of his childhood home in Hannibal, Missouri, were similar to those in the small town of Memphis, a few hundred miles down the Mississippi River.

In novels, travel accounts, and commentaries, Twain covered mumps, measles, cholera, consumption (tuberculosis), scarlet fever, whooping cough, "yaller janders" (jaundice), patent medicine, and medical charlatans. He wrote about medical fads and the changing treatments of the time. In *Huckleberry Finn*, Twain tells of books generally found in homes of the period: the big family Bible, *Pilgrim's Progress*, *The Life and Speeches of Henry Clay*, and *Dr. Gunn's Domestic Medicine*, "which told you all about what to do if a body was sick or dead."

Two of the doctors that Clemens remembered, Dr. Hugh Meredith and Dr. Thomas Jefferson Chowing, were traditional medical practitioners of the day. Their principal treatments were castor oil used as a laxative, calomel, a mercury compound used as a diuretic and purgative, (which sometimes caused the loss of teeth), jalap, another cathartic to stimulate intestinal action, bleeding by incision, and mustard plasters applied to the chest or back.

Clemens's boyhood in the small river town of Hannibal was the basis for the fictional adventures of his protagonists Tom Sawyer and Huckleberry Finn. Without their medical references, they would be poorer indeed. Imagine Tom and Huck without spunk-water or dead cats to cure warts, without grave robbers to dig up cadavers for medical study, without Aunt Polly's health periodicals, and without the cold water treatment, hot baths, blister plasters, and especially the painkiller that Tom fed to Peter the cat.

"...Tom pried [Aunt Polly's cat's] mouth open and poured down the Pain-killer. Peter sprang a couple of yards in the air, and then delivered a war-whoop and set off round and round the room, banging against furniture, upsetting flower-pots, and making a general havoc. Next he rose on his hind feet and pranced around, in a frenzy of enjoyment, with his head over his shoulder and his voice proclaiming his unappeasable happiness. Then he went tearing around the house again spreading chaos and destruction in his path. Aunt Polly entered in time to see him throw a few double summersets, deliver a final mighty hurrah, and sail through the open window, carrying the rest of the flower-pots with him."

After seeing the cat's frenzied reaction, Aunt Polly no longer forced Tom to take the painkiller. Tom Sawyer's story of the cat's wild gyrations when fed Perry Davis Pain Killer skewered patent medicines in a way no scientific paper could.

Despite the humor in Mark Twain's stories, they had a basis in fact. In the major cholera epidemic of 1849, Jane Clemens, Samuel's mother and the model for Tom Sawyer's Aunt Polly, gave Perry Davis Pain Killer to young Sam Clemens. Typical patent medicines were largely alcohol and flavoring. Davis's remedy also included laudanum, camphor, and cayenne pepper. In his autobiography, Clemens later wrote: "it was a most detestable medicine, Perry Davis Pain-Killer. Mr. Pavey's… man, who was a person of good judgment and considerable curiosity, wanted to sample it and I let him. It was his opinion that it was made of hell-fire," and so thought the thousands who annually took the medicines of that period.

ABOVE: *An ad for Perry Davis' Vegetable Pain Killer, published around 1860, describes it as a panacea for all types of ailments from the common cold to cholera. Mark Twain (opposite page), who was given the medicine as a child, described it as "a detestable medicine."*
RIGHT: *An ad for Wolcott's Instant Pain Annihilator. The ad claims that this preparation will instantly relieve the demons of pain associated with such conditions as headache, weak nerves, consumption, burns and scalds, toothache, catarrh, and neuralgia.*

A BOOMING ECONOMY

this military expedition. "As early as May last [1846] in anticipation of a call for volunteers for the war just breaking out with Mexico, the stirring notes of the fife and drum were heard in Memphis; in a few weeks, five companies were made up and remained in organization [however] only three were admitted into service. Three companies arrived at Camp Carroll, the place designated for the encampment, two miles east of Memphis. Our camp, so far as could be observed, was free from local causes of disease…and both men and horses were furnished with excellent water from a single large spring."

In August, the Eagle Guards set out on their long overland journey. "From Memphis to Little Rock the road passes through a low flat marsh country pregnant with local causes of disease and our troops suffered more perhaps in performing that distance of one hundred and fifty miles than on any other portion of the route… most severely in the Mississippi [river] swamp, a distance of forty-five miles." Robards described fevers as *remittent, intermittent,* or *continued* and noted them to be more malignant along the riverbanks. The Eagle Guards returned to Memphis, arriving on June 5, 1847, to a rousing welcome. They had not seen much fighting but were said to be in good health and had delivered "plenty of good hard service." A skilled surgeon, Robards later joined the reorganized faculty of the Memphis Medical College in 1852.

In 1846, the state of Tennessee issued charters for two medical schools in Memphis. The Memphis Medical College curriculum was that of traditional medicine, while the Botanico Medical College program was based on an alternate treatment philosophy employing herbal remedies and steam. Both schools opened in November, each with a 16-week course of instruction emphasizing diseases believed to be of Southern origin, such as malarial fever, yellow fever, and related ills. Students of both schools did their clinical work at the Memphis Hospital on Union Avenue. Elsewhere in Tennessee, medical departments were added to existing colleges. The Medical Department of the University of Nashville was organized in October 1850. In the late 1880s, the University of Tennessee Medical Department (a merger of earlier schools), also located in Nashville, later became the main unit of the University of Tennessee Medical College.

The Botanico Medical College

The Botanico Medical College faculty included Drs. James Weaver, anatomy and surgery; William Byrd Powell, physiology and pathology; Michael Gabbert, theory and practice of medicine; J. M. Hill, materia medica, therapeutics and pharmacy; Hugh Quinn, midwifery and diseases of women and children; and William Hyer, chemistry and botany, the latter reflecting the focus on natural herbal medicines developed in opposition to harsh chemical remedies. The school, located on Beale Street east of Lauderdale, was successful enough to add a three-story brick addition in 1854. The faculty operated an outpatient clinic, which provided free medical care for the poor. Dr.

TOP: *Physicians in Memphis, as listed in the first city directory, 1849.* ABOVE: *Dr. Howell R. Robards' article, "The Diseases of the Army of Occupation in the Summer of 1846," appeared in the March 1847 issue of* The Western Journal of Medicine and Surgery.

Botanico Medical College

William Byrd Powell edited the school's journal, *Southwestern Medical Reformer*, from its founding in 1846 until 1850 when he moved to Lexington, Kentucky.

The Memphis Medical College

The Memphis Medical College, which opened in a hall on Main at Exchange, conducted three sessions between November 1846 and February 1849, when it ceased operation because of financial difficulties and disagreements within the faculty. Not willing to see the school close permanently, Dr. Lewis Shanks reorganized the school from 1851 to 1852 and gathered a faculty with outstanding credentials. The college opened in the Exchange Building (Front Street north of Poplar), a commodious building which included the city hall, the mayor's office, a lecture hall, and the medical school.

Lewis Shanks was dean of the faculty and professor of obstetrics and the diseases of women and children. Other faculty were Ayres P. Merrill, professor of materia medica and therapeutics; Hardy V. Wooten, professor of principles and practice of medicine; Edward H. Leffingwell, professor of chemistry and toxicology; Howell R. Robards, professor of surgery; Arthur K. Taylor, professor of anatomy; Charles Todd Quintard, professor of physiology and pathology; and Daniel F. Wright, demonstrator of anatomy.

When the reorganized college opened in November 1852, it included a free clinic, which provided care for the poor and educational benefits for the medical students.

ABOVE AND LEFT: *The Botanico Medical College was located on Beale Street across from the Hunt-Phelan home. The two images, the title and ribbon, were part of a 17" x 22" diploma earned by Dr. Starke Dupuy, the great-great-grandfather of Dr. Jerre Freeman. He received it on March 1, 1847, about one year after the school was chartered. The gold medallion contains the name, "The Botanico Medical College of Memphis." Within its circle is a physician at his desk with a book, paper, and a potted herb.*
LEFT: *An ad for the Memphis Medical College from the 1855 Memphis* City Directory *which lists some of the pioneer physicians in Memphis. The Botanico Medical College also advertised its courses and faculty in the same directory.*

Patients were examined, treated, and prescribed for without charge by the medical faculty, and the city paid for medicines and surgical dressings. A number of surgeries were also performed in the college infirmary.

Ayres P. Merrill, a champion of public health, education, and civic betterment, served on the board of aldermen, the school board, and board of health. He urged improvement of public sanitation, including installation of a sewer system. Merrill was an organizer and charter member of the Elmwood Cemetery Association. He also revived efforts to secure a marine hospital for Memphis. In addition to his teaching, Merrill, along with Shanks and Quintard, wrote and edited the *Memphis Medical Recorder*,

published from July 1852 until financial difficulties forced suspension in May 1858.

Arthur K. Taylor was the son of Dr. William Vannah Taylor, a native of North Carolina who had moved to Holly Springs, Mississippi, and then to Memphis in 1842. Arthur had three brothers, William V., Jr., John, and Richard, all doctors, who practiced medicine in Memphis. The January 1854 issue of the *Memphis Medical Recorder* includes a detailed report of an operation Dr. Arthur Taylor performed on a young man suffering from an enormous osteosarcoma of the jaw. Dr. Shanks administered the chloroform and Dr. Wright assisted. The young man made a full recovery, regaining the ability to chew and swallow.

Charles Todd Quintard, Doctor and Bishop

Charles Todd Quintard, a native of Connecticut, received his medical education at New York University and interned at Bellevue Hospital. He came to Memphis in 1851 from Athens, Georgia, to join the faculty of the Memphis Medical College, where he gained recognition as an outstanding teacher and physician. Quintard was especially interested in public health and compiled detailed statistics on the city's health for the three-year period from 1851 to 1853, later reported in the *Memphis Medical Recorder*. He concluded

ABOVE: *The Exchange Building, located on Front north of Poplar, housed the Memphis Medical College during the 1850s. It also contained city hall. As seen in Henry Lewis's painting of the Memphis waterfront around 1854, the three-story building sits on the bluff directly above the sail of the small boat.* OPPOSITE PAGE: *Dr. Charles Todd Quintard became the Episcopal Bishop of Tennessee in 1865. The image is from the 1880s.*

that the college dispensary had significantly reduced mortality because patients came for treatment in the early stage of illness or injury, allowing the doctors to provide more effective treatment. He added that all surgical operations at the clinic, in addition to relieving suffering, had been successful. In 1855, following private study with James Hervey Otey, Episcopal bishop of Tennessee then residing in Memphis, Quintard was ordained a priest in the Episcopal Church and resigned from the faculty of Memphis Medical College. During the Civil War, Dr. Quintard served both as chaplain and surgeon in the Army of Tennessee. In 1865, Quintard was chosen bishop of Tennessee following Bishop Otey's death in April 1863. He reestablished the University of the South at Sewanee, which also had a medical school until the early twentieth century. The university still maintains a hospital.

John Millington, who followed Edward Leffingwell as Professor of Chemistry and Toxicology, was born near London. He was a student of the English scientists Michael Faraday and Sir Humphry Davy and later taught chemistry at Guy's Hospital in London for 14 years. Millington came to America in 1829 and was engaged in a number of engineering and scientific projects. In 1836, he began a 12-year tenure at the College of William and Mary. In 1838, he received his M.D. degree from Jefferson Medical College in Philadelphia. Millington's next position was at the University of Mississippi, where he was professor of chemistry. In 1853, he joined the faculty of the Memphis Medical College. At the outbreak of the Civil War, Millington retired and moved to LaGrange, Tennessee. When federal military units gained control of LaGrange in 1862, they converted Millington's home into a hospital, destroyed his possessions, and burned all his scientific papers and research material. Millington later went to Philadelphia and after the war moved to Richmond, Virginia. He is buried in Bruton Parish Churchyard in Williamsburg. John Millington's career successfully combined chemistry, engineering, and medicine.

Daniel F. Wright, who had studied chemistry and toxicology under Dr. Leffingwell, was a keen student of chemistry and pharmacy. His articles in the *Memphis Medical Recorder* stressed the need for reform in the dispensing and administering of drugs. He analyzed many prescriptions, some of which had appeared in medical journals, and pointed out their ineffectiveness or, worse, their conflicting ingredients. His careful analyses led him to conclude that the deaths of many young children suffering with dysentery and other gastrointestinal complaints were the result of injudicious use of opium, a powerful component in various patent medicines administered to children. He urged establishment of colleges of pharmacy, which would educate ethical, trained pharmacists. Following medical service in the Civil War, Dr. Wright moved to Clarksville, Tennessee. He served for many years on the State Board of Health and was particularly interested in maintaining the health of schoolchildren.

Growth of Patent Medicines

Two developments in the first half of the nineteenth century greatly accelerated the use of patent medicines. The spread of cheap newspapers aided the advertising of patent medicines, and the flood of immigrants to this country provided a ready market for these remedies. Most consisted of alcohol combined with vegetable (botanic) ingredients and flavoring. Quinine, which varied in effective strength, was taken for fevers, and laudanum, a preparation of opium and morphine in alcohol, was used as a painkiller. With no standard formulation of the principal ingredient and no regulatory control, addiction to patent medicines was common.

In 1888, The Cotton Bale Medicine Company, Helena, Arkansas, manufactured eight patent medicines for a variety of complaints. They included a tonic for chills and fevers, a blood elixir, cough syrup, liniment, liver pills, healing balm, worm and liver shot, and "twins."

The American Pharmaceutical Association was founded in Philadelphia in 1852 in an effort to regulate the practice of pharmacy and establish a standard drug formulary. It would take another 50 years to achieve this goal.

When the Memphis Medical College again experienced financial difficulties, the school was sold in 1859 to Dr. Lunsford P. Yandell, Sr., who, with his son, Lunsford P. Yandell, Jr., came to Memphis from the University of Louisville College of Medicine. Lunsford Yandell, Sr., was recognized by his medical colleagues as one of the preeminent physicians of the nineteenth century, a doctor gifted as a writer, lecturer, and practitioner. Before Yandell was able to restructure the Memphis Medical College, the school was forced to close at the outbreak of the Civil War as students and faculty enlisted in the Confederate Medical Service. After the war, Dr. Yandell returned to the University of Louisville. He later served as president of the American Medical Association.

For years, the *Appeal* had urged the city to provide a charity hospital for Memphis residents, as admission to the Memphis Hospital, funded by the state, was restricted to nonresidents and river boatmen. Through the determined efforts of doctors such as Ayres P. Merrill, and Robert W. Mitchell, the city council approved the opening of a charity hospital in December 1860. Dr. Mitchell, secretary of the board of health, was appointed Charity Hospital physician and Dr. Merrill named chairman of the board of trustees. Opened in a former cotton warehouse in the suburb of Chelsea, the hospital reported the admission of 60 patients in January 1861. Patients were moved to the old navy yard hospital in April 1861 when the cotton warehouse was reclaimed for the manufacture of gun wadding.

During the 20-year period when Memphis moved from a small, rowdy town on the Mississippi to a position of importance as the largest river port between St. Louis and New Orleans, disease was an ever-present fact of life. Numerous appearances of cholera, the annual cases of malarial fevers, typhoid, dysentery, an epidemic of smallpox in 1850, an outbreak of yellow fever in 1855, and other contagious illnesses were health issues addressed by the medical profession. Lacking any understanding of the bacterial and viral origin of disease, treatment focused on symptoms; nevertheless, some surgeries and a few medicines, especially quinine, provided positive benefits.

Between 1840 and 1860, the medical profession in Memphis had grown substantially and included men whose education was the best available at that time. The legislature had provided funds for the first hospital in the state, two medical schools were organized, and charity clinics conducted. In addition, a new charity hospital for local residents had been authorized by the city council. By 1860, the 1840 population of 1,799 had expanded to 22,623, of whom twenty-six percent were immigrants, mostly Irish and German, with some English, Scots, and French. As the decade of the 1860s began, Memphis was a thriving river port, the center of the cotton trade, and the terminus for a major railroad. Neither the city's businessmen nor members of its medical profession could have foreseen that the next two decades would see the national tragedy of the Civil War and the yellow fever epidemics of the 1870s, in which disease brought catastrophic death and devastation to Memphis.

LEFT: *Dr. Lunsford P. Yandell, Sr. purchased the Memphis Medical College in 1859, but the Civil War interrupted his plans to reorganize the college. He had come to Memphis from the University of Louisville College of Medicine, where he was professor of chemistry and pharmacy.* BELOW: *The levee at Memphis, around 1860. With its position at the center of the cotton trade, and as the terminus of a major railroad, Memphis had become a thriving river port.*

A BOOMING ECONOMY

CHAPTER THREE

A CIVIL WAR HOSPITAL CENTER

1861–1865

IN 1860, IN SPITE OF SECTIONAL CONTROVERSY over the question of slavery, Memphis was a pro-Union city, largely the result of its economic ties with the upper Midwest. The election of Lincoln in late 1860 and the resulting secession of South Carolina and other cotton states shifted public opinion in support of the Confederacy. When President Lincoln called for troops after the firing on Fort Sumter in April 1861, Memphians overwhelmingly ratified Tennessee's secession from the Union.

The call to arms by military leaders of the United States and the Confederate States created problems of a vast and complex nature that neither side fully comprehended nor was prepared to address. The logistics of providing food, uniforms, arms, and combat training for men with no prior military experience taxed the best administrative and military minds. Military leaders had not anticipated the need for medical care during the early months of the war and contagious diseases swept through hastily organized military camps.

Confederate General Gideon Pillow, commander at Memphis, waged a constant battle with the Tennessee legislature for funds to provide food and uniforms for troops under his command. The city's strategic location made it a major recruiting center for military units organized in West Tennessee and North Mississippi. Outbreaks of disease in quickly organized military camps caused serious illness. Rural recruits who had not been vaccinated and had not acquired immunity to childhood diseases contracted

ABOVE: *At Petersburg, Virginia, federal troops were issued rations of whiskey and quinine twice a day. It was thought that both were "...very efficient safeguards against diseases of the camp."*

LEFT: *The Irving Block was used by the Confederacy as a hospital prior to the federal occupation of Memphis. After the Battle of Memphis on June 6, 1862, it was used as a federal prison. In August 1864, General Nathan Bedford Forrest raided Memphis. One objective was to free the prisoners in the Irving Block, which was unsuccessful.*

During the Civil War, the Overton Hotel was converted to a general hospital. At the time of this photograph, in 1863, it had 1,000 beds.

measles, chicken pox, mumps, and smallpox. Contaminated water and poor sanitation resulted in typhoid, dysentery, and other gastrointestinal illness.

Measles caused great havoc in both federal and Confederate camps, as soldiers often suffered from the equally debilitating aftermath of the disease, such as pneumonia, hearing loss, ophthalmia, and other infections. Southern doctors agreed that measles and its often-serious aftermath had cost more lives and invalidism than any other disease. While the number of cases diminished, measles remained a significant disease during the course of the war. In the South, insect-vectored diseases, especially mosquito-transmitted malaria, filled the hospitals. Other causes of disease included pneumonia and respiratory infections, scurvy caused by dietary deficiency, and the inevitable sexually transmitted syphilis and gonorrhea. Disease accounted for sixty-six percent of federal deaths and more than seventy-five percent of Confederate deaths.

Confederate Hospitals in Memphis

The need for hospital facilities to care for the seriously ill was soon apparent to local authorities. In June 1861, the state of Tennessee requisitioned Memphis Hospital on Union Avenue, and designated it as the Army Hospital. This single hospital was not adequate for the needs of sick soldiers brought to Memphis from military camps in the area. Memphis women, organized as the Southern Mothers Society, acquired the Irving Block building on Second Street across from Court Square, and by September 1861 had organized a hospital with facilities for some 400 soldiers. Generous Memphians donated bedding, food, and medicines.

When large numbers of wounded were brought to Memphis following the Battle of Belmont, Missouri, which took place on November 7, 1861, military authorities took possession of the Overton Hotel on Main at Poplar, which was ideally suited for use as a hospital, adding an operating room, surgeon's office, and pharmacy. Designated as an official Confederate States of America hospital, the Overton was placed under the command of Dr. C. H. Mastin, supervisor of hospitals, and a prominent Mobile physician. By the end of 1861, Memphis had more than 1000 hospital beds for the care of sick and wounded Confederate soldiers.

From the outset of the war, a major objective of the federal government was full control of the Mississippi River, which divided the eastern and western halves of the Confederacy. Because Tennessee was in the upper South, Union strategy called for control of the Cumberland and Tennessee rivers, which would provide easy access to the trans-Appalachian Confederacy. In February 1862, General Ulysses S. Grant moved first against Fort Henry, an earthen work fort guarding access to the Tennessee River at the Kentucky-Tennessee border. Nearly under water from heavy winter rains, Fort Henry surrendered on February 6. Grant then moved ten miles over land and attacked Fort Donelson, which controlled access to the Cumberland River, also near the Tennessee-Kentucky line. Fort Donelson's inept commander surrendered on February 16. Some wounded Confederate soldiers who escaped capture by Grant's forces made their way to Memphis and were admitted to Overton Hospital.

SHILOH—A BLOODY FIELD OF BATTLE, APRIL 6-7, 1862

Grant moved his troops up the Tennessee River, stopping at Pittsburg Landing, Tennessee, on the west side of the river across from Savannah. In an effort to halt Grant's move into Mississippi, a large Confederate army commanded by General Albert Sidney Johnston staged a surprise attack on federal troops just after dawn on Sunday April 6, 1862. The Battle of Shiloh was the first immense battle in the western theatre of the war, and the carnage of the two-day conflict left the battlefield strewn with the mangled bodies of men and animals. The 24,000 men killed or wounded shocked the nation. It was a staggering loss for both armies and made clear the war would not be quickly won by the North.

In his report on the battle, Dr. Nelson Derby, USA, pointed out the complete lack of facilities and trained personnel available to treat the wounded. Derby described the struggle to save lives in the pouring rain and morass of mud and concluded, "the circumstances attending the battle of Shiloh were fearful, and the agonies of the wounded were beyond all description." Derby was also critical of the excessive number of amputations performed by volunteer doctors anxious to operate. With no facilities to care for the wounded, Major B. J. D. Irwin, USA, appropriated tents and medical equipment and organized a large tent field hospital. Dr. Irwin later wrote that "the first tent field hospital of any magnitude [was established] for the reception and treatment of the wounded on the field of battle." The value

Appalled at the lack of facilities to treat the wounded following the Battle of Shiloh, Major B. J. D. Irwin and his medical officers set up a large, well organized tent hospital near the town of Hamburg to care for the thousands injured in that horrific battle.

Leg amputation among federal troops at Fortress Monroe, Louisiana. Amputations were performed "excessively" at Shiloh, according to Dr. Nelson Derby, USA. Those lucky enough to survive the surgery faced the risk of postoperative sepsis or gangrene.

of tent field hospitals was developed into a system used by both armies.

Confederate General Albert Sidney Johnston, leading his men on horseback in the thick of battle, was shot in the right leg—a minie ball went through his boot and severed the popliteal artery. In the excitement of combat, Johnston was unaware of the serious nature of his injury because blood collected in his boot. Johnston had sent Dr. David W. Yandell, medical chief of staff, to care for wounded prisoners. In shock from loss of blood, Johnston died shortly after being taken off his horse and laid on the ground. He had bled to death in a matter of minutes.

Johnston's death placed General P. G. T. Beauregard in command of the Confederate troops. With the arrival of strong Union reinforcements from Nashville, the Confederate forces lost the advantage of a surprise attack on Sunday. Hoping to save the important rail junction at Corinth, Beauregard ordered the evacuation of all troops on Monday afternoon, April 7. Trains of army wagons transported more than 5,000 seriously wounded to Corinth. With no hospitals to receive these men, wounded soldiers were placed anywhere they could be laid. Shattered arms and legs resulted in wholesale amputations. The diary of Confederate nurse Kate Cumming records her horror at the gruesome sight of piles of arms and legs stacked like cordwood. Several hundred of the Confederates wounded at Shiloh were transported to Memphis. Doctors worked long hours caring for these men, many of whom were dying when they arrived. By May 2, 1862, there were more than 1,200 patients in Memphis hospitals.

As Confederate forces withdrew from Shiloh to Corinth, the South suffered another strategic military defeat in the loss of Island No. 10, a fortified sandbar in the Mississippi River south of Cairo that had prevented federal gunboats from gaining access to Memphis and points south. In an effort to hold the rail center at Corinth, Beauregard pulled all troops from the forts in West Tennessee, including the garrison at Memphis. Confederate sick and convalescent soldiers were ordered back to duty and the wounded evacuated to Grenada.

NAVAL BATTLE AT MEMPHIS—JUNE 6, 1862

Expecting Memphis to fall into federal hands, Beauregard ordered businesses to send all supplies south or destroy what could not be moved. Hundreds of barrels of molasses were dumped in the Mississippi River and thousands of bales of burning cotton lit the night sky. The anticipated loss of Memphis to federal forces took place on June 6, 1862, when gunboats commanded by Colonel Charles Ellet, Jr., defeated the Confederate River Defense Fleet commanded by Captain James Montgomery. With no military garrison in Memphis, Mayor John Park surrendered the city to naval officers who came ashore. Victory, with much smoke and little bloodshed, belonged to the federal forces.

Confusion prevailed as federal officers established control of the city and quickly took possession of Overton Hospital. Dr. G. W. Curry, who had been in charge of the Southern Mothers' Hospital on Second, returned to Memphis under a flag of truce to care for 50 Confederate

The Battle of Memphis, one of the largest naval engagements of the Civil War, took place on June 6, 1862. It lasted for 90 minutes and left Memphis in federal hands for the remainder of the war.

soldiers left unattended in the state Army Hospital. The federal Medical Department did not immediately occupy this hospital (Memphis Hospital on Union), and it was used as a contagious disease hospital during the smallpox epidemic in the winter of 1862 to 1863. The Southern Mothers' Hospital on Second Street was converted to a military prison, which became so notorious for the harsh treatment of civilian prisoners, including women, that President Lincoln ordered an investigation into the prison operation.

When General William T. Sherman took command of the federal garrison at Memphis on July 20, 1862, he notified General Grant that the city had good hospitals, including regimental tent hospitals in the camps. Sherman preferred sending the sick to field hospitals because men were quickly returned to active duty. He had Overton Hospital enlarged to 1,000 beds under the command of Dr. Nelson R. Derby, USA.

As part of Grant's objective to gain full control of the Mississippi River, Sherman launched a major attack against heavily fortified Vicksburg on December 28-29, 1862. As soldiers attempted to climb the bluff at Vicksburg, they were cut down by canon and rifle fire. The attack was a disaster, and the Northern press called for Sherman's removal. Hundreds of wounded were brought back to Memphis.

Federal Hospitals in Memphis

In early 1863, Major B. J. D. Irwin, USA, was sent to Memphis, where as chief surgeon and superintendent of hospitals, he directed the operation of all facilities, appointing the doctors who managed the Memphis hospitals. A university graduate and career officer, Dr. Irwin had received his medical education in New York followed by service as a military surgeon in the Arizona territory.

To support Grant's military activities in the Southwest, federal authorities appropriated large mercantile buildings in Memphis for use as hospitals. A massive effort was required to organize and equip these hospitals. The Quartermaster's Department took over large warehouses to hold the enormous quantities of cots, bedding, clothing, medicines, instruments, surgical dressings, and ancillary supplies for the Memphis hospitals. Grant's post hospitals in West Tennessee and his field hospitals also were supplied from Memphis. As hospital supplies moved downriver from St. Louis and Cincinnati, medicine smuggled to the

RIGHT: *An etching by John A. Volck, showing the smuggling of barrels, boxes, and bundles of medicine by Confederate troops. Memphis, out of the war early, was the center of a lucrative smuggling trade on river and land.* BELOW: *Dr. B. J. D. Irwin with his wife, 1880s. In 1863, Major Irwin was appointed head surgeon and superintendent of federal hospitals in Memphis. As such, he appointed the doctors who managed the Memphis hospitals and directed the operation of all the facilities.*

Confederacy also came south. It was a lucrative trade for those willing to take the risk.

Federal hospitals included the Gayoso (500 beds); Washington (600 beds); Webster (500 beds); Jackson (500 beds); and Union (700 beds), all located on Main Street between Beale and Union. The 1000-bed Overton on Main at Poplar remained the principal hospital. The Jefferson (500 beds) and Adams (600 beds) hospitals were located on Second Street across from Calvary Episcopal Church. An officers' hospital with 100 beds was organized in an ornate cotton office building on Front Street. Salaried federal officers paid for their medical care and were treated in separate facilities. A 50-bed hospital to treat cases of gangrene was set up in the First Baptist Church on Second near Adams.

The state Army Hospital on Union, which the federal Medical Department used for contagious diseases, was enlarged with the addition of four wooden barrack buildings able to accommodate 400 patients. The Commercial Hotel on Jefferson between Front and Main was appropriated in March 1863 for use as a marine hospital to serve federal naval personnel on the Mississippi River. The Navy Yard Hospital, which had served as the location of the Memphis Charity Hospital after 1861, was commandeered for federal use, and the

LEFT: *The Federals adapted several commercial buildings as hospitals. Photographs of these hospitals were gathered by Dr. B. J. D. Irwin and bound in a ledger. Remarkably, they show rare glimpses of Memphis buildings. The Jefferson Block, built in the Greek Revival style, has on its left exterior wall, handbills for minstrel shows and plays. The business on the left is the Cafe Cosmopolitan. Next door is a law office.*
BELOW: *The Officers' Hospital located on Front Street with a two-story wrought iron overhang.*

A CIVIL WAR HOSPITAL CENTER

ABOVE: *The U.S. Sanitary Commission lodge at a convalescent camp near Alexandria, Virginia, 1863. The commission was the forerunner of the American Red Cross.* OPPOSITE PAGE: *Instruments for resection used by the federal Army, illustrated in* The Medical and Surgical History of the War of the Rebellion, *published by the U.S. War Department in 1879.*

civilian patients were moved to the Botanico Medical College on Beale. In 1864, the Memphis Charity Hospital was moved again, this time to the Exchange building on Front Street. After the war, city aldermen refused to provide adequate funding to operate the Charity Hospital, and it was permanently closed in 1867.

The United States Sanitary Commission, the forerunner of the American Red Cross, was formed in Washington, D.C. in 1861. An official agency of the U.S. government, the commission recruited civilian nurses, gathered personal supplies for soldiers, and provided assistance to federal hospitals and soldiers' homes. The Western Sanitary Commission was organized in St. Louis to provide a base of operation closer to the western theatre of war, and a branch was established in Memphis. The Sanitary Commission also outfitted steamboats as floating hospitals, sparing wounded men brought to Memphis and other federal hospitals the trauma of being transported in jolting land vehicles. The *Red Rover*, captured from the Confederates, operated on the Mississippi River and the *City of Memphis* on the Ohio River. The Sanitary Commission operated a fleet of hospital steamers on all major rivers.

On June 20, 1862, Flag Officer Charles H. Davis, commanding the Western Flotilla, wrote the secretary of the navy from Memphis advising him that Sisters of the Holy Cross, St. Mary's Notre Dame, Indiana, were providing nurses for the hospital boat *Red Rover* and other hospital steamers. In Memphis, the Holy Cross Sisters also worked at the Overton and Naval Hospitals. Other nurses who served in Memphis hospitals included the Sisters of Charity (Cincinnati), Dominican Sisters from St. Agnes School, Memphis, and local women. During the Civil War, more than 5,000 women, religious and lay, served in federal and Confederate base and field hospitals in the South.

Although scientific proof of the bacterial origin of disease had yet to be established, by 1863, military doctors had learned that strict sanitation and the generous use of disinfectants such as carbolic acid and chlorine reduced mortality. Wounds were cleansed with iodine, vinegar, or bromine. Separating the sick from the wounded improved the recovery rate of the injured. The Gayoso and the Jefferson hospitals were designated for wounded only. Surgical procedures were determined by the kind of ammunition used. Soft lead bullets smashed bones, crushed tissues, and impacted dirt and clothing into wounds increasing the likelihood of serious infection. Cone-shaped minie balls caused a bursting type of wound that shattered bones and increased the fatality of abdominal injuries. Most major injuries to the extremities resulted in amputation. If a soldier survived the trauma of amputation, he faced the postoperative complication of sepsis or gangrene.

Dr. F. Noel Burke, USA, surgeon in charge of Jefferson Hospital and later the Gayoso, was an able surgeon who carried out surgical procedures that saved a few soldiers from amputation of an arm. One of the most successful efforts in treating wound complications took place at the Gangrene Hospital. Drs. Charles H. Cleaveland and George B. Weeks had worked with Dr. Middleton Goldsmith, USA, surgeon in charge of the 5000-bed Jefferson Hospital near Louisville. Dr. Goldsmith developed a method of treating gangrene, the most feared of all hospital complications, used by Cleaveland and Weeks. While the patient was under anesthesia, the sloughing tissue was dissected out with forceps, the wound edges were painted with bromine, and the wound was then packed with cotton lint soaked in a mild bromine solution which burned away the dead tissue and promoted healing through the granulation of healthy tissue.

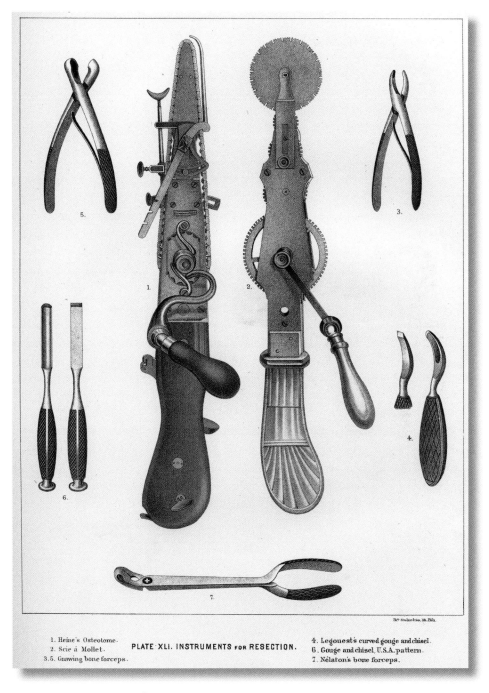

1. Heine's Osteotome.
2. Scie à Mollet.
3,5. Gnawing bone forceps.
4. Legouest's curved gouge and chisel.
6. Gouge and chisel, U.S.A. pattern.
7. Nélaton's bone forceps.

PLATE XLI. INSTRUMENTS FOR RESECTION.

GRANT'S SIEGE OF VICKSBURG—MAY 19 – JULY 4, 1863

Grant launched the siege of Vicksburg on May 19, 1863. The battered, starved-out city surrendered on July 4, 1863, a fact eclipsed by the enormous Battle of Gettysburg, which took place July 1-3, in which some 46,000 men were killed or wounded. Federal losses at Vicksburg were heavy, and disease was as devastating to Grant's troops as the gun and canon fire rained down by Vicksburg's Confederate defenders. The summer heat, malaria, dysentery, and typhoid caused great morbidity. When the Confederate stronghold at Port Hudson, south of Vicksburg, surrendered on July 9, federal army and navy units controlled the Mississippi River from its source to the Gulf of Mexico. From May through August, hospital steamers filled with sick and wounded landed almost daily at Memphis. In the first nine months of 1863, records document that 63,800 men passed through Memphis hospitals.

General Sherman left Memphis in October 1863 to relieve General William S. Rosecrans, besieged in Chattanooga. After securing Chattanooga, Sherman began his move toward Atlanta in May 1864 and had the city under federal control by September 3, 1864. Losses following the Battle of Atlanta were enormous—nearly 32,000 federal troops were killed, wounded or captured, and Confederate losses numbered some 35,000. By the summer of 1864, specially equipped hospital trains took badly wounded federal soldiers to hospitals in the North, although Sherman kept 25,000 backup hospital beds in Memphis, Nashville, and Chattanooga.

Although the major focus of the war had moved to the Southeastern theatre, small battles in West Tennessee, along with seasonal diseases, provided patients for Memphis hospitals. The circumstances of the last large-scale care provided by the military hospitals in Memphis were especially tragic. On April 21, 1865, shortly after Lee's surrender at Appomattox Courthouse, the steamboat *Sultana* left New Orleans with some 180 passengers and crew. Docking at Vicksburg on April 24, repairs were made to leaking boilers. Waiting at Vicksburg for transport north were hundreds of men who had been released from Confederate prisons in Andersonville, Georgia, and Cahaba, Alabama, and marched overland for embarkation north. Anxious to get the men moving homeward, federal officers at Vicksburg ordered some 2000 men to board the *Sultana*, dangerously overloading the boat.

The *Sultana* Horror

On the evening of April 26, 1865, the *Sultana* docked briefly at Memphis to discharge a few passengers and then moved across the river and took on coal at Hopefield, Arkansas. At about two a.m. on April 27, the *Sultana* moved upriver against a flood stage current. Under full pressure, the hastily repaired boilers exploded in an Armageddon of flames, boiling water, and flying debris, hurling passengers into the cold, dark waters of the Mississippi. Many drowned in the swirling current. The explosion could be seen in Memphis and boats were dispatched to rescue survivors. Approximately 600 soldiers and a few civilians were brought to federal hospitals. More than half of those rescued died of injuries, primarily burns. Although the exact number of soldiers boarded at Vicksburg was not recorded, it is estimated that between 1500 and 1700 of the *Sultana* victims were Union soldiers. A historic marker on the Memphis riverfront commemorates the greatest marine disaster in the nation's history.

In 1862 and 1863, the federal Medical Department had organized more than 5,000 hospital beds in Memphis and

Reinforcements for Grant's army leave Memphis for Vicksburg aboard steamboats. Steamboats were also used as hospital ships, such as the R. C. Wood. *It was remodeled and made into a hospital ship to transport the sick and wounded to and from various cities, including Memphis. Between April 1863 and April 1865, it carried 11,024 soldiers*

ABOVE: *Overloaded with federal troops released from Confederate prisons, the Sultana's boilers exploded just north of Memphis.* LEFT: *After the Civil War, Dr. Gustavus B. Thornton became the local surgeon for the Illinois Central and Southern Railroad and president of the Memphis Board of Health.*

demonstrated skill in their arrangement and operation. The last patient was discharged from Overton Hospital on August 3, 1865. Major Irwin, promoted to the third highest rank in the federal Medical Department, was transferred to Louisville prior to assignment in the west. He later rose in rank to assistant surgeon general. Strategically located on the Mississippi River, Memphis served as an important hospital center throughout the Civil War, gaining its first national recognition for medical care.

In 1866, the military government operating Memphis returned control of the state-owned hospital to Tennessee (Memphis Hospital on Union built in 1841). The war left the state impoverished, and the legislature did not appropriate funds to support Memphis Hospital. Title was transferred to local government, and the city assumed support of the hospital, which was thereafter identified as City Hospital. Fixtures from the Memphis Charity Hospital, closed in 1867, were sent to the newly named City Hospital. Dr. Gustavus B. Thornton was appointed hospital administrator. In 1872, the state relinquished title to the ten-acre tract (now Forrest Park) on which the hospital was located.

Southern doctors who had been in charge of military hospitals became leaders in the public health movement in the United States. In Memphis, Robert W. Mitchell, Richard B. Maury, Dudley D. Saunders, and Gustavus B. Thornton were such leaders. By 1880, the research of Louis Pasteur in France and Robert Koch in Germany proving the bacterial origin of disease would make significant contributions to medicine and sanitary reforms during the last quarter of the nineteenth century.

CHAPTER FOUR

A DECADE OF DISEASE

1870–1879

THE TWO MOST SIGNIFICANT DISEASES OF African origin transferred to the New World were malaria and yellow fever, which reached North America in the mid-seventeenth century. By the nineteenth century, epidemics of yellow fever in the port cities of America made this the most feared disease in the country. Ships arriving from Africa brought the infected mosquitoes in their water casks. In coastal areas where yellow fever was endemic, or regularly present, many people had the fever and thus acquired immunity to this viral disease.

Memphis had its first documented outbreak of yellow fever in 1855, which was confined to an area south of the city. Estimates of the number of cases vary. Memphis Hospital reported 128 deaths and the county recorded 150 deaths between August 14 and November 31. Among those who died were three doctors—Reuben Berry, Zeno T. Harris, and Michael Gabbert, a founder of the Botanico Medical College.

The Memphis chapter of the Howard Association, a benevolent service organization, started in 1855, was based on the model which originated in New Orleans in 1837. The Howard Association was composed of young businessmen dedicated to care of the sick. The name honored English reformer John Howard (1726-1790), who committed his life and fortune to reforming hospitals, prisons, orphanages, and other institutions of incarceration.

In the decade following the Civil War, Memphis suffered from financial difficulties, some of which were the consequence of Reconstruction, while others were the result of reckless spending by the city council during the boom decades. Pleas from public health officers and the medical community to address sanitary reforms went unheeded or, if temporarily addressed, lacked authority and funding to carry out needed measures. The yellow fever epidemic of 1867 had a reported 2,500 cases and 250 deaths. The reactivated Howard Association applied to the state legislature for a charter of incorporation, granted in January 1869.

During the winter of 1873, Memphis experienced a series of unrelenting calamities. The first, an epidemic of infectious equine disease, killed hundreds of horses and mules, crippling business and public transportation. The freezing of the Mississippi suspended all river traffic creating great hardships, including

ABOVE: *The adult female of* Aedes aegypti *is the carrier of the yellow fever virus, its male counterpart unable to bite humans because of the structure of its mouth parts. The* Aedes *mosquito bites with its abdomen held parallel to the skin, a factor which distinguishes it from* Anopheles, *the carrier of malaria.* OPPOSITE PAGE: *An officer of the Howard Association finds the bodies of a dead mother and child during the yellow fever epidemic of 1878.*

A Decade of Disease

a lack of coal, the city's primary source of fuel. This was followed by an epidemic of smallpox. An outbreak of cholera in the spring claimed some 276 victims.

Yellow Fever Strikes in 1873

The city had a few weeks of respite in the summer, but in September the Board of Health confirmed yellow fever had been present in the city since mid-August. The announcement resulted in a mass exodus of all who had the means to seek safety farther north. Medical historian and University of Tennessee professor Dr. Simon R. Bruesch points out that the delay in reporting the presence of the disease arose from the difficulty of diagnosing yellow fever until there had been a sufficient number of cases to confirm that it was not one of the related febrile diseases such as dengue or malaria. The yellow fever of 1873 did not exhibit the severe symptoms which accompanied the more virulent epidemic of 1878. When the killing frost came in late November, there had been 2,000 deaths out of 5,000 known cases.

Memphis Appeal editor John M. Keating rebuked city authorities for negligence in public sanitation. On November 23, 1873, he wrote, "We fear the apathy and feeling of false security, which an abatement of the disease induces. Another epidemic and the blow, which has prostrated our trade will annihilate it, to say nothing of the uncounted victims doomed to death under the shadow of pestilence."

BELOW: *Arrest of yellow fever refugees by the safety patrol of Memphis. Memphians brought the disease to other towns as they escaped the scourge in Memphis.*
OPPOSITE PAGE: *Dr. William C. Cavanaugh, former health officer, was the physician in charge of Camp Father Matthew, which cared for some 400 residents.*

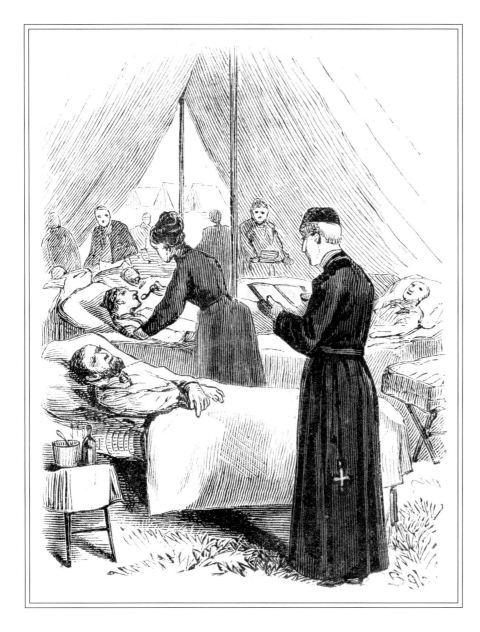

More concerned with the effects of the disastrous 1873 financial panic and the bankruptcies of large eastern financial institutions, politicians and businessmen failed to heed Keating's warning of a devastating future epidemic. His prediction would become a terrifying reality.

THE 1878 YELLOW FEVER DISASTER

The summer of 1878 was extremely hot, and the newspaper reported an especially heavy mosquito infestation. Although there were sporadic cases of yellow fever in July and early August, it was not until August 23 that the Health Department officially confirmed the presence of fever in the city. In the ensuing panic, some 25,000 people fled by train, steamboat, or other conveyance seeking a place of safety. Many who fled were already infected and took the virus to other towns and communities. Germantown, Collierville, Brownsville, Somerville, Jackson, and other towns on railroad lines out of Memphis experienced heavy yellow fever mortality in spite of efforts to quarantine their communities. A notable example of disease transfer was the epidemic in Chattanooga, a city believed beyond the reach of yellow fever, which was infected as a result of Memphis refugees who went by train to the East Tennessee city. The first death of a Memphis refugee in Chattanooga occurred on August 21. The fever spread rapidly in September and, while not as severe as in other Tennessee towns, lasted well into 1879.

The frenzied exodus of more than half the Memphis citizenry in less than a week resulted in a suspension of government services and a paralysis of business; nevertheless, many of the city's best citizens remained to provide basic services for the stricken city. The Citizens' Relief Committee, headed by cotton factor Charles G. Fisher, took over the functions of government, organized refugee camps outside the city, contacted the federal government for assistance, and arranged for the distribution of food, tents, and other supplies. Of some 20,000 who remained in the city, about 14,000 were black and 6,000 were white, many of them immigrants. The committee sent 5,000 people to tent camps located outside the city. These camps were organized under the direction of Drs. R. B. Nall, Paul Otey, and Dudley D. Saunders who had organized military tent hospitals during the Civil War.

HOWARD SOCIETY REACTIVATED

The reactivated Howard Society chose Dr. Robert W. Mitchell, experienced in treating yellow fever, to head the Howard Medical Corps and Dr. William E. Rogers to direct the Howard Infirmaries. The Court Street Infirmary cared for sick doctors and nurses, while the Market Street Infirmary received local citizens. The Howard Nursing Corps, directed

DR. ROBERT WOOD MITCHELL

The life and medical career of Dr. Robert Wood Mitchell spanned a period that included the Civil War, devastating yellow fever epidemics, the establishment of scientifically based medicine, and the first medical specialties. Born in 1831 in Madison County, Tennessee, he was the son of General Guilford and Mary Wood Mitchell. In 1836, the family moved to Grenada and later to Jackson, Mississippi, where Mitchell attended Centenary College.

Moving to Vicksburg, Mitchell entered the drug business, which aroused his interest in medicine. He apprenticed with his brother-in-law in Yazoo City, receiving his formal education in the medical department of the University of Louisiana (now Tulane) at New Orleans. His M.D. degree was awarded in 1856. The next year, Mitchell was placed in charge of a Vicksburg hospital, where he had his first experience with yellow fever and may have acquired immunity to the disease. Dr. Mitchell moved to Memphis in 1858 and was elected secretary of the Board of Health in 1859. During the Civil War, he served as a surgeon in several military units, finally advancing to division surgeon in the Army of Tennessee. Dr. Samuel Hollingsworth Stout, medical director of the Army of Tennessee, named five Tennesseans who were distinguished for their work and service in general and field hospitals during the war: Drs. Robert W. Mitchell; Frank Rice; Gustavus B. Thornton; J. M. Keller; and Benjamin W. Avent, all Memphis doctors.

When the Memphis Medical College was revived in 1868, Mitchell was named professor of *materia medica*. In spite of the excellent faculty, which included Drs. Alexander Erskine; Benjamin W. Avent; Richard B. Maury; Gustavus B. Thornton; Dudley D. Saunders; Alfred H. Voorhies; Felix McFarland; and Robert Thummel,

by former Mayor John Johnson, functioned as a separate unit and hired some 3,000 nurses during the course of the epidemic, including many who came from Southern cities, which had experienced numerous outbreaks of yellow fever.

The City Hospital on Union, directed by Dr. Gustavus B. Thornton, admitted 460 yellow fever patients during the epidemic. Volunteer doctors T. J. Lynn and E. T. Easley assisted Thornton in the care of the sick. The doctors themselves had severe cases of fever. Drs. Thornton and Lynn recovered, but Dr. Easley died. Regarding treatment, Dr. Thornton observed, "active medication to arrest [yellow fever] is not only useless, but positively injurious…. success-

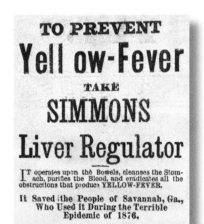

An ad for Simmons Liver Regulator claims to prevent yellow fever. The contents of this patent medicine were not stated, but another, named Bedford Alum and Iron Springs Mass, claimed to prevent and cure yellow fever. Its ingredient was a combination of alum and iron which was supposed to be taken often and in large quantities.

ful treatment should alleviate suffering and assist nature throw off or eliminate this poison from the system." Dr. Robert W. Mitchell recommended a mustard foot bath and light blanket for the chill, followed by castor oil to cleanse the bowel, no further medication, and complete bed rest.

Opposite page: *Dr. Robert Wood Mitchell, c. 1887. Dr. Mitchell served as a surgeon in the Confederacy and later became medical director of the Memphis Howard Association.* Above: *The Mitchell and Maury Sanitarium for the diseases of women, 111 Court, between 3rd and 4th Streets. This is the second location of their practice, in a brand new building built in 1886. The Sanitarium, which opened at 73 Court in 1885, had the first organized training school for nurses in Tennessee.*

the school closed in 1873, largely the result of a devastated post-Civil War economy. In 1872, Mitchell married Rebecca Park, daughter of William and Rebecca Fox Park, a prominent Memphis family. They had no children.

At the outbreak of the 1878 yellow fever epidemic, the Howard Association was called into service, and Mitchell, an experienced hospital administrator, was appointed director of the Howard Medical Corps. He coordinated the work of all local and nonresident Howard physicians under unbelievably difficult circumstances. During both epidemics, Rebecca Mitchell remained in Memphis to work at the side of her husband. The surgeon general of the United States acknowledged Mitchell's leadership as worthy of the highest commendation. In 1878, President Rutherford B. Hayes appointed Mitchell to the first National Board of Health. Working at both national and local levels, Mitchell was a leader in the field of public health and sanitation.

Robert W. Mitchell and Richard B. Maury became medical partners in 1867, an association that continued until Dr. Mitchell's death in 1903. A partnership of such length assumes a bond of common experiences and interests, which included their active participation in medical education and their support of public schools and colleges. From 1870 to 1876, Mitchell was a member of the Board of Education, serving terms as president and vice president.

Mitchell was an active member of the American Medical Association, the National Board of Health, the American Public Health Association, the Tennessee Medical Society, and the Memphis and Shelby County Medical Society. One of the giants of Memphis medicine, Robert Wood Mitchell maintained a busy medical practice until shortly before his death on November 2, 1903.

In July 1878, when the New Orleans public health officer confirmed that yellow fever was present in that city and moving upriver, Dr. Robert W. Mitchell, president of the Board of Health, urged the city council to establish quarantine stations at points outside Memphis, which he believed essential to prevent those infected with fever from entering Memphis. Quarantine, which the business community considered a handicap to trade, was denied by the city council. Dr. Mitchell resigned from the Board of Health and Dr. Dudley Dunn Saunders was appointed president. Others on the board were Health Officer Dr. John Henry Erskine; Board Secretary Dr. R. F. Brown; Dr. Benjamin W. Avent; Dr. Richard B. Maury; Mayor John R. Flippin; and Chief of Police A. Phillip Athey.

The board purchased great quantities of carbolic acid for use as a disinfectant, as well as tons of lime to spread on streets. Hundreds of barrels of pine tar were burned to disperse the deadly miasmas and cannons were shot off to clear the air. As the fever daily claimed more victims, meetings ceased. All Board of Health members had yellow fever; Dr. Benjamin W. Avent died on September 12 and Dr. John H. Erskine on September 17.

September Death Rate 200 Per Day

The number of fever cases increased with frightening rapidity. At the peak of the epidemic in early September, the daily death rate exceeded 200. On September 2, Charles G. Fisher reported 400 new cases. On the 7th, he wrote his sister in Covington, "I have never seen such a fearful time in all my life. Rev. Charles Parsons died last night Dr. Saunders, Dr. Avent are both down with the fever. Dean George Harris and Mayor Flippin and Chief of Police Athey are all

A Decade of Disease

Active members of the Howard Association of Memphis during the Yellow Fever Epidemic of 1878. Of the 32 members who served, ten died. Those who died are pictured within the center oval.

very low. God help us! Where will the end be?"

On September 20, Fisher telegraphed a New York business associate. "Deaths to date 2,250. Number sick now about 23,000. Average deaths 60 percent of sick. We are feeding some 10,000 persons, sick and destitute in camps and in the city. Our city is a hospital. Fifteen volunteer physicians have died, two others sick. Many nurses have died [including] many that had the fever before and thought themselves proof. We are praying for frost. It is our only hope."

The frost came too late for Charles G. Fisher—he died the morning of September 26th. General Luke Wright, executive member of the Citizens' Relief Committee (he was later governor general of the Philippines) wrote Dr. Charles Fisher, Sr., "Charlie was taken with a desperate case of fever. He had been so brave, so indefatigable in his labors for the suffering and sick that it seemed he could not be reached by the destroyer. He was our friend, our counselor and our guide in all our troubles. In those moments when we were faint and broken by the terrible burden upon us and appalled by death all around us, his words of hope, cheer, and counsel restored us."

Heroic Work of the Sisters

The Episcopal Sisters of St. Mary, who conducted St. Mary's School for Girls, also managed the Church Home Orphanage. The Citizens' Relief Committee asked the Sisters to take charge of Canfield Orphan Asylum, an agency organized for black children after the Civil War. When Sisters Constance, Superior at Memphis, and Thecla died in early September, other sisters came from New York to help minister to the sick and care for the orphans. After arriving in Memphis, Sister Clare wrote the Mother Superior at Peekskill, "Sister Helen and I came to Canfield Orphanage this morning. We take chil-

Top: *Dr. Dudley Dunn Saunders, president of the Memphis Board of Health, 1878. He was also a Confederate surgeon and a founder of the Memphis Hospital Medical College.* Above: *Dr. John Henry Erskine was appointed Memphis's health officer by Mayor J. R. Flippin in 1876. He died on September 17 during the 1878 yellow fever epidemic.*

dren from parents' deathbeds and keep them here until it is safe for them to go to the Church Home. There are forty children here now, including six babies, and the numbers increase daily. We are living on soda crackers and water. I see only a desolate waste and hear only pitiful voices. Pray for us."

Sisters Ruth and Frances, working at the Church Home Orphanage, succumbed to the fever and died, exhausted from caring for so many sick children.

Sisters of the Dominican Order suffered heavy losses in their Memphis community. Mother Alphonsa, Superior of the Dominican community at St. Agnes School for Girls, was in Illinois when she learned that yellow fever was raging in Memphis. Hurrying back, she found four of her sisters very sick. Among the Dominican Sisters who died were the school infirmarian, the superior of LaSallette Academy (a day school), and the sister in charge of St. Peter's Orphanage. Mother Alphonsa died on September 6. Two groups of Franciscan sisters came from St. Louis to work as volunteer nurses. Only one of their number escaped yellow fever. Dominican and Franciscan priests lost local clergy, and volunteer priests who came to take the place of those who had died also succumbed.

Clergy of all faith groups gave without reserve to provide help of every kind—and paid a heavy price for their faithful service. The Reverend Charles Carroll Parsons, Rector of St. Lazarus Episcopal Church, died on September 6 while working with the sisters at St. Mary's Cathedral. Dr. Edward C. Slater, pastor of First Methodist Church, died on September 9, the week of heaviest mortality. His daughter Mary said, "Pa did right. I would rather he had stayed and died as he did, than desert his people in the hour of their greatest need." Mary Slater, her sister, and mother died before the epidemic ended.

A Decade of Disease **41**

None worked harder than the Reverend Dr. W. E. Boggs of Second Presbyterian Church, whose effort in behalf of orphans was prodigious. After recovering from a serious case of yellow fever himself, he continued to minister to the sick and dying. Rabbi Ferdinand Sarner died while serving members of his congregation.

Annie Cook—Madam and Heroine

On September 12, the *Appeal* reported the death of a volunteer who, although not affiliated with any religious body or medical organization, had taken in the sick and dying. Annie Cook, madam of the Mansion House on Gayoso east of Main Street, died nursing those stricken with the fever. When thousands fled, Annie turned her elegant bordello into a hospital and, with the help of Emily Sutton, nursed many who had no home or place of refuge. The *Appeal's* florid Victorian obituary was nonetheless true, "After a life of shame, she ventured all she had of life and property for the sick."

Dr. William James Armstrong, a Howard physician, sent his wife and eight children to Columbia, Tennessee, when

ABOVE: *Sister Hughetta, of the Episcopal Sisters of St. Mary, survived the yellow fever epidemic. She is shown with four girls at St. Mary's School, Sewanee, Tennessee, around 1890.* LEFT: *A painting of Dr. William James Armstrong, c. 1870s. Dr. Armstrong wrote daily to his wife about the epidemic. On August 28th, he described its virulence. "...in its general symptoms is like Yellow Fever, but in its spread, in its tendency to break out after exposure for a short time & in its tendency to spread in new districts, is unlike any other fever ever known before...." He signs the letter, as he always did, "your Husband."*

yellow fever was declared epidemic in Memphis. From his office on Alabama near St. Mary's Episcopal Cathedral, Dr. Armstrong assisted the sisters and nurses at the cathedral in caring for fever cases in the Eighth Ward, which suffered the heaviest mortality. He was also attending physician at the two orphanages managed by the sisters. A collection of letters written to his wife, Louisa Hanna Armstrong, reveals a sense of helpless frustration because of his inability to treat the yellow fever, as well as his fear as the death toll mounted daily. On September 1, he wrote, "Gloom impenetrable, through which there is no view to mortal eyes, overhangs our dear Memphis. The sights that now greet me every hour in the day are beyond the . . . 1873 [epidemic]. Our best citizens are going by the dozens, and we poor doctors stand by abashed at the perfect uselessness of our remedies. It is appalling, and makes the very bravest quake. I never was, in all my life, so full of sympathy and sorrow for suffering humanity. I feel sometimes that I must run away [but] I see that anyone who runs away is taken sick and dies—so that if I remain someone will be near to attend me. God grant that I may be able to administer to the sick throughout."

Worn out from his labors tending the sick, Dr. William Armstrong died on September 20. His nurse said that even in his delirium, Armstrong expressed concern for his patients.

The Hoped-for Hard Freeze—October 19, 1878

The ice and hard frost of October 19 killed off the mosquitoes and the cases of fever began to diminish, although deaths continued through December. Many of the later deaths occurred after an individual seemed to have recovered only to suffer a fatal relapse. On October 26, the epi-

Food distributed through the Citizens' Relief Committee was provided by the federal government as farmers from outlying areas were afraid to come into town.

A Decade of Disease 43

demic having been declared ended, Dr. Robert W. Mitchell gathered the Howard physicians together in a solemn farewell dinner to pay tribute to their fallen comrades. The epic battle against yellow fever had claimed 34 resident and volunteer physicians, including five volunteers from Ohio. One, Dr. Richard H. Tate of Cincinnati, the only black doctor to work in Memphis, died on September 21. The fall 1878 issue of the *Cincinnati Lancet* carried an extensive account of the Memphis epidemic and paid tribute to those Ohio physicians who gave their lives working in the fever-stricken city of Memphis.

Pathologists and epidemiologists attribute the virulence of the 1878 epidemic to a mutation in the yellow fever virus that caused great mortality. Dr. Greensville Dowell of Houston, Texas, a noted authority who worked in Memphis, declared the yellow fever of 1878 "virulent and violent and particularly fatal." Symptoms included high fever, chills, cramps, and severe aching. The liver and kidneys shut down and the skin turned yellow. Internal bleeding mixed with stomach acid caused the so-called black vomit. Delirium and convulsions were common. Death followed in two to five days. Sometimes, the patient appeared to have recovered only to sink into a fatal relapse. Entire families came down with the fever and parents and children died within hours of one another. Nurses died beside their patients.

The doctors and nurses, clergy and religious, undertakers and druggists, newspapermen and telegraphers, police and militia who worked in Memphis during the 1878 epidemic labored under conditions which beggar description. A journalist who came to the city at the height of the epidemic wrote, "A stranger in Memphis might have though he was in hell—the glare of the barrels of burning tar, the eerie silence broken only by wagons loaded with coffins that rattled through the streets day and night, the stench, the smoke of burning bedding, the overpowering sense of death and gloom."

In the weeks after the epidemic was declared ended, most who left Memphis returned to the city and to homes they had abandoned as panic-stricken residents fled. They found houses that had been ransacked and streets full of burned bedding and household furnishings. The stench of death still hung over the city, but as the wheels of commerce began to turn, life returned to some degree of normalcy.

In a city severely chastised for its years of sanitary neglect, Memphis had not recovered from the 1878 yellow fever when the disease made another comeback in 1879. The winter of 1878 to 1879 was mild, and epidemiologists believe that enough infected mosquitoes survived through the winter to fuel this further round of yellow fever. Of 2,000 fever cases reported in 1879, there were 600 deaths. Most of these victims had fled the city in 1878 and provided a pool of nonimmune people for the mosquito-transmitted disease.

An Analysis of the 1878 Disaster

Historians agree that the 1878 yellow fever epidemic in Memphis was one of the worst urban disasters to strike an American city. The 5,150 deaths in the Memphis area represented about twenty-five percent of the total mortality in the entire Mississippi River valley. What made this epidemic so devastating? Ignorance of the cause was certainly foremost. Scientific research in Europe had only recently established the germ theory of disease, but proof of insect-transmitted disease was still in the future. Although opinions varied, it was the commonly held belief among Southern physicians that miasmas, or foul air arising from open sewers, filthy streets, and decaying plant and animal material caused the disease. Based on observation of the disease, other doctors argued for contagion as the mode of transmission. Whatever the cause of yellow fever, city government had failed to establish a health department with authority and funding to enforce even minimal sanitation.

Opposite page, above: *"Near the End" from the* Memphis Daily Appeal, *1878, describes the abatement of the yellow fever epidemic.* Opposite page, below: *In 1879, the* Memphis Daily Appeal *reported on the state of cleanliness in the city. "More Filth and Nastiness" was printed on February 11th, part of which appears here. "Poisonous vapors," seventh line from the top, is used to describe a link to disease.* Below: *During the summer of 1879, Memphis suffered another yellow fever epidemic. An August 9, 1879, report in* Harper's Weekly, *mentioned that the population was already fleeing the city. On August 16, the magazine printed this sketch of African Americans en route to Kansas because of the epidemic.*

A member of the Tennessee Board of Health, Dr. Richard B. Maury reported: "The epidemic of 1878 is characterized by two important features—unusual malignity and unusual rapidity of march." He concluded that the devastation had to be attributed "not to man's ignorance, but his wanton disregard and willful neglect of well known sanitarian laws."

The 1878 epidemic brought the full force of the recently organized National Board of Health to bear on public health and sanitation in Memphis. Dr. Robert W. Mitchell was appointed to the National Board, which met in Memphis early in 1879. George Waring, a former army engineer, presented a plan to city authorities for the installation of an underground sanitary sewer system. Reforms demanded by national and state health departments were initiated under the direction of Dr. Gustavus B. Thornton, president of the Memphis Board of Health. The loss of tax revenue left the city of Memphis unable to make payments on its bonded indebtedness and the city council declared bankruptcy. The city's charter was revoked in 1879 and Memphis was made a taxing district of the state. In the next decade, capable officials appointed by the state would restore fiscal stability and make sanitation and public health a priority.

CHAPTER FIVE

THE GILDED AGE

1880–1899

When a new decade began in 1880, Memphis had been devastated by three yellow fever epidemics and the population reduced by almost half to 33,592. The state legislature revoked the city charter and made the municipality a taxing district of the state. After years of fiscal mismanagement, the outlook was bleak for the bankrupt city. The Gilded Age, marked by lavish displays of wealth and the rapid growth of industry, railroads, and banking in the Northeast, had a limited impact in Memphis and its agricultural surroundings. The immediate problem facing city leaders was sanitation and public health.

The first of two far-reaching benefits to public health was the installation of a sanitary sewer system designed by George E. Waring, a military engineer. Waring's system of underground clay pipe sewers had been successfully installed in a few large cities, and his sewer plan was approved for installation in Memphis. More than 30 miles of sanitary sewers were laid in 1880, and by 1889 another 20 miles had been installed. A new board of health with enforcement power, directed by Dr. Gustavus B. Thornton, initiated garbage collection, cleaning of streets and alleys, and stone paving of major streets to replace rotted wooden block paving. Dr. Thornton presented a summary of this work, which was entitled "Six Years of Sanitary Work in Memphis," at a meeting of the American Public Health Association, which was held in Toronto, Canada, from October 4 to 8, 1886.

The city's greatest health benefit was the discovery in 1887 of the Memphis aquifer, a vast reservoir of artesian water under the Memphis sands. From its founding, the city's water supply had come from cisterns and shallow wells. In the 1870s, a private company built water lines and pumped water from the polluted Wolf River and later the Mississippi River, both a source of disease. By the end of the 1890s, the availability of pure water from the aquifer had reduced water-borne disease by seventy percent and this plentiful source of water proved a long-term asset to the city's economy as well as its health.

In spite of the benefits provided by sanitary sewers and a pure water supply, diseases such as typhoid, tuberculosis, diphtheria, and scarlet fever remained significant causes of death in the Bluff City. Much of this mortality was attributable to poverty, contaminated milk, and the continued use of polluted water. Raw milk was the major source of tuberculosis, and it required strenuous efforts by public health officials and the city press to force compliance with local ordinances requiring the pasteurization of milk and certification that dairy herds were disease free.

Opposite page: *The original Memphis Hospital Medical College was located on the south side of Union across from City Hospital. The first enlargement in 1892 doubled the size of the college. In 1901, a large facility named Rogers Hall was built on the north side of Union, near Marshall. An image of Rogers Hall can be found in Chapter Six.*

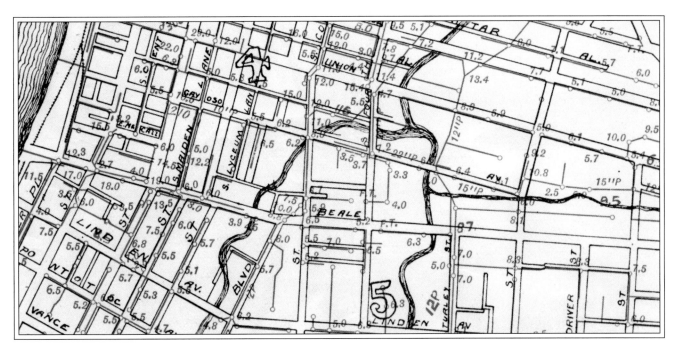

ABOVE: *Portion of a map showing the Bayou Gayoso and the sewer system in the vicinity of Union and Beale streets, 1908. The bayou is represented by wide black lines, while the numbers refer to the sewer system, shown as thin lines. The Mississippi River is at the extreme left. The sewer system, made of vitrified pipe six inches in diameter, was started January 20, 1880. As of 1902, the system covered 176.96 miles (from Map of Memphis, Tennessee, Showing Its Sewer System).* LEFT: *The pumping station of the Memphis Artesian Water Works, located on Auction Street, 1904. Built in 1890, it is now on property next to St. Jude Children's Hospital.*

In 1880, authorities of the taxing district petitioned the state for permission to issue bonds for construction of the Shelby County Poor and Insane Asylum. Permission to issue $30,000 in bonds was granted in 1881. The asylum was located on a large tract of land on the Raleigh Road (later Jackson Avenue). The second quarter report for 1886 states, "Patients remaining from the first quarter, 196; sane admitted, 110; insane admitted, 200; number treated, 326." Dr. Wilford Gragg, for whom Gragg School was named, was supervising physician from 1891 to 1896. Dr. Newton Ford Raines (father of Dr. Samuel Raines) served two terms from 1897 to 1901.

At the 1899 annual meeting of the Tri-State Medical Association, Dr. Raines presented a paper on "Management of the Insane." He objected to the word "asylum," preferring "hospital for the insane," as it is "my ambition to make this a place of active treatment and not one of simple safe-keeping—a hospital in fact." Dr. William Krauss commended the excellent work done by Dr. Raines, affirming that he deserved the greatest credit for efficient administration and many improvements in operating the county asylum.

Memphis Hospital Medical College, 1880

Organization of the Memphis Hospital Medical College reestablished medical education in the Bluff City. The Memphis Medical College, which had closed at the outbreak of the Civil War, reopened in 1867 but met with postwar financial difficulties and closed permanently in 1873. Dr. William E. Rogers took the lead in organizing a new medical school but the planned opening in the fall of 1878 was another victim of the yellow fever epidemic. Dr. Rogers was himself seriously ill with fever.

The college opened in November 1880 in a new building on Union Avenue at Dunlap across from the City Hospital. Dr. Rogers was appointed dean of the college. The faculty was reorganized several times in the early years. After Dr. Rogers's death in 1885, the trustees named Dr. A. G. Sinclair dean. Following Sinclair, Dr. Francis L. Sim and other capable men led the school until its merger in 1913 with the University of Tennessee College of Medicine. The faculty included prominent physicians who established themselves as leaders of early specialty practice.

In 1899, the two-year course of study became a four-year program requiring a high school diploma for admission. Many of the city's medical leaders graduated from the Memphis Hospital Medical College during its 33 years of operation: Richmond McKinney (1884); E. A. Neely (1886), later associate editor of the *Memphis Medical Monthly*; William Krauss and Matthew DePass (1889); and Eugene J. Johnson and Battle Malone (1899). Malone's senior paper on surgery won first place. A significant number of graduates of Memphis Hospital Medical College obtained postgraduate training in European centers or in New York.

New Medical Societies

The Memphis Medical Society, reorganized as the Shelby County Medical Society in 1876, suffered from the 1878 and 1879 yellow fever epidemics. In a history of the Medical Society written for the *Memphis Medical Journal* in 1925, Dr. Edward C. Ellett states, "In March 1884 new

Announcement of commencement, Memphis Hospital Medical College, March 1, 1888. The portrait insert is of Dr. William E. Rogers, one of the school's founders and its first dean.

officers were elected, and the affairs of the Society took a turn for the better." Reviewing existing minutes and other print sources, Ellett wrote, "The interest of the Society was largely in medical and obstetrical cases. In the 1880s there was not much surgery being done and hospitals were few and poorly equipped. There was much discussion of the fevers—typhoid, typho-malarial, continued fevers, and the various forms of malaria." In June 1887, the name was changed to Memphis Medical Society; however, in 1902 to meet representation requirements adopted by the American Medical Association and the Tennessee Medical Society, the name was changed again to Memphis and Shelby County Medical Society.

The Tri-State Medical Association of Mississippi, Arkansas, and Tennessee was organized on September 23, 1880, and the first meeting held in Memphis in November of that year. A change in the constitution, which became effective in 1883, provided for a single annual meeting to be held in Memphis. Attendance at these meetings grew as excellent programs featured speakers from leading medical centers throughout the country. In 1932, reflecting the teaching aspect of the annual meetings, the organization name was changed to Mid-South Postgraduate Medical Assembly.

Efforts to obtain laws regulating the practice of medicine in Tennessee finally met with success in 1889 when the General Assembly enacted legislation requiring the licensing of all physicians and the regulation of medicine and surgery.

BEALE STREET: MEDICINE AND MUSIC

By DR. RICHARD RAICHELSON

Originally a wealthy residential neighborhood, Beale Street survived as a historic district because of its connection to the blues and Memphis music. Robertson Topp, entrepreneur, lawyer, politician, and businessman, owned 414 acres of land in South Memphis that included Beale Street. Topp's elegant mansion, constructed in 1841, was located east of the William Richardson Hunt residence, the present Hunt-Phelan house. The Hunt residence, built in 1842 of red brick in the Federalist style, was enlarged in 1855 with a two-story, Ionic-columned portico on the Beale front. It is the only remaining antebellum home in the area.

The Botanico-Medical College, built in 1846, was located east of Lauderdale across from the Hunt mansion. It closed in 1861 and did not reopen after the Civil War. White physicians resided in the area for several decades after the Civil War. Dr. Francis L. Sim, who also had an office on Beale Street for years, was a professor at the Memphis Hospital Medical College as were Drs. William B. Rogers, Alexander Erskine, and Richard B. Maury, whose homes were on upper Beale Street.

During the Civil War, large numbers of African-American federal troops were stationed at Fort Pickering, located on the bluff overlooking the Mississippi River, and following the war, many of them remained in the city. By 1885, Beale Street began to emerge as the business and entertainment center for blacks in the Mid-South, although other ethnic groups lived and worked in the community well into the twentieth century.

Memphis native John C. Walker, an 1882 graduate of Meharry Medical College in Nashville, was the first African-American physician to practice on Beale Street. Others during this early period were Alexander Burrell, Thomas A. Brown, Reverend Jacob P. Jay, and Albert Sidney J. Burkett, a prominent physician, who practiced on Beale after graduating from Meharry in 1884. In the twentieth century, Drs. Jacob C. Hairston; A. N. Kittrelle; Frances "Fannie" Kneeland; Arthur T. Martin; L. G. Patterson; and Cleveland A. Terrell, among others, had offices on Beale Street. For ten years, Terrell and Patterson operated an infirmary at 159 Beale prior to the opening of Jane Terrell Hospital in 1909.

Dr. A. N. Kittrelle, c.1908. Born in Georgia, he received training as a bricklayer from Tuskegee, ending up as a master mason before graduating with a degree in medicine in 1905.

For African Americans, medicine around Beale Street had several faces. Blues musicians, including Gus Cannon, Furry Lewis, and Jim Jackson, entertained at medicine shows and hawked patent remedies such as Jack Rabbit Liniment. Shows operated in Memphis and traveled through the rural South. In 1913, Gus Cannon, billed as "Banjo Joe," entertained as a banjoist, juggler, and comedian with the Dr. Stokey Medicine Show.

Graduate pharmacist George R. Jackson and his sister Flossie opened the first black-owned drugstore on Beale in 1893. Over Jackson's New Era Pharmacy was the office of Dr. L. R. Ross, a dentist who advertised as "My Tooth Doctor." Fahlen and Kleinschmidt Drug Store at the northwest corner of Beale and Second was in a building that was a prime location for medical offices in the 1880s and 1890s. Much later, the building housed Lansky Brothers Clothiers made famous by Elvis Presley. The Randolph Building constructed at Beale and Main in 1891 was a favorite location for physicians, but none were black.

In 1899, black millionaire Robert R. Church, Sr., built Church's Park and Auditorium on Beale at Fourth, one of the largest auditoriums in the South, where W.C. Handy directed the house orchestra. Beale Street was more than the blues. It was the center for a vibrant cultural exchange. The offices of African-American doctors, pharmacies, mercantile businesses such as A. Schwab and Goldsmith Brothers, the Beale Street Market, Church's Solvent Savings Bank, nightclubs, restaurants, and the Beale Street Baptist Church provided a unique mix like no other street in Memphis.

Above: *The U.S. Marine Hospital as it looked in 1898. The central portion of the original hospital, with its cupola, was moved 300 feet to a northwest exposure and remodeled as junior officers' quarters when the facility was rebuilt in 1937.* Right: *Dr. William Bodie Rogers, c.1900. He was professor of surgery at the Memphis Hospital Medical College.*

Ninety-two Memphis doctors registered with the Health Department as medical practitioners: 81 were regulars (allopaths), two eclectics, one a Thomsopath (botanic), four homeopaths, and four unspecified. Five of the 92 were African-American doctors, including three graduates of Meharry Medical College—Drs. T. C. Cottrell, A. S. J. Burchett, and Y. S. Moore. Two had not graduated from a medical school.

U. S. Marine Hospital, 1884

The need for additional hospital facilities in the last quarter of the nineteenth century was a cause of concern to the medical community. As early as the 1830s, city and state authorities had petitioned Congress to authorize construction of a marine hospital in Memphis. This long-sought facility became a reality in 1884 when the U.S. Marine Hospital Service appropriated $46,000 to build a hospital at Fort Pickering. For years, merchant seamen had been sent to City Hospital on Union. Dr. Henry Rose Carter, noted epidemiologist and one of the first researchers to identify the mosquito as the yellow fever vector, was in charge of the Marine Hospital at Memphis from 1885 to 1886.

Organized in 1798, the U.S. Marine Hospital Service was the first federally funded health care and disease prevention agency established in this country. Merchant seamen were initially charged 20 cents per month for this care. The network of hospitals was reorganized in 1879 along military lines under a uniformed service, directed by a supervisory surgeon, later given the title of surgeon general. To better define its expanded services and research programs, in 1912 the agency was renamed the United States Public Health Service.

New Medical Infirmaries

As the science of bacteriology was incorporated into medical practice, infirmaries, which were small hospitals, opened in the 1880s and 1890s, providing medical and surgical

The Gilded Age   51

RICHARD BROOKE MAURY: PHYSICIAN, TEACHER, LEADER

Richard Brooke Maury was born in Georgetown, D.C. on February 5, 1834, the son of Richard Brooke Maury, Sr. and Ellen Magruder Maury. Educated in Virginia schools, Maury received his medical degree from the University of Virginia in 1857. While working as an intern at Bellevue Hospital, Maury earned a second medical degree at the University of New York. In 1859, Maury moved to Port Gibson, Mississippi, where he established his first medical practice. Two years later, he married Jane Seeley Ellett, daughter of the Honorable Henry Thomas Ellett and Rebecca Seeley.

During the Civil War, Maury was in charge of hospitals in Mississippi and Alabama. After the war, he moved to Memphis, and in 1867 became associated in practice with Robert W. Mitchell. When the Memphis Medical College was revived in 1867, Maury was named professor of physiology and Mitchell, professor of *materia medica*. In spite of an excellent faculty, the school suffered from postwar financial reverses and closed in 1873. In the yellow fever epidemics of the 1870s, Maury and Mitchell provided critical medical leadership to the stricken city.

Maury's surgical skills were enhanced through work with leading doctors in Europe, including study with Lawson Tait, England's premier gynecological surgeon. Maury was recognized internationally when elected a fellow of the British Gynecological Society in 1886. Earlier named a fellow of the American Gynecological Society (1883), Maury served as president of the society in 1906. A skilled surgeon and gifted teacher, he was appointed professor of

ABOVE: *Dr. Richard B. Maury (1834-1919), c.1888.*
OPPOSITE PAGE: *A portion of those attending Dr. Maury's 80th birthday celebration in 1914. He is second from the left at the head table.*

facilities for private patients. Rogers Surgical Infirmary, opened in 1883 by Dr. William E. Rogers and his son Dr. William Bodie Rogers, was an early surgical infirmary. After the senior Dr. Rogers's death, Shepherd A. Rogers became a partner with his brother; later Drs. Bennett Graves Henning and Frank Williford joined the partnership.

In 1885, two of the city's leading physicians, Dr. Robert W. Mitchell and Dr. Richard B. Maury, opened the first infirmary for the diseases of women. Dr. Edward Dana Mitchell (he was not related to Robert W. Mitchell) and Dr. William Wood Taylor became associated in this practice, which established the city's first training program for nurses in 1887. Lena Angevin, who received the first diploma in June 1889, later organized the training school for nurses when the new Memphis City Hospital opened in 1898.

While Dr. Mitchell was directing construction of a larger infirmary in 1886, Dr. Maury spent several months in Europe observing the work of prominent gynecologists, primarily Lawson Tait, Britain's leading surgeon and pioneer in pelvic surgery. At the February 1888 meeting of the Memphis Medical Society, Maury presented a summary of his work over the prior six months, including surgery for gynecological surgery at the Memphis Hospital Medical College in 1888.

Of Maury's diagnostic ability, his colleague Alexander Erskine said, "Dr. Maury could more forcibly and clearly explain a puzzling and difficult pathological situation ...than any man I ever knew; presenting etiology, diagnosis, and prognosis, plainly and succinctly." A former medical student affirmed, "Dr. Maury never failed to command the instant attention and respect of his students. He prepared his lectures with great care. Like everything else he did, he taught seriously and well."

In addition to his accomplishments in medicine, Maury's abilities were employed in other interests. He served for a number of years on the Memphis School Board, work recognized in the naming of Maury School in his honor. In 1907, Maury organized the City Club and served as first president of this civic group, which worked to secure responsible, ethical government. A faithful churchman, Maury was honored for his 27 years of service on the vestry (lay board) of Calvary Episcopal Church, ten as senior warden.

The Egyptians, an elite cultural organization, was formed in 1913 with the object of providing a forum for the presentation of papers on topics of interest in science, religion, economics, and literature. The limited membership included Drs. Benjamin F. Turner and Richard B. Maury. Rabbi William Fineshriber selected the name Egyptians, acknowledging that Memphis was named for an ancient center of culture in Egypt. At the January 1915 meeting of the Egyptians, Maury's paper, "The Practice of Medicine from the Standpoint of the Physician, Patient, and Public," broadly sketched the history of medicine. Noting the many life-saving, scientific discoveries made in his lifetime, Maury concluded he was living in the "golden age of medicine."

On the occasion of Dr. Maury's eightieth birthday in 1914, Memphis doctors honored him with a dinner at the Peabody Hotel (the original Peabody on Main at Monroe), presenting him a silver loving cup in recognition of his long and distinguished career in medicine. Richard Brooke Maury, physician, teacher, leader, died on March 17, 1919, the first of three generations of Memphis physicians. The family name lives on in the Maury-Turley Building at the Church Health Center on Peabody Avenue.

peritonitis, ovarian cysts, tumors, adhesions, and a ruptured tubal pregnancy. He performed the first surgery in Memphis for an extra-uterine pregnancy. Maury credited his success to careful observation of the surgical work done by Lawson Tait and the assistance of his partners. Dr. Mitchell developed an improved inhaler for administering anesthesia, which was an important aid in surgery.

Following the death of Dr. Robert W. Mitchell in 1903, Dr. Maury retired, and ownership of the infirmary was transferred to his son, Dr. John M. Maury, and to Dr. Edward C. Ellett. (Dr. Edward Dana Mitchell died suddenly in 1896, and Dr. William W. Taylor joined another practice group.) The infirmary cases now included both ophthalmological and gynecological surgery.

Dr. Thomas J. Crofford, who had come to Memphis to join the faculty of the Memphis Hospital Medical College as professor of gynecology, opened an infirmary for the diseases of women in 1886. Reorganized in 1891 as the Memphis Sanitarium, his enlarged facility occupied a new brick building near First Methodist Church at Second and Poplar. Drs. William Bodie Rogers (surgery) and Bennett Graves Henning (proctology) joined Dr. Crofford. The

new sanitarium was equipped for aseptic surgery and organized into specialty floors. A highly regarded physician and teacher, Dr. Crofford was elected vice-president of the American Association of Obstetricians and Gynecologists in 1899.

In 1885, Dr. G. W. Overall, a former professor of physiology and diseases of the nervous system at the Memphis Hospital Medical College, and Dr. Robert L. Knox, opened the Overall-Knox Electric Infirmary for the treatment of diseases of the nervous system. Electrical treatment for nervous disorders was much in vogue at the end of the nineteenth century.

LEFT: *In a painting presented to the Tennessee Nursing Association in 1938, Lena Angevin Warner wears her Spanish-American war medals and uniform. In 1889, she became the first nurse to receive a diploma in Tennessee and later founded the Tennessee Nursing Association.* BELOW: *Graduates of the Memphis City Hospital Training School for Nurses, 1903. The school was organized by Lena Angevin Warner when the hospital was opened in 1898.*

St. Joseph Hospital, 1889

St. Joseph Hospital, the first private general hospital in Memphis, was an undertaking of Father Francis Moening, O.F.M., and Dr. E. Miles Willett, Sr. A generous gift from Mrs. Catherine Hamilton, with additional support from leaders of the city's religious communities, provided funds for a hospital to serve the northern part of the city. Opened in early 1889 in a house on Jackson Avenue, the hospital was staffed by the Poor Sisters of St. Francis, Lafayette, Indiana. In November, the hospital moved into a new two-story frame building with accommodations for 65 patients. The

Below: *St. Joseph Hospital opened a modern 200-bed facility in 1895. This photo was made after construction of the third addition in 1926.* Right: *Dr. Heber Jones (1848-1916), a highly respected physician, was on the staff of both the St. Joseph and Lucy Brinkley hospitals, and the College of Physicians and Surgeons. He was also president of the Memphis and Shelby County Medical Society from 1888-1889 and secretary of the Medical Society from 1876-1879.*

The Gilded Age 55

> **THE LUCY BRINKLEY HOSPITAL for WOMEN**
>
> INCORPORATED 1891   REORGANIZED 1906
>
> STAFF
>
> R. B. MAURY, M.D.
> *Consulting Surgeon and Gynecologist*
>
> HEBER JONES, M.D.
> *Consulting Physician*
>
> SURGEONS AND GYNECOLOGISTS
>
> W. W. TAYLOR, M.D.   M. GOLTMAN, M.D.
> J. M. MAURY, M.D.   E. M. HOLDER, M.D.
>
> OBSTETRICIANS
>
> J. L. ANDREWS, M.D.   MOORE MOORE, M.D.
>
> PATHOLOGIST
>
> WM. KRAUSS, M.D.
>
> This is a surgical, gynecological and obstetrical Hospital; situated on high grounds in the resident district in Memphis and overlooking beautiful Forrest Park.
>
> The building is new, large and commodious; it is elegantly furnished, fire-proof in construction, and has been designed especially for the purposes for which it is to be used. It is heated by the hot-water system, with provision for the admission of fresh air to each room separately. It is lighted by gas and electricity.
>
> The rooms for patients are large, all outside, and each one is provided with hard-wood floor, electric lights and bells, and every other arrangement which seems adapted to secure the comfort, seclusion and safety of the patients. Besides the private rooms, there are wards intended for patients whose circumstances are such that they can not afford a private room.
>
> Well-trained nurses will give the patients the best care and attention.
>
> The Hospital is reached by the Madison Avenue cars, getting off at Dunlap Street and going south two blocks.
>
> For further particulars, address
>
> THE LUCY BRINKLEY HOSPITAL
> 855 Union Ave., Memphis, Tenn.

ABOVE: *An ad for the Lucy Brinkley Hospital from the August 1907 issue of the* Memphis Medical Monthly. *The 3-story brick building had just opened in its new quarters on Union Avenue.* RIGHT: *Dr. Thomas J. Crofford, c.1900, was on the faculty of the Memphis Hospital Medical College and operated a sanitarium for women. He wrote on gynecology for the* Memphis Medical Monthly.

attending medical staff included Heber Jones, C. T. Peckham, A. G. Sinclair, and E. Miles Willett, Jr., "to whom is due the credit of planning and supervising the construction of the building." (Dr. Willett, Sr. died on Feb. 6, 1888, a year before the first hospital opened.) Annual reports from the hospital's first decade confirm that half the patient admissions were charity cases. Care of the poor would remain an important part of the hospital's mission. In 1895, a new St. Joseph Hospital opened in a modern 200-bed facility. Large additions to this structure were made in 1901, 1910, and 1926.

The Women's Hospital Association, a philanthropic organization chartered by the state, was formed to provide hospital care for poor women and children. In 1893, with the assistance of Dr. Richard B. Maury, consulting surgeon, and Drs. William W. Taylor and Eugene E. Haynes, attending physicians, the first hospital for women opened in a building at 106 Washington east of Third Street. A generous gift from businessman Hu L. Brinkley, made in memory of his wife, provided funding for the Lucy Brinkley Hospital. In 1907, the hospital moved into a new building at 855 Union Avenue.

MEDICAL JOURNALS

An important adjunct to the exchange of medical information came from local journals published in the last two decades of the nineteenth century. The first issue of the *Mississippi Valley Medical Monthly* appeared in January 1881. Edited by Dr. Julius Wise, Memphis Hospital Medical College, the journal's stated purpose was the study of endemic and epidemic diseases of the Mississippi Valley, especially yellow fever and malaria, as well as sanitation, quarantine, and medical legislation.

A member of the Howard Medical Corps, Wise served in the yellow fever epidemics of 1873, 1878, and 1879. When Dr. Wise married and moved to St. Louis in 1882, Dr. Francis L. Sim assumed editorship of the *Mississippi Valley Medical Monthly*, published under that title through 1888, when the name was changed to *Memphis Medical Monthly (MMM)*, which was published from 1889 through 1921. No issues were published in 1922 and 1923. Renamed the *Memphis Medical Journal*, publication resumed in 1924 under the sponsorship of the Memphis and Shelby County Medical Society.

The *Memphis Journal of the Medical Sciences* was published from March 1889 until 1893. Its editorial staff included Drs. Alexander Erskine; Thomas J. Crofford;

Shepard A. Rogers; Bennett G. Henning; William Bodie Rogers; and James L. Minor. Crofford wrote on gynecology, Rogers on general surgery, Henning and Shepard Rogers on medicine and therapeutics, and Minor on diseases of the eye and ear. When Dr. William Krauss joined the staff in 1890, he wrote on bacteriology and also translated articles from German medical journals.

Of shorter duration was the *Memphis Lancet*, published from July 1898 through April 1900. The editorial board included Drs. Richard B. Maury; M. B. Herman; William W. Taylor; William Krauss; John M. Maury; Max Goltman; Edward C. Ellett; and Stephen. E. Rice, leaders in the development of early medical specialties. The *Memphis Lancet* published articles on research in yellow fever and malaria, papers by Drs. Goltman and Krauss on microscopic studies and the diagnostic use of x-ray, improvements in anesthesia, advances in surgery, progress in drug reform, and the state of public health.

Dr. Francis L. Sim was editor of the Memphis Medical Monthly *and on the staff of the Memphis Hospital Medical College. He was dean of the faculty from 1889 until his death in 1894.*

Ophthalmology—A New Specialty

In addition to obstetrics and gynecology, ophthalmology was one of the earliest specialties in Memphis. The first ophthalmologist of record, Dr. Alfred H. Voorhies, was a military surgeon who came to Memphis at the end of the Civil War. When the Memphis Medical College reopened in 1867, Voorhies was named professor of aural and ophthalmic surgery. After the school closed in 1873, Voorhies continued in private practice until 1880 when he moved to San Francisco.

Dr. A.G. Sinclair, professor of ophthalmology and otology at the Memphis Hospital Medical College, was the second recognized ophthalmologist in Memphis. His training included graduate work in New York and Detroit hospitals. From the time he arrived in Memphis in 1880, Sinclair performed successful eye surgery. He removed the left eye of his colleague Dr. Francis L. Sim, who suffered from a detached retina and choroiditis. Sinclair was the first Memphis ophthalmologist to use cocaine as a local anesthetic in cataract surgery.

Dr. James. L. Minor, a graduate of the University of Virginia Medical College, was the third ophthalmologist in Memphis. He did extensive clinical work in New York hospitals, including six years at the New York Eye and Ear Infirmary. In 1884, The *New York Medical Recorder* wrote that Dr. Minor was the first surgeon in America to use cocaine as a local anesthetic in the extraction of cataracts. He left New York and established his practice in Memphis in 1887. In 1899, as a public service, Dr. Minor examined the eyes of 682 schoolchildren attending the Market Street and Linden Street schools. He emphasized the importance of good eyesight in the learning process and the need for proper light and ventilation in the classroom. Committed to the prevention of blindness, Minor wrote that ophthalmia neonatorum, responsible for ten percent of all cases of blindness, could be prevented by sterile techniques at delivery.

Pediatrics, taught as a separate course at the Memphis Hospital Medical College, was in its beginning stage in Memphis. The *Memphis Medical Monthly* took note of the untiring efforts of Dr. Abraham Jacobi of New York to establish pediatrics as a separate field of medicine.

Women Doctors in the 1880s

Little information is available on the few women in medical practice in Memphis between 1880 and 1899. Dr. Rachel Gowling was in practice by 1891, and in 1895 she and Dr. Louise Drouillard had offices in the Randolph Building at Main and Beale, where other physicians were located in the late nineteenth century. By 1897, Dr. Mary O'Driscoll had joined Drouillard and Gowling. Two years later, the *Memphis Lancet* reported the death of Dr. Mary O'Driscoll, who died in St. Louis on April 8, 1899. Drs. Drouillard and Gowling continued their practice in Memphis into the 1920s.

Dr. Georgia Lee Patton, the first African-American woman to practice medicine in Memphis, was an 1893 graduate of Meharry Medical College, an educational agency of the Methodist Church. Following graduation, Dr. Patton paid her own passage to Liberia, where she worked for two years as a missionary doctor in Monrovia. Suffering from tropical illnesses, she returned to the United States in 1895 and located in Memphis, opening an office at 282 North Second. On December 19, 1897, Georgia Patton married David Washington, a postal employee who later accumulated extensive real estate holdings. Dr. Georgia Lee Patton Washington died on November 8, 1900, the cause of death given as tuberculosis, a widely prevalent disease in Memphis at that time. She and her two baby sons are buried in historic Zion Cemetery on South Parkway East.

The Great Bridge and the New City Hospital

The construction of the Great Bridge at Memphis, the first bridge across the Mississippi River below St. Louis, provided significant economic benefits to Memphis. Construction of the bridge, the third longest in the world, was a major engineering feat. A great celebration marked the opening of the bridge on May 12, 1892, as some 50,000 Memphians lined the bluff to watch 18 steam locomotives cross the bridge. Prior to construction of the bridge, east-west railroad traffic was moved across the Mississippi River by railroad barges, a difficult and time-consuming procedure. Later named the Frisco Bridge, the railroad bridge extended the Memphis trade area and brought tons of cotton from Arkansas and points west to the Memphis market. By World War I, Memphis was the largest rail freight center in the South. For residents of Arkansas, the bridge also facilitated access to medical care in Memphis.

ABOVE: *Dr. Georgia Esther Lee Patton was the first African-American woman to receive a license to practice medicine and surgery in Tennessee and the first to practice in Memphis.* BELOW: *The City of Memphis Hospital as it looked in the 1920s.* OPPOSITE PAGE, ABOVE: *Dr. Alexander Erskine was mainly responsible for the design of the new City of Memphis Hospital.*

58 MEMPHIS MEDICINE: A HISTORY *of* SCIENCE AND SERVICE

As early as the immediate post-Civil War period, Dr. Gustavus B. Thornton, City Hospital superintendent, joined by local doctors, urged Memphis authorities to build a new city hospital. The hospital on Union Avenue built in 1841 and located on the site of the present Forrest Park was outdated, without aseptic surgical facilities, and lacked a professional nursing staff. It was not until 1895 that the legislature authorized the city to collect a special *ad valorem* tax for construction of a new hospital. The city council retained Dr. John Shaw Billings of New York, designer of the Johns Hopkins Medical School and Hospital, as consulting architect. Organization of the hospital was largely the work of Dr. Alexander Erskine. Work began in July 1897 at a site on the north side of Madison just east of Dunlap and was completed a year later. On July 2, 1898, under supervision of City Hospital physician Pope M. Farrington, 75 patients were transferred from the old hospital to the new facility. The Mitchell Maury Training School for Nurses, directed by Lena Angevine Warner, was moved to the new City Hospital.

During the Gilded Age, cotton remained the economic lifeblood of Memphis, followed closely by the hardwood lumber and cottonseed oil industries. With the expansion of railroads, the city continued its dominant position as the transportation center of a large regional area. Honest and efficient government provided by Taxing District officials brought fiscal reforms, which enabled Memphis to regain its charter in 1891 and its power to tax in 1893. Major reforms in sanitation, including the Waring sanitary sewer system and a pure artesian water supply, provided long-term health benefits. By the end of the nineteenth century, the Memphis Hospital Medical College, three hospitals, and several infirmaries testified to the city's growth as a medical center. The new century would bring long-needed food and drug reforms, commission government, strict measures to control tuberculosis and smallpox, mandated changes in medical education, and new medical specialties.

THE GILDED AGE   59

CHAPTER SIX

THE PROGRESSIVE ERA

1900–1915

When Memphis entered a new century, medicine had been established as a scientific discipline, the Reed Commission was in Cuba unraveling the connection between the mosquito and yellow fever, and social changes, notably woman suffrage, were underway. Significant reforms initiated during President Theodore Roosevelt's administration included pure food and drug legislation, conservation of public lands, and creation of national parks.

The progressive movement, which championed greater efficiency and fiscal responsibility in city government, was the impetus behind the city's adoption of commission government on January 22, 1909. Edward Hull Crump, a young, red-haired Mississippian who had moved to Memphis in 1894, was elected mayor and headed the new city government. In 1909, Crump appointed Dr. Max Goltman as superintendent of the reorganized Department of Health (previously the Board of Health), which was given increased powers and responsibilities under the new commission government. Dr. Goltman undertook a major campaign to reduce deaths from tuberculosis and typhoid fever and continued aggressive efforts to obtain legislation requiring pasteurization of milk. Contaminated milk was for years the source of much gastrointestinal illness in addition to being the chief cause of tuberculosis transmission from infected dairy cattle.

With pride, the Chamber of Commerce announced that the city's population in 1900 was 102,320, placing Memphis in the upper ranks of Southern cities. With eastward growth, the city limits were expanded to the Parkways in 1908. Other improvements included streets, sidewalks, and the street railway system. The city purchased properties that created one of the best public park systems in the country: Overton Park in the heart of the city and Riverside Park overlooking the Mississippi River. And for all the children, and not a few adults, the opening of the zoo in 1907 was a great delight. As modern buildings appeared on the downtown skyline, doctors moved their offices to newer locations. The elegant Exchange Building, constructed in 1910 at Second and Madison, was for many years a prominent location for medical offices.

Opposite page: *Surgeon Dr. Max Goltman with his three sons, Jack S., Alfred M., and David W. Goltman, 1929. Each of his sons became physicians. Dr. Goltman was superintendent of the Memphis Health Department from 1910 to 1914. He later became professor of surgery in the Medical College, University of Tennessee, and chief surgeon at the Memphis Baptist Memorial and General Hospitals.* Below: *To accommodate record enrollments, the Memphis Hospital Medical College built Rogers Hall, which opened for the 1902-03 session.*

Congressional approval of the Pure Food and Drug Act on June 30, 1906, provided important protection of public health. Although activists had worked for years to obtain this legislation, Upton Sinclair's book *The Jungle*, which exposed the filth and terrible working conditions in the meat industry, was the catalyst that brought a great public outcry for reform. The law provided for the inspection of meat and prohibited the manufacture, sale, or transportation of adulterated food and patent medicines. In spite of the new legislation, the patent medicine business did not suffer a fatal blow. Dr. William Krauss worked for enactment of legislation in Tennessee that would protect the public from reckless advertising of patent remedies that contained dangerous or useless ingredients.

New Medical Schools

Medical education in Memphis made substantial gains during the Progressive Era. The Memphis Hospital Medical College, having experienced record enrollment for the ses-

The Progressive Era

ABE PLOUGH— ENTREPRENEUR AND PHILANTHROPIST

In 1908 an enterprising young man named Abe Plough opened a small drug business in one room of a building located at 93 North Second Street. His first product was Plough's Antiseptic Healing Oil. The business expanded, and in 1915 he purchased the Battier Pharmacy at Beale and Hernando. Later drug store acquisitions included the Pantaze chain and the Peabody Hotel drug store.

Incorporated in 1918, the company name was changed from Plough Chemical to Plough, Inc. The purchase of the Gerstle Medicine Company of Chattanooga in 1922 included rights to the trade name St. Joseph. During the influenza pandemic of 1918 to 1919, the widespread use of aspirin established acetylsalicylic acid (ASA) as an effective medication to reduce fever and aching. When Bayer's U.S. patent on the copyrighted name aspirin expired in 1920, Plough decided to make aspirin under the St. Joseph label. The name and special cellophane packaging made St. Joseph aspirin one of the company's most successful products, the first to be nationally advertised. In 1947 the company developed St. Joseph Aspirin for Children, which quickly won acceptance from the medical profession and the public.

In the 1930s, Plough added a popular line of inexpensive cosmetics and the Penetro products—nose drops, cough syrup, and cold tablets. New lines of skin care products, including internationally known Coppertone and Solarcaine, and the Dr. Scholl's foot-care products were leading sellers.

Plough Laboratories, a wholly owned subsidiary of the parent company, was formed in 1961 to develop new drugs. To meet space requirements for greater manufacturing capacity in Memphis, Plough, Inc., acquired large proper-

ties on Jackson Avenue in 1969 and began construction of the administrative and research center, which grew to more than one million square feet under a single roof, making it the largest manufacturing-administrative facility in Memphis at that time.

In 1967, Plough, Inc., purchased the Maybelline cosmetics firm. In addition to the manufacturing of drugs, cosmetics, and health care items, Plough, Inc., acquired companies that made a wide range of household utility products. Initiatives by the Plough and Schering Corporations to merge the two companies were completed on January 16, 1971, forming the Schering Plough Corporation, combining Schering's international pharmaceuticals with Plough's proprietary drugs, toiletries, cosmetics, and household products. In 2009, Merck purchased Schering Plough for $41 billion dollars, as giant pharmaceutical companies sought mergers that would result in cost savings through combined research and reduction in the workforce.

In 1973, Plough gave a one million dollar gift to the Memphis Community Foundation in memory of his parents, Moses and Julia Plough. The organization was renamed Plough Community Foundation and the gift designated to fund programs for health, youth, education, and civic improvement. The Foundation provided countless scholarships, as well as endowed programs for pharmacy schools. Mr. Plough often said his philosophy of philanthropy was based on "helping people help themselves," and many of his gifts were challenge grants to enlist the active participation of recipients.

Although he retired in 1976, Abe Plough continued to work every day at his office, where he managed the Plough

Foundation. When he died on September 14, 1984, at the age of 92, Plough had turned a one-room drug manufacturing operation into a multibillion dollar international business. His philanthropies earned him accolades as the city's most generous man and he was honored as the first citizen of Memphis. Today, the Plough Foundation continues the philanthropic work begun by its generous founder.

OPPOSITE PAGE: *Plough's Antiseptic which could be used externally or internally as the situation demanded. Note that Plough's first patent medicine contained 70 percent alcohol. It was also recommended for children.*
ABOVE: *Abe Plough (seated) looks at the progress thermometer for United Way with Bob Healey, 1976. Plough made contributions to the fund each year.*

side of Madison Avenue opposite Memphis City Hospital. The college catalog noted a "free hospital dispensary located in the basement of its building ...where everyday advice and medicines are given to a large number of patients."

Faculty for the fourth academic session of 1909 to 1910 included: Heber Jones, dean and professor of clinical medicine; Edward C. Ellett, ophthalmology; Richmond McKinney, otolaryngology; Arthur G. Jacobs, diseases of children; John M. Maury, principles of gynecology; J. L. McLean, clinical gynecology, Percy W. Toombs, physiology; J. S. Toombs, clinical medicine, Louis Leroy, principles and practice of medicine; Willis C. Campbell, orthopedic surgery; Oswald S. McCown, obstetrics; James A. Crisler, anatomy and surgery; Marcus Haase, dermatology; W. H. Pistole, pharmacology and therapeutics, George G. Buford, diseases of the nervous system; Eugene M. Holder, principles and practice of surgery; Maximillian Goltman, clinical surgery; George R. Livermore, genitourinary diseases; Edwin D. Watkins, chemistry; Edward Clay Mitchell, bacteriology and histology; William T. Pride, pathology; and Clarence L. Sivley, LL.B., medical jurisprudence.

Dr. William Krauss—Pathology and Radiology

Dr. William Krauss, who pioneered the development of clinical pathology, taught at various times in all the Memphis medical schools. He is credited with doing more to advance

ABOVE: *An ad for the second session of the College of Physicians and Surgeons of Memphis,* Memphis Medical Monthly, *August 1907. When the college merged with the University of Tennessee, the building was renamed "Lindsley Hall." Note that Dr. Heber Jones was the dean and professor of clinical medicine.* RIGHT: *Dr. William Krauss flanked by the microscope he introduced to the field of pathology.*

sions from 1899 to 1900 and from 1900 to 1901, undertook a major enlargement of its facilities. Rogers Hall, named in honor of founder Dr. William E. Rogers, was ready for the fall session of 1902. The building was later incorporated in the University of Tennessee College of Medicine campus as the Dental College.

In 1905, a group of the city's leading physicians organized the College of Physicians and Surgeons, which offered a graded four-year curriculum and modern laboratory facilities. Faculty included doctors who were in the forefront of developing medical specialties. Opened for the 1906-1907 term, the college occupied a handsome new building on the south

ABOVE: *Dr. Louis Leroy received his medical degree from the University of Pennsylvania. After moving to Nashville in 1899, he became bacteriologist for the state of Tennessee. Leroy was the first physician in Nashville to use x-rays. He moved to Memphis in 1906.* RIGHT: *An ad for the University of West Tennessee, 1908. Most noted African-American physicians in Memphis served at one time on the faculty.*

scientific medicine in Memphis than any other practitioner. His introduction of the oil immersion microscope in laboratory work was of critical importance to the emerging field of pathology. Dr. Louis Leroy, who came to Memphis from Vanderbilt in 1906 to teach at the College of Physicians and Surgeons, was also a trained pathologist and bacteriologist, having taught those subjects at medical schools in Chicago and Nashville. The reorganized University of Tennessee College of Medicine introduced its first formal course in clinical pathology in the opening session of 1911 to 1912.

Three Memphis physicians were among the first in this country to understand and utilize the new x-ray technology discovered November 8, 1895, by German physicist Dr. Wilhelm Conrad Roentgen, who held the chair of physics at the University of Wurzburg. The early work of Drs. Max Goltman and William Krauss has been noted in the previous chapter. Dr. Walter Sibley Lawrence, who came to Memphis in 1905, was one of the first to use x-rays. In "Present State of Radiotherapy," (1906) Lawrence wrote, "A few months after Roentgen's paper, I assisted Dr. Eugene R. Corson of Savannah, Georgia, in doing some of the first experimental work in this country." Dr. Corson was a prominent Savannah physician recognized for his early work in radiography. Thus, Goltman, Krauss, and Lawrence stand as Memphis pioneers in the early use of x-rays. While Goltman and Krauss later gave up their work with x-rays, Lawrence continued his leadership in this field, heading the University of Tennessee Department of Radiology for some 25 years. Other Memphis physicians who worked with x-rays shortly after the turn of the century were Drs. George G. Buford; John L. Jelks; Jesse Cullings; Louis Leroy; Lyman Chapman; and T. R. Montgomery.

THE UNIVERSITY OF WEST TENNESSEE—1907

The University of West Tennessee, a small medical school organized in Jackson, Tennessee, by Dr. Miles V. Lynk (1871-1956), moved to Memphis in 1907 and located at 1190 South Phillips Place. Its importance in providing an educational facility for training African-American doctors, dentists, nurses, pharmacists, and lawyers far exceeded its size. From the outset, the University of West Tennessee had a four-year medical curriculum and a three-year program

for nurses. During its 16 years of existence, its graduates established practices in Memphis and other large cities.

Dr. Lynk, an 1891 graduate of Meharry Medical College in Nashville, became a protégé of Dr. Robert F. Boyd. In 1895, Lynk and Boyd, with ten other African-American physicians, organized the National Medical Society in Atlanta. Lynk was also a founding member of the Bluff City Medical Society. He edited the first journal for African-American doctors, and during his long career wrote on medicine, black history, and literature, culminating with his autobiography, *Sixty Years in Medicine*.

The University of West Tennessee faculty were all African Americans, and in 1907 to 1908 included Drs. Robert G. Martin; Jacob C. Hairston; Cleveland A. Terrell; A. M. Kittrell; L. G. Patterson; Lucius Samuel Henderson; A. D. Byas; A. L. Thompson; E. E. Nesbitt; E. C. Craigen; J. L. DeLoney; Felix R. Newman; John W. Winchester; and Frances (Fannie) M. Kneeland, the second African-American woman to practice in Memphis. Dr. Kneeland, an 1898 graduate of Meharry Medical College, whose private practice was in obstetrics and gynecology, conducted the training program for nurses. Kneeland's first office was located on Beale Street; later she moved to 825 Walker Avenue near LeMoyne College. Drs. J. C. Clark, H. H. Kennedy, and John H. Steward taught dentistry. George R. Jackson, the first African-American graduate of the University of Michigan School of Pharmacy, was professor of pharmacy. Beebe Steven Lynk, also a graduate in pharmacy, taught chemistry and medical Latin.

OPPOSITE PAGE, ABOVE: *Dr. Miles V. Lynk, center, with faculty from the University of West Tennessee, 1922. His staff included physicians B. F. McCleave, C. A. Terrell, J. C. Hairston, R. L. Flagg, S. W. Prioleaux, T. E. Cox, B. D. Harrell, N. M. Watson, and F. A. Moore.* OPPOSITE PAGE, BELOW: *Dr. Frances Kneeland (Fannie) graduated from Meharry Medical College in 1894 and was on the staff of the University of West Tennessee, Memphis.* BELOW: *An ad for the University of Tennessee Medical School, 1923, points to the consolidation and merger of those schools from which it was born.*

THE UNIVERSITY OF TENNESSEE MEDICAL COLLEGE COMES TO MEMPHIS—1911

Great reforms came from recommendations embodied in the famous Flexner Report, *Medical Education in the United States and Canada* (1910). Trustees of the Carnegie Foundation for the Advancement of Teaching had commissioned Abraham Flexner to investigate medical schools in the United States and make a formal report of his findings. His scathing criticism of the large number of schools that lacked adequate funding, laboratory equipment, or an affiliated teaching hospital brought sweeping changes to medical education. Critical of all but one of the ten medical schools in Tennessee (six white, four black), Flexner recommended closing all schools except Vanderbilt Medical School in Nashville.

Following publication of Flexner's report, presidents and chancellors of the medical schools in Nashville and Knoxville reached agreements to combine assets into a single institution, the University of Tennessee College of Medicine.

UNIVERSITY OF TENNESSEE

COLLEGE OF MEDICINE, SCHOOL OF PHARMACY AND COLLEGE OF DENTISTRY, Memphis, Tenn.

Baptist Memorial Hospital, capacity 150 beds, 40 beds under control of this College.

150 feet south is site of new Methodist Hospital soon to be built.

Across the street is the Memphis City Hospital. Capacity 250 beds, under Clinical control of this college.

Alongside is the Municipal Hospital for Contagious Diseases to be erected. All autopsies in city hospital in presence of and with the assistance of students of Pathology—40 to 60 per year.

Lindsley Hall, the main building; four stories, 34 halls and rooms. Office of Registrar-Bursar, General Library and Museum here. Fourth and third year subjects, Organic and Physiological Chemistry and half of Free Dispensary instruction are offered in this building.

The entire fourth floor is fitted up as a large laboratory, with a research laboratory and a professor's office adjoining, hereafter to be used in instruction in Pharmacy. In animal house in yard are kept animals for experimental purposes.

Eve Hall, new four-story Laboratory building completed in 1912. Five halls and 12 rooms. Office of Dean, the all-time Professor of Pathology, laboratories Bacteriology, Pathology and Physiology, two departmental libraries and three research laboratories are on second, third and fourth floors. First floor set apart for part of Free Dispensary instruction. Clinical Microscopy taught in this building.

Rogers Hall, across Forrest Park from Lindsley Hall, four stories, 37 halls and rooms, including beautiful Auditorium and gallery seating 1000 persons. Here are the laboratories for the departments of Anatomy; Chemistry; Histology and Embryology; and Practical Pharmacy. The College of Dentistry also has ample space in this large building.

Most of the first and second year medical subjects are taught in Rogers Hall, while third and fourth year medical subjects are offered at the other two buildings on account of their proximity to hospitals now maintaining more than three hundred free beds.

Four medical colleges, united by consolidation and mergement, form one college embodying all essentials of a thoroughly equipped medical school. **Ten all-time teachers. Ten separate well-equipped laboratories for fundamental instruction** besides several research and private laboratories for full-time professors.

Three practically new college buildings, erected in 1901, 1906 and 1912, respectively, with nearly 100 halls and rooms. More than 100 in combined faculties of the three Memphis departments. 170 in faculties of ten departments of the University, Knoxville and Memphis together.

For copies of the University of Tennessee Bulletin, address the **Registrar-Bursar** or the Dean of that department about which information is desired.

BIG ORANGE—THE UNIVERSITY OF TENNESSEE DOCTORS

The consolidation of Memphis medical schools into the University of Tennessee College of Medicine, which moved to Memphis in 1911, formed the foundation of an outstanding teaching and research institution. The school's early years were not without financial difficulties and staffing problems, although one program was uniquely successful. Few, if any, medical schools have ever had a football team, especially one that racked up powerhouse victories. The University of Tennessee Doctors' 1922-1924 seasons, including an undefeated year in 1922, gave them "bragging rights" to some big wins such as the November 17, 1922, victory reported by the *Commercial Appeal*, "Ole Miss. Proves Easy for Doctors, Who Win 32-0." National recognition garnered by the team's success was a recruiting tool that brought an increase in enrollment providing important financial benefits at a critical time for the medical school.

The University of Tennessee College of Medicine football team was reorganized in 1921, and Bill Brennan, a former major league umpire and coach at Tulane, was hired as head coach, with George Tandy as line coach. Financial backing was provided by Memphis surgeons Louis Levy and Jim Bodley; Charlie Campbell, Memphis dentist; Robert S. Vinsant, dean of the Dental School; and Robert L. Crow, dean of the Pharmacy School. The financial support of these backers provided both great football and scholarships for team members. Dr. Levy continued to fund scholarships for medical students long after the team disbanded, and when he died, sportswriter

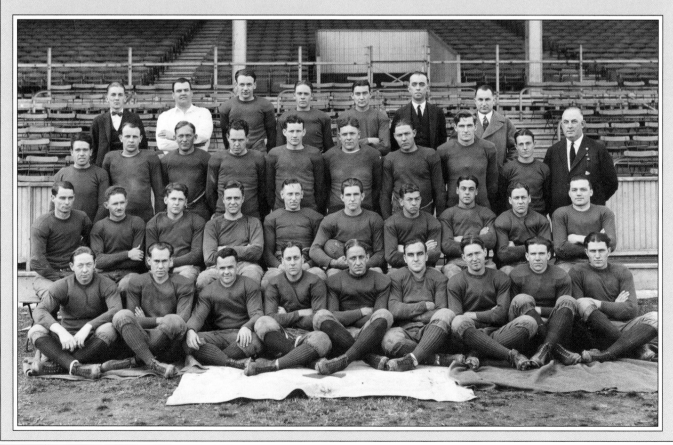

Walter Stewart said, "Dr. Levy was the godfather of one of the greatest football teams time has sired, a cleated horde [that] was a fearsome engine of destruction."

Phil E. White, All-American fullback from the University of Oklahoma, played several seasons for the New York Giants after graduating from medical school. Sam H. Sanders, a Southwest Conference star halfback from Texas A&M and Doctors team member, rose through the academic ranks at the University of Tennessee, where he was appointed professor in the Department of Ear, Nose and Throat Surgery, followed by chairmanship of the Department of Otolaryngology. Dr. Sanders was the last surviving member of the Doctors team. Sam L Raines, first string Washington & Lee player and fullback on the Doctors team, chaired the Department of Urology at Tennessee.

The University of Tennessee Doctors team had two pairs of brothers: the Gotten twins, Nicholas and Henry, and Julian "Big" Sullivan and D. A. "Little" Sullivan. Nicholas Gotten, a pioneer in the field of neurosurgery, was professor of neurology at the University of Tennessee and head of the Neurosurgery Department at Methodist Hospital. His twin brother Henry was a leading Memphis internist. Julian Sullivan, M.D. and his brother D. A. Sullivan, D.D.S., took their medical skills home to Cleveland, Tennessee. Virgil Payne, who played for Tulane and then for the Doctors, practiced nose and throat surgery in Pine Bluff, Arkansas.

Other members of the Doctors team included physicians Maurice "Bully" Doak; R. R. Swindell; Jeff Hanna; Shed Davis; and J. B. Futrell. Also on the team were dentists Hobart "Hobie" Ford; C. R. "Monk" McLaughlin; Jake Plesofsky; N. T. "Ned" Underwood; Ortie King; A. B. Moorehead; George Graham; R. C. "Jitney" Ford; Malcolm Prewitt; and William B. Cockroft, who later founded United Inns. Keeping the Doctors organized were team managers Clyde Croswell, Memphis pediatrician; Troy Bagwell, Knoxville orthopedic surgeon; A. B. Moorehead, Memphis dentist; and C. L. Green, Clinton, Tennessee, dentist. Although the team played only a few years, it blazed a bright trail and left an outstanding record of accomplishment.

OPPOSITE PAGE, BELOW: *The U.T. Docs football team, 1924. Dr. Louis Levy is in the last row, far right.*
BELOW: *U.T. Docs reunion held February 11, 1960, at the Hotel Peabody. Standing, left to right: J. B. Futrell, Louis Leggett, Jake Plesofsky, Clyde Croswell, R. T. (Tarzan) Holt, Sam Sanders, Richard Ching, W. L. Wilhelm, Sam Raines, A. B. Morehead, E. K. Munn, N. T. Underwood, A. M. McArthur. Seated, left to right: Otis Beck, Robert C. Ford, Joe Palermo, Virgil Payne, Roy Gullett, D. A. Sullivan (front), C. R. McLaughlin (back of Sullivan), R. G. Laird, Frank Weinel, Charles Campbell, Malcolm Prewitt, S. J. Sullivan, John Metz.*

With financial support from local leaders, the school moved to Memphis in 1911 and occupied the College of Physicians and Surgeons building, renamed Lindsley Hall. By contract, the University of Tennessee agreed to "take over the property, goodwill, and equipment of the College of Physicians and Surgeons, retain its faculty, and establish the same as the Medical Department of the University of Tennessee." The following year, Eve Hall, a four-story laboratory building, was constructed behind Lindsley Hall.

In 1913, at the end of its thirty-third academic year, the Memphis Hospital Medical College merged with the University of Tennessee, an action approved by the American Medical Association Council on Medical Education. The final merger completing the University of Tennessee College of Medicine took place in 1914 when the medical department of Lincoln Memorial University, located in Knoxville, transferred its students to Memphis. The merger of five institutions into the University of Tennessee College of Medicine formed the foundation of the medical center with the City Hospital as the primary clinical teaching facility.

WOMEN DOCTORS—KANE AND YORK

Elizabeth C. Kane, a native of Memphis, was one of the few women who established a medical practice in Memphis prior to World War I. Denied admission to the Memphis Hospital Medical College, she attended the Women's Medical College of Baltimore for two years then transferred to the University of Nashville Medical Department, graduating in April 1898. In 1900, Kane opened an office in the Memphis Woman's Building at 295 Second Street and in 1902 was elected to the City Hospital staff as a specialist in gynecology and obstetrics. A member of local and state medical societies, Kane was a fellow of the American Medical Association. During her career, she actively supported health education for women. Dr. Kane died in 1932 at the age of 56, lauded for her competency in practice, splendid character, and generosity in putting humanity above private gain. Her estate was left in trust to provide care for poor women and children.

A major step forward in medical education was the admission of women students to the University of Tennessee College of Medicine. In 1913, Sara Conyers York was the first woman to graduate from the University of Tennessee and did so with honors at the head of her class. The first woman to intern at Baptist Memorial Hospital, York also completed residencies at three other hospitals before returning to take up practice in her home community of Ripley, Tennessee. In 1963, on the fiftieth anniversary of her graduation from the University of Tennessee, Dr. Sara

ABOVE: *Dr. Sara Conyers York Murray, 1963, when she was honored at a University of Tennessee commencement for her 50 years of medical practice.* OPPOSITE PAGE: *A 1938 ad shows children who were receiving care for TB from the City Health Department at Oakville Sanatorium. Prior to the opening of Oakville, TB patients were housed in wooden, screened buildings containing only bare necessities. The medical staff consisted of a visiting physician who looked after 35-40 patients.*

Conyers York Murray was recognized at the commencement ceremony for her five decades of medical practice. On her ninetieth birthday in 1968, Dr. Murray was honored by the state for a lifetime of service to her West Tennessee community.

IMPORTANT SMALL HOSPITALS

Between 1900 and 1915, several small hospitals and one large institution opened in Memphis. The new Lucy Brinkley Hospital on Union at Dunlap received its first patients on June 4. 1907. Dr. Richard B. Maury was consulting surgeon and president of the staff and Heber Jones was consulting surgeon and vice president. From 1918 to 1921, the hospital was owned and operated by Methodist Hospital while their new facility was under construction. After the new Methodist Hospital on Lamar was sold to the Veterans Bureau in 1922, the staff moved back to the Brinkley location. The building was later sold to a private medical group, enlarged, and opened as the Woman's Hospital in 1925. It closed in the mid-1930s, a victim of the Depression.

In the early years of the twentieth century, tuberculosis was the leading cause of death in Memphis and Shelby County and Memphis experienced the highest tuberculosis death rate of any municipality with a population of more than 100,000. In 1909, the Health Department established a hospital for tubercular patients in two small wooden buildings located near the riverfront. As the death rate from the "white plague" increased, the Tennessee legislature authorized construction of the Oakville (Tennessee) Memorial Sanatorium for the treatment of tuberculosis. Opened in

July 1921 and later known as Oakville Memorial Hospital, funding came from city and county governments. Several additions to the Oakville facility provided care for both white and black patients, including children. In December 1948, the new West Tennessee Chest Disease Hospital at Jefferson and Dunlap assumed care of tubercular patients. With advances in chemotherapy, the cases of tuberculosis decreased, and the Oakville hospital became a treatment center for non-tubercular pulmonary disease.

HOSPITALS FOR THE AFRICAN-AMERICAN COMMUNITY

Two hospitals provided years of care for the African-American community. Jane Terrell Hospital, funded by the Negro Baptist Association, opened in 1909 in a former private residence at 698 Williams Avenue, just east of Mississippi Boulevard. Dr. Cleveland A. Terrell, a graduate of Meharry Medical College, was superintendent of the hospital, which was enlarged several times and included a training school for nurses. Most of the black registered nurses who worked for the Health Department were graduates of this school. When Dr. Terrell died in 1938, the hospital was renamed Terrell Memorial to honor his years of service. Leadership passed to Dr. Norman M. Watson, surgeon-in-chief, who continued in that position until the hospital closed in 1963.

The medical staff of the Jane Terrell Hospital, 1940s. Back row, left to right: William O. Speight Sr., Jacob Hairston, E. M. Wilkins, unknown, J. H. Gilton, unknown, C. J. Covington (standing to right of post). J. Brawner, P. W. Bailey, unknown. unknown; Middle row: unknown nurse, unknown nurse, unknown, unknown, Wheelock Bisson, possibly E. E. Burt, unknown, unknown nurse; James Byas (in front of post), possibly H. H. Johnson (in front of Byas), unknown nurse; Front row, 4th from left, Dr. Miles Lynk, 7th from left, Dr. N. M. Watson, rest unknown.

Collins Chapel CME Church, located at 678 Washington Avenue, founded the Collins Chapel Hospital, a small hospital and old folks' home. Opened in 1910, the hospital at 418 Ashland was operated by a group of African-American physicians but closed after a few years for lack of operating funds. In 1919, Dr. William S. Martin took over operation and remodeled the hospital. In 1928, he built an annex doubling the number of beds. Martin continued to finance the operation of the hospital until 1930 when the CME Church began regular support. Collins Chapel trained its own nurses and maintained an ambulance service.

Dr. Martin, a 1907 graduate of Meharry Medical College, interned at Mercy Hospital in Nashville and Bellevue Hospital in New York. He did postgraduate work at the

University of Minnesota Hospital at St. Paul, and was a member of several medical organizations, including the Bluff City Medical Society (he was president in 1908), the Volunteer State Medical Association, and the National Medical Association. In 1955, completion of a new 44-bed Collins Chapel Hospital on Ayers was the fulfillment of Martin's years of effort to provide a modern health care facility for African Americans. After the segregation of Memphis hospitals ended, Collins Chapel Hospital closed in 1971.

Dr. Martin and his brothers were known nationally as owners of the Memphis Red Sox, one of the teams in the Negro Baseball League. In 1947, Dr. Martin built Martin Stadium on E. H. Crump Boulevard at Wellington where the National Baseball League games were played to a packed stadium. Dr. Martin died on May 17, 1958. He was survived by his wife, Eva Cartman Martin, long-time superintendent of Collins Chapel Hospital, and his brothers, Dr. A. T. Martin, a physician, and Dr. B. B. Martin, a dentist, both of Memphis, and Dr. J. B. Martin, a pharmacist, of Chicago.

The Gartly-Ramsay Hospital—1910

In 1910, Dr. George Gartly and Mr. R. G. Ramsay, Sr., opened a small hospital in the old McDavitt home on Jackson Avenue. Dr. Gartly incorporated physical therapy treatment based on the system devised at Battle Creek, Michigan. A new fireproof wing in 1929 added operating rooms and obstetrical facilities. Dr. Bryce W. Fontaine was the first chief of staff. For more than 30 years, the hospital conducted a training school for its nurses. Gartly-Ramsay Hospital operated as a general hospital until 1950 and then as a psychiatric facility until it closed in 1974. William Faulkner, acclaimed

Top: *Gartly-Ramsay Hospital, 1930s, with its fireproof wing added in 1929. The hospital was located at 669 Jackson Avenue.* Above: *Baptist Memorial Hospital, 899 Madison, c. 1920. The center section, with its double curved stairway, is the original building built in 1912.*

novelist and resident of Oxford, Mississippi, was a sometime patient at the hospital during its last years.

Baptist Memorial Hospital—1912

The opening of Baptist Memorial Hospital in July 1912 created an institution that became the largest hospital facility in the medical center complex. An imposing building with a sweeping Italianate staircase, the hospital was funded by the Tri-States Baptist Hospital Association and built on Madison Avenue on land that was given by the College of Physicians and Surgeons in 1907 for the purpose of having

MOSQUITOES, MALARIA, AND COLONEL JOSEPH A. LePRINCE

A BRIEF NOTE IN THE FALL ISSUE OF THE 1881 *Mississippi Valley Medical Monthly* reported the increase of malaria in states such as New York, New Jersey, Delaware, and Pennsylvania where malaria previously had not been present. Thousands of Union soldiers infected with the malaria parasite Plasmodium during the Civil War provided the source for the spread of this disease. Mosquitoes bite humans infected with the malaria parasite and spread the Plasmodium, which the mosquito harbors in its stomach and salivary glands. On August 20, 1897, Dr. Ronald Ross, a British physician attached to the Indian Medical Service, who had spent years investigating how humans became infected with malaria, identified the mechanism by which mosquitoes, the intermediate host, transmitted the malaria Plasmodium to humans

The conquest of malaria in the South provided enormous benefits to both public health and the regional economy. The work done in Cuba by pathologists and sanitarians of the Walter Reed Commission laid the groundwork for eliminating yellow fever and malaria in this country. After the *aedes aegypti* was identified as the vector of yellow fever, Dr. William C. Gorgas, U.S. Army Medical Corps, and Joseph A. LePrince, a young graduate engineer, developed an eradication program that rid Cuba of mosquito species that transmitted yellow fever and malaria.

In 1905, when the United States government obtained the rights to build the Panama Canal, Gorgas and LePrince, with other medical officers and army engineers, went to Panama and prepared a detailed plan for the eradication of mosquitoes and other disease-bearing insects in the Canal Zone, incorporating methods developed in Cuba. After the Panama Canal opened in 1914, the government sent LePrince to New Orleans to train American sanitary engineers in malaria control techniques perfected in Panama.

During World War I, LePrince organized mosquito control programs at army and navy installations in the South. He was also sent to Memphis to undertake a campaign to

OPPOSITE PAGE: *Colonel Joseph A. LePrince stands next to a malaria poster, as he reads an issue of the Arkansas Vector Control Bulletin. The plaque on the poster reads, "Memphis and Shelby County Health Department, Malaria—Mosquito Control."* ABOVE: *Ditching used in the control of malaria was pioneed in Panama. Ditches were dug to drain standing bodies of water, or to increase the flow of water, in order to remove mosquito breeding sites. In the South, the method was very successful in the control of malaria during the early twentieth century. In Memphis, during the 1930s, abandoned ditches were filled in, as well as cleaned and straightened. Straightened ditches and bayous were lined with concrete and other materials to prevent erosion, a very effective approach to mosquito control.*

rid the city of mosquitoes. In 1918, Memphis became the permanent home of the LePrince family. (Dr. Lester R. Graves is a grandson of Colonel LePrince.)

During the Great Flood of 1927, LePrince, on special assignment to the American Red Cross, undertook an extensive program to treat flooded land with oil and insect larvicide. Working from the U.S. Public Health Office in the Federal Building, LePrince directed malaria control programs that included draining swamps, filling low areas, constructing concrete ditches, oiling standing water, using minnows in fresh water to eat mosquito eggs and larvae, and spraying Paris green larvicide along river and lakeshore edges. One of the most effective controls was proper screening of houses. From 1935 to 1942, the Works Progress Administration (WPA) put thousands of men to work building concrete ditches and draining swamps throughout the South. In Memphis, some 80 miles of concrete ditches has been constructed by 1942.

LePrince was alert to the importance of educating the public concerning the cost of malaria. This economic burden included wages lost by workers suffering from recurring attacks of malarial fever, the cost of medical services and prescription drugs, money wasted on useless patent medicine chill tonics, and the economic loss to families from unnecessary deaths. In the rural South, there were more than 5,000 malaria deaths annually.

When World War II erupted in 1941, the Japanese quickly gained control of quinine- producing areas in Southeast Asia. LePrince worked with sanitarians to develop effective controls for tropical diseases, while researchers at the University of Tennessee developed an effective substitute for quinine.

In 1951, the American Society of Tropical Medicine and Hygiene established the Joseph Augustin LePrince Medal to recognize outstanding work in malariology. Honored for his pioneering work in eradicating malaria, LePrince was the first recipient of the medal. Whether lecturing students in Memphis high schools, studying new ditching techniques in Shelby County, or wading in hip boots in Tennessee lakes, Colonel Joseph A. LePrince inspired and encouraged others through a lifetime dedicated to the elimination of mosquito-borne diseases.

ABOVE: *Dr. Marcus Haase, secretary of the Memphis Board of Health, sits at his desk in the board's laboratory, 1903.* RIGHT: *Dr. Joseph Augustus Crisler, Sr., c.1910, was a partner of Dr. Eugene Johnson. Around 1907, Crisler joined with Dr. E. M. Holder to provide surgical and gynecological care. He later organized the Crisler Clinic in an old Overton home just east of Methodist Hospital.*

a hospital next to the medical school. When the hospital opened, Drs. John M. Maury and Edward C. Ellett moved their practice to the hospital, occupying an entire floor.

To help fill hospital beds, directors of Baptist Memorial Hospital went to Yazoo City, Mississippi, and persuaded Dr. Eugene J. Johnson to move his surgical practice to Memphis. A graduate of the Memphis Hospital Medical College and a well-known Southern surgeon who had served as president of the Tri-State Medical Association, Johnson moved his office and staff to Memphis in 1912. He quickly established a surgical practice that may have been the largest in the South. For a few years, Johnson and James A. Crisler were surgical partners.

Practicing during the era of the surgical giants, Johnson's record of operative procedures at Baptist Memorial Hospital is impressive by any standard. His speed and stamina in the operating room were legendary, and he often worked for more than 12 hours at a time. When he died on February 18, 1938, the *Commercial Appeal* reported that he had performed more than 40,000 surgeries at Baptist Hospital alone, and his total surgeries exceeded the number of any living doctor of that time. Newspaper obituaries noted Johnson's kindness and generosity in caring for many patients without charge. Colleagues estimated that half his surgical procedures were charity cases. His generosity also provided scholarship assistance for medical students. A skilled surgeon, generous, and hardworking, Dr. Eugene J. Johnson touched the hearts and lives of an enormous number of Mid-Southerners during his medical career.

POSTGRADUATE TRAINING IN EUROPE

During the first decades of the twentieth century, many of the city's practitioners went to clinics and hospitals in New York and Europe for additional training. In 1906, Dr. Marcus Haase went to London and Paris to study diseases of the skin. Drs. Richmond McKinney, W. Likely Simpson, and Archibald Lewis went to Europe in 1908 for advanced work in EENT. Dr. Simpson returned to Vienna in 1912 to receive the degree of doctor of ophthalmology.

Dr. Arthur G. Jacobs, the city's earliest pediatric specialist, also availed himself of the excellent postgraduate training offered in Berlin and Vienna. Dr. Edward C. Ellett and Dr. John M. Maury made frequent trips to European medical centers. Dr. Willis C. Campbell went to London in 1909 to work at the Royal National Orthopaedic Hospital and then went to Vienna to train with world-famous orthopedic surgeon Adolph Lorenz. Whatever their specialty field, Vienna was the favorite city of Memphis doctors for postgraduate training, perhaps due in part to the glamorous life of the city.

Prior to leaving for Europe in 1908, Dr. Richmond McKinney resigned as editor of the *Memphis Medical Monthly*, a job he had filled with great credit since 1896. In announcing his retirement, McKinney said he was turning the work over to Dr. Eugene Rosamond, who had the literary and professional qualifications to assume the editorship. In his incoming salutatory, Dr. Rosamond promised "to help further the recognition of our own magnificent city as the great medical center...she is destined to become," a statement both oratorical and prophetic.

In 1911, Dr. Rosamond resigned, citing poor health, and the editorship was passed to Dr. James Lindsey Andrews. When Dr. Willis C. Campbell was appointed professor of orthopedic surgery at the UT College of Medicine, the *Memphis Medical Monthly* added a regular section on orthopedic surgery written by Campbell. Early articles covered joint disease, surgery for congenital hip dislocation, and the importance of a careful history when examining children for orthopedic procedures.

Floods of 1912 and 1913

While the courageous work of doctors and nurses during the yellow fever epidemics of the 1870s has been extensively covered, the humanitarian service of the medical community during four devastating floods in the 25 years from 1912 to 1937 is not as well known. In March 1912, the lower Mississippi River valley experienced the first of four major floods, followed by those of 1913, 1927, and 1937, the most destructive of the four. Although the city's location on the bluff protected it from serious flood damage, high water backed up the Wolf River into low-lying areas of North Memphis and, in the South, water backed up in the Nonconnah flood plain.

In 1912, the Citizens' Relief Committee, with the help of local and national service agencies, provided food, shelter, and medical care for more than 20,000 people driven from their homes in the Mid-South. Some 2,000 refugees were housed in a tent city at the Fairgrounds. Memphis doctors vaccinated thousands of individuals. The 1912 floodwaters deposited great quantities of mud on a sandbar at the mouth of the Wolf River, creating the appropriately named Mud Island, officially named City Island. In April 1913, the

The 1912 flood in North Memphis. Note the sign, "Boil Water Before Drinking," a concern for public health.

press reported that North Memphis was again under water. Levees on tributary rivers broke, sending thousands to Memphis for food, shelter, and medical attention.

Society of Ophthalmology and Otolaryngology

As the Progressive Era was drawing to a close, one of the city's leading specialty groups was given birth. The Memphis Society of Ophthalmology and Otolaryngology was formally organized on September 23, 1915, with the following members: Dr. Julian B. Blue; Edward C. Ellett; Robert Fagin; Pope M. Farrington; J. F. Hill; Rufus W. Hooker; William Howard; Louis Leroy; Louis Levy; Archibald C. Lewis; James L Minor; H. F. Minor (nephew of J. L.); Richmond McKinney; Bruce F. Moore; G. H. Savage; John J. Shea; W. Likely Simpson; and James Blue Stanford.

The first examination for any American specialty board, the American Board of Ophthalmology, the first specialty board established in the United States, was given at Lindsley Hall in December 1916. Ten candidates sat for the examination; five were Memphians: Drs. J. B. Blue; Louis Levy; Archibald Lewis; W. Likely Simpson; and James Blue Stanford. All passed the examination for board certification.

As the year 1915 ended, the city had benefited from reforms in government and public health, the opening of new hospitals, the consolidation of several medical schools into the University of Tennessee College of Medicine, and the continued development of specialization in medical practice, which would be greatly accelerated by World War I.

CHAPTER SEVEN

WORLD WAR I AND THE INFLUENZA PANDEMIC

1917–1919

THE OUTBREAK OF World War I in 1914 in Europe had little impact in this country until the German U-boat sinking of the *Lusitania* on May 7, 1915, brought the war to the attention of the American public. Outrage at the sinking of Britain's civilian luxury liner with great loss of life and the U-boat threat to American shipping turned public opinion against Germany. Although President Wilson was reelected in 1916 on his promise to maintain America's neutrality, when Germany resumed unrestricted submarine warfare early in 1917, Wilson asked Congress for a declaration of war on April 2, 1917.

In Memphis, young men were quick to enlist. Park Field was constructed near Millington as a training base for pilots, helping pave the way for the establishment of a strong aviation center in Memphis. During World War II, Park Field became part of the U.S. Naval Air Station.

The Tennessee State Medical Association reported in September 1917 that thirty-one percent of the Tennessee physicians who had received commissions in the army medical reserve corps were from Memphis and Shelby County. Months before the United States officially entered the war, leading American doctors such as George Crile, Harvey Cushing, and Elliott G. Brackett took medical teams to England and France to organize military hospitals and centers for specialty care. They were in the vanguard of American doctors, dentists, and nurses who volunteered for military service before this country was officially at war. To expedite getting trained medical teams to Europe, Dr. Crile (founder of the Cleveland Clinic, 1921) recommended to Surgeon General William C. Gorgas that fully staffed medical units for military service come from large general hospitals and medical schools in the United States, an undertaking facilitated by the American Red Cross.

BELOW: *The hospital unit of Memphis physicians organized by Dr. W. Battle Malone, Camp Grenleaf, Fort Oglethorpe, Georgia. Top row, left to right: J. O. Boala, Hiram Mann, L. L. Keller, unknown, J. E. Johnson, C. D. Blassingame, Lewis Virdell, J. P. Owens. Middle row, left to right: Charles Bender, K. M. Buck, Robin Mason, L. L. Meyer, Lynn Anderson, S. E. Frierson, A. F. Cooper, C. K. Summers, unknown, unknown, unknown. Bottom row, left to right: T. F. Coughlin, Edward Clay Mitchell, Thomas Coppedge, W. T. Swink, W. Battle Malone, S. N. Brinson, Joel J. Hobson, Jack Henry, Ed Thompson, W. A. Carnes, Louis Levy.* OPPOSITE PAGE: *A recruitment poster for the American Red Cross, 1918. For a wounded soldier, the symbol of the red cross rising like the sun was a sign of hope.*

MALONE AND MEMPHIS GENERAL HOSPITAL
UNIT P

Such a fully staffed unit was formed in Memphis. Dr. W. Battle Malone, professor of surgery and clinical surgery at the University of Tennessee, organized Memphis General Hospital Red Cross Unit P. By June 1917, Unit P was staffed with 12 volunteer medical officers, 20 volunteer nurses, and 50 enlisted corpsmen. Memphis doctors in Unit P included Ernest L. Anderson; S. N. Brinson; Kinsey Mansfield Buck; Arthur F. Cooper; Thomas Nelson Coppedge; Benjamin F. Dunnavant; Samuel Evander Frierson; Joel Hobson; Robin F. Mason; Lucius McGehee, Jr.; Walter T. Swink; and Edward G. Thompson. Chief Nurse Myrtle Archer, later superintendent of nurses at Baptist Memorial Hospital, directed the selection and training of nurses. Surgical instruments and supplies to equip Unit P were provided by public and private subscription.

Unit P, the South's first volunteer group to go overseas, left Memphis on November 17, 1917. From London, the unit went to Chaumont, France, near Paris, headquarters of the American Expeditionary Force commanded by General John J. Pershing. There, Unit P became part of Base Hospital 15, which had been formed at Roosevelt Hospital in New York. Some Base Hospital 15 staff were sent to the front, where Memphis doctors and nurses worked in field hospitals within sound of cannon fire. Dr. Malone received the Distinguished Service Cross for military valor.

Morgan and Tennessee Ambulance Company One

In late summer 1917, Dr. J. Logan Morgan of Memphis organized Tennessee Ambulance Company Number One. Ambulance drivers were noncombatants trained to provide immediate care for the wounded at field dressing stations and evacuate the injured to field hospitals Ambulance units used motorized vehicles, as well as horse-drawn ambulances, which were able to move the wounded from rough terrain where there was no road. It was dangerous duty and seven Memphis men in this unit were killed.

Ambulance organizations such as those affiliated with the Red Cross and the American Field Service attracted many volunteers—some looking for excitement, most for humanitarian reasons, others because they were pacifists. Ambulance company volunteers included well-known American and British writers Ernest Hemingway; John Dos Passos; W. Somerset Maugham; E. E. Cummings; Malcolm Cowley; and Dashiell Hammett. Hemingway and Dos Passos used their wartime experiences as themes for novels.

Dr. Percy A. Perkins of Memphis was promoted to the rank of major and placed in charge of the 117th Sanitary Train of the 42nd Division, assisted by Captain John R. Drake, also of Memphis. Sanitary trains were fully equipped to handle surgery and care for the wounded. These special trains moved wounded soldiers from the front to large general hospitals and delivered medical supplies to field hospitals.

Mitchell, Smythe, and Base Hospital 57

Base Hospital 57 was organized at Camp Greenleaf, Georgia, on April 2, 1918, and placed under the command of Colonel Edward Clay Mitchell, University of Tennessee faculty member and a former West Point student. Lt. Col. Frank David Smythe was named chief of the surgical service and Major David M. (Max) Henning, assistant chief. Both were University of Tennessee faculty members. Among the 29 Memphis doctors who served in this unit were Shields Abernathy; Hugh Boyd; James Parvin Carter; William F. Clary; J. A. McIntosh; Hugh Nash; Robert H. Pegram; John J. Shea; James Blue Stanford; and James A. Vaughan.

Margaret E. Thompson and Georgia Holmes recruited the 27 volunteer nurses assigned to Base Hospital 57. Soon after its arrival in France, the unit was ordered to Paris in late September 1918 to organize a 2,000-bed hospital in the Preparatory Department of the University of Paris, a school for boys. Dr. Smythe, chief of surgery, reported on the successful use of the Carrell-Dakin method of wound treatment. French surgeon Alexis Carrell and British chemist

Opposite page: *Medical staff unloading wounded soldiers, France. Note the gun carriage to the right used as a litter.* Above: *U.S. Army Base Hospital 57, Paris, France, 1918.* Right: *Dr. Charles Decatur Blassingame, 1930s, when he was assistant professor of otology, laryngology, and rhinology at the University of Tennessee College of Medicine.*

Henry Dakin developed a solution of sodium hypochlorite (0.485%) and boric acid (4%) in sterile water for irrigating wounds, which prevented infection, as well as tissue damage caused by harsh antiseptics such as carbolic acid. During the peak of the influenza epidemic in October 1918, Base Hospital 57 cared for more than 2,000 sick and wounded. The last American military hospital operating in Paris, Base Hospital 57, closed in August 1919, and the staff sailed for the United States.

Blassingame and Evacuation Hospital No. 49

Charles D. Blassingame, an otolaryngologist, was among a number of Memphis physicians who headed military hospitals. Captain Blassingame was placed in charge of Evacuation Hospital No. 49, also organized at Camp Greenleaf, Georgia. Men recruited in late summer 1918 to staff Evacuation Hospital 49 fell victim to the influenza then moving rapidly through military installations in the United States and fighting units in Europe. Influenza, called Spanish Flu, was so named because Spain was the only European country reporting outbreaks of the disease. Other countries suppressed the news on the basis of military security.

When Evacuation Hospital 49 arrived at the port of Brest, France, on October 26, 1918, the unit was immediately sent to the Meuse-Argonne front. The Meuse-Argonne offensive (September 26-November 11, 1918) was the largest World War I military operation in which American troops participated. In combat, the evacuation hospital was the vital link in the chain of medical service, and its staff performed life saving operations under the most difficult and dangerous conditions. The wounded were brought in by ambulance or on litters and their condition assessed by a triage nurse. If not too critical, they were x-rayed before surgery. The organizational flow was such that a team of doctors could do 40 or more operations in a 24-hour period. In January 1919, Evacuation Hospital 49 was sent to Coblenz, Germany, with the army of occupation. Medical personnel treated American troops and examined discharged German soldiers who were being sent home. Evacuation Hospital 49 staff left Germany for America in May 1919.

Ellett, Ophthalmology, and Base Hospital 115

Dr. Edward Coleman Ellett, internationally recognized eye surgeon, was placed in command of Base Hospital 115, a unit whose medical officers were specialists in head and face surgery. The hospital was installed in the Hotel Ruhl in Vichy, on September 6, 1918. Vichy, a spa town located in mountainous central France, was the World War I hospital center of France. The American Expeditionary Force leased 86 hotels for use as hospitals. Doctors assigned to Base Hospital 115 were specialists in ophthalmology, otolaryngology, neurosurgery, and oral plastic surgery. A companion

BELOW: *With a sign on the back wall reading, "Silence—No Smoking," three operations take place in this makeshift operating room.* OPPOSITE PAGE: *American soldiers in a field hospital, France. Special attention is being given the patient at the rear, left.*

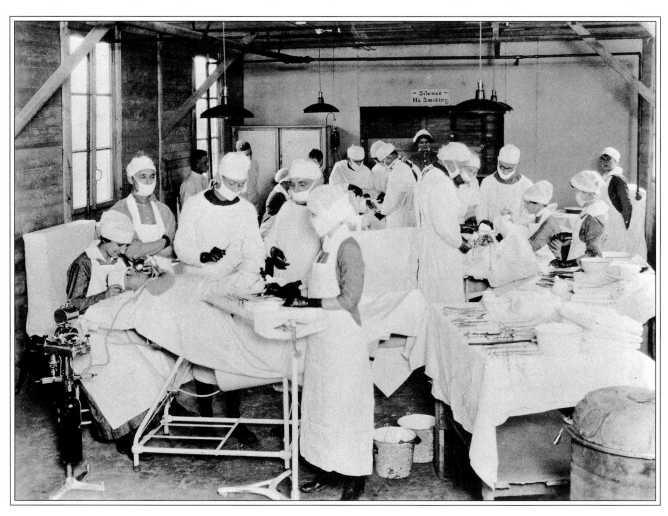

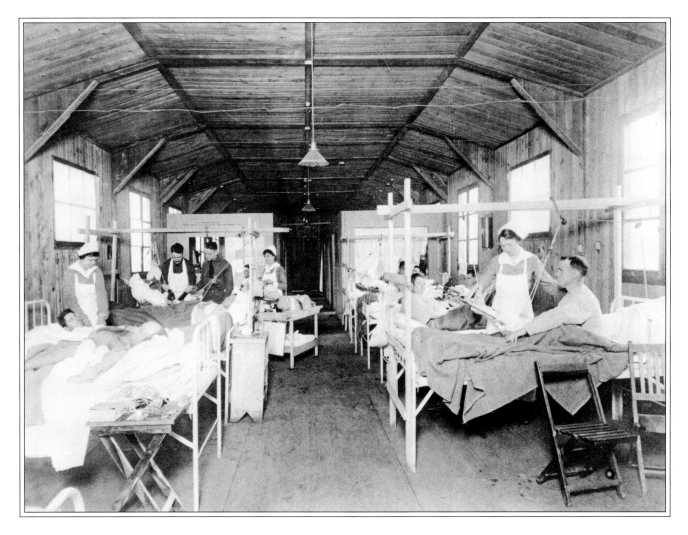

hospital for head and face surgery had been organized at Cape May, New Jersey. Both hospitals reported directly to the surgeon general.

The surgeon general's report on Base Hospital 115 states, "From this time on [September 6, 1918] all the head, face, and neuro-surgery was transferred to this hospital. Hotel Ruhl was a large and handsome building, so in addition to head and face, many general surgical cases of every type were admitted and placed in separate wards." Medical services provided at Base Hospital 115 included rehabilitation training for soldiers who had been blinded by battle injuries. The number and diversity of cases treated by Dr. Ellett provided a unique surgical experience. After the war, he limited his practice to ophthalmology. His association with specialists from medical centers in the United States served him well during his leadership of national organizations.

In 1915, Dr. Harvey Cushing spent time with the French Army at the American Ambulance Hospital in Paris, organized by Americans living in France. He returned to Boston where he formed Base Hospital 5, which went overseas in 1916 as a Red Cross hospital unit that served in France treating British and French wounded. When America entered the war, the operation of Base Hospital 5 was transferred to the American Expeditionary Force, and Cushing was named senior consultant in neurosurgery. In October 1918, Cushing contracted influenza, then a raging epidemic filling military hospitals throughout Europe. He was bedridden for some months with a polyneuropathy affecting his upper and lower extremities, later believed to be Guillain-Barre Syndrome.

Semmes, Neurosurgery, and Base Hospital 87

One of Cushing's ablest students, Raphael Eustace Semmes, did his student and postgraduate work at Johns Hopkins Hospital. His close association with Cushing was important in establishing Semmes as an early pioneer in neurosurgery. During World War I, Surgeon General Gorgas reorganized the medical services of the American Expeditionary Force and made neurosurgery a separate subspecialty of general surgery. When General Gorgas asked for volunteers trained in neurosurgery, Semmes was one of a select group sent for

EDWARD COLEMAN ELLETT—SURGEON, TEACHER, EXEMPLAR

THE OUTSTANDING SURGICAL ABILITIES OF Memphis's pioneering specialists established the city's reputation as a treatment and research center. Such a physician was Edward Coleman Ellett, skilled surgeon, gifted teacher, and internationally renowned ophthalmologist.

Ellett, the son of Judge Henry T. and Katherine Coleman Ellett, was born in Memphis on December 8, 1869. His father was a prominent member of the Memphis Bar. Ellett attended Southwestern College (now Rhodes College) and received his A.B. degree from the University of the South at Sewanee, Tennessee. His M.D. degree was earned at the University of Pennsylvania Medical School in 1891, followed by an internship at St. Agnes Hospital. Ellett was senior resident and house surgeon at the Wills Eye Hospital in Philadelphia, an institution which played a vital role in establishing ophthalmology as a separate branch of medicine.

Ellett returned to Memphis in 1893 and entered his first partnership with James L. Minor. Their practice was limited to eye and ear diseases. With William Krauss and Max Goltman, Ellett organized the Memphis Pathological Society in 1898. He was the first doctor in Memphis to own a microtome used for sectioning eye tissue.

A leader in medical education, Ellett was an organizing member of the College of Physicians and Surgeons and professor of ophthalmology. When this school merged with the University of Tennessee College of Medicine in 1911, he continued to head the Department of Ophthalmology until 1922 except for his absence during World War I. His distinguished service as commanding officer of Base Hospital 115 is described on pages 82-83. Ellett was a founding member of the American Board of Ophthalmology organized in Memphis in 1915, the first specialty board established in this country.

Ellett held memberships in leading medical societies, including the American Academy of Ophthalmology and Otolaryngology, serving as president in 1926. He was president of the American Ophthalmological Society (1932), the International Ophthalmological Association, the Tennessee State Medical Association, and The Memphis and Shelby County Medical Society. In addition, he chaired the American Medical Association Section on Ophthalmology and the Southern Medical Association Section on EENT.

When eye specialists from the Mid-South gathered in Memphis on December 10, 1935, to recognize one of the

Opposite page: *Dr. Edward Coleman Ellett, ophthalmologist and a founding member of the American Board of Ophthalmology.* Right: *An ad for Dr. E. C. Ellett's Private Hospital,* Memphis Medical Monthly, *August 1907.* Above: *The Memphis Eye, Ear, Nose and Throat Hospital under construction. It opened in August 1926 at 1052/1060 Madison Avenue.*

world's leading ophthalmologists, Ellettt was uniquely honored. It was "Ellett Day" and skilled surgeons operated on patients with impaired vision. The doctors gave their service, and the Memphis Eye, Ear, Nose and Throat Hospital and Baptist Memorial Hospital donated the use of their facilities. In the Depression, it was a generous way to recognize an outstanding physician. At the dinner, which climaxed this extraordinary day, Ellett recalled his early days of training (1891-1893). "We had cocaine as a local anesthetic, general anesthesia was in use, and we knew there was such a thing as antisepsis, though I shudder now to think how it was practiced. In the Wills Eye Hospital we did not have a single gown or mask or cap; chemical antiseptics were the rule. Nothing was boiled; in fact, we had no autoclaves and no rubber gloves."

The Leslie Dana Gold Medal was another honor awarded in recognition of Ellett's "meritorious service in the conservation of vision." Sponsored by the National Society for the Prevention of Blindness, the presentation was made in St. Louis on December 14, 1939. The Memphis and Shelby County Medical Society honored Ellett in May 1943 on the fiftieth anniversary of his entry into medical practice. Friends and medical colleagues gathered at the University Club dinner to celebrate a life of extraordinary accomplishments. Equally important to Ellettt were the visits and letters from patients expressing gratitude for his vision-saving work.

Dr. Ralph Rychener, Ellett's long-time associate in practice said: "Probably no other man in the Mid-South has exerted such an influence for good in the field of medicine by precept and integrity. Thousands benefited from his surgical skill." Edward Coleman Ellett died on June 8, 1947, while attending the annual meeting of the American Medical Association in Atlantic City.

The inscription on the Leslie Dana medal testified, "Edward Coleman Ellett, inspiring teacher, skilled surgeon, understanding and sympathetic clinician, and friend." "I was blind, but now I see."

fensive numbered 26,277 killed and 95,786 wounded. The peak of the influenza epidemic occurred at the same time, filling thousands of hospital beds in Toul. For meritorious service, Semmes was promoted to the rank of captain on May 5, 1919.

Dr. Neuton Stern enlisted in the Army Medical Corps in 1917 following his internship at Massachusetts General Hospital and initial military training at Fort Sill, Oklahoma. From there, Lieutenant Stern was assigned to a unit in Paris responsible for purchasing medical supplies for the great number of American hospitals

ABOVE: *Dr. Neuton S. Stern talks with Mrs. W. B. Rosenfield, 1941. Dr. Stern, one of the first internists and cardiologists in Memphis, introduced the first electrocardiograph machine to the city.* RIGHT: *Orthopedic surgeon Dr. Alphonse H. Meyer, Sr., with his wife, 1938. His son, Dr. Alphonse H. Meyer, Jr., who died in 1983, was a surgeon and assistant professor of surgery at the University of Tennessee Center for Health Sciences. His father died in 1973 at the age of 91.*

advanced training at the New York Neurological Institute, the first hospital organized in the United States for the treatment of neurological disorders. On completion of this training, Semmes was commissioned a 1st Lieutenant in the Army Medical Corps and ordered to report to Camp McArthur, Texas, on August 1, 1918, for assignment to Base Hospital 87.

When the personnel of this hospital unit embarked from Hoboken, New Jersey, on September 15, 1918, influenza broke out among the ship's crew and quickly spread to the Army men on board, causing serious illness. On arrival at Brest, France, the receiving port for American troops, 90 men and officers had to be hospitalized; of these, five died. Base Hospital 87 men left Brest and arrived in Toul, France, on October 8, 1918, where the staff took charge of the Gas Hospital and the Neurological Hospital. Because of the critical surgical needs resulting from the massive Meuse-Argonne offensive, Semmes performed mostly general surgery and some neurosurgery. Casualties from this major of-

Memphis's first plastic surgeon, Dr. Joseph E. Johnson, in his World War I uniform.

in France. When he mustered out of service, Stern went to England to study with Sir Thomas Lewis, then the world's leading cardiologist.

Meyer, Orthopedics, and Base Hospital 32

Memphian Dr. Alphonse Meyer was part of a team headed by Boston's noted orthopedic surgeon Elliott G. Brackett that went to Europe before the United States entered the war. Originally serving with the British Expeditionary Force, the unit was transferred to the American Expeditionary Force and assigned to Base Hospital 32 located at Contrexville, in the northeast of France near Toul. Close to the front lines, Base Hospital 32 had one of the largest orthopedic services in France. The terrible injuries causes by exploding bombs, shrapnel, and close range machine gun fire required skilled surgery. Compound fractures represented ninety percent of the work done by the orthopedic surgeons of this unit. Wound infections were a serious problem; all of the bomb wounds, eighty percent of the high explosive wounds, and ten percent of the machinegun wounds became infected. Returning to Memphis after his military service, Dr. Meyer joined the University of Tennessee faculty as an associate professor of orthopedics.

The radiology service in military hospitals made significant contributions to successful surgery. As part of the reorganization of the medical service, the surgeon general had established the Division of Roentgenology in 1918. The important contribution of radiologists during the war hastened the development of radiology as a medical specialty.

Joseph E. Johnson, Reconstructive Surgery, and Base Hospital 52

The disfiguring injuries of World War I required skilled plastic surgery to restore faces and bodies. Dr. Joseph Edward Johnson, the first plastic surgeon in Memphis, was appointed director of plastic surgery at the Army School in Philadelphia in 1917. Commissioned a major, Johnson was attached to the medical staff of Base Hospital 52 at Rimacourt in northeastern France. The hospital was one of many barrack-type hospitals where thousands of surgical procedures were carried out. After his wartime experience in plastic and reconstructive surgery, most of Johnson's postwar practice was in plastic surgery. For some years, he was professor of plastic surgery at the University of Tennessee. When he died in 1931, a *Commercial Appeal* editorial noted that he was one of the preeminent plastic surgeons in the nation, a fact not well known in Memphis. The editorial writer concluded, "Only those who have witnessed what medical science accomplished for the war mutilated can fully appreciate what those in Dr. Johnson's field have accomplished. Plastic surgery is creative art, combined with medical science."

Where Memphis Doctors Served

According to records compiled by the Memphis and Shelby County Medical Society in 1924, a total of 105 doctors then practicing in Memphis served in World War I military units. Three Memphis physicians, Conley H. Sanford, D. H. Anthony, and Norwin B. Norris, served with the U.S. Navy. Lieutenant Norris was killed on September 30, 1918, when a German U-boat torpedoed the U.S.S. *Ticonderoga*. Dr. Casa Collier served on the British transport *Waimana*. Four Memphis physicians were assigned to the Aviation Branch: Frank Ward Smythe, Percy Wood, Louis Levy, and R. D. Henderson.

Hubert Sage was the first Memphis physician to lose his life in World War I. In September 1917, he volunteered for service with the British Expeditionary Force and received additional surgical training in an English hospital. Dr. Sage was killed in May 1918 in a field hospital in France that came under German aerial attack, inflicting heavy casualties on the patients and hospital staff.

Dr. Grover Carter enlisted in Memphis on August 4, 1917, was attached to the British Expeditonary Force, and sent to the War Hospital in Dartford, England. On April 19, 1918, Captain Carter was transferred to the 121st Brigade, Royal Field Artillery. He was fatally wounded on October 15, 1918. Two weeks earlier, he had met his brother in Paris, where Dr Parvin Carter was on the staff of Base Hospital 57.

Late in 1918, Captain Robert B. Underwood died of pneumonia in Rouen, France. Pneumonia was a deadly bacterial complication of influenza causing great mortality. The influenza pandemic of 1918 to 1919 caused more military deaths than all battlefield casualties and injuries combined. There were 385 known deaths of Memphis soldiers in World War I: 76 were direct battle casualties; 242 died of wounds; 40 died of influenza-pneumonia; and 27 deaths resulted from accidents or other causes. United States military deaths through December 31, 1918, included 54,402 battle casualties and 63,114 disease-related deaths. Influenza would continue to claim lives through 1919. Respiratory illness led all other diseases as the cause of death and postwar disability.

Influenza—A Deadly Killer

As American troops were shipped back to the United States, the influenza virus traveled with them. On a global basis, the 1918-1919 influenza pandemic is estimated to have killed 50 million people, making it the deadliest epidemic in recorded history. Influenza traveled from central Europe eastward via Turkey into Asia, India, and China. The death rate in Africa was overwhelming. As the disease crossed the Pacific Ocean, whole villages in the Pacific Islands and Alaska were wiped out. No country in the world escaped the influenza virus. The social effects of the disease were enormous, as the greatest number of people died in the six-month period between October 1918 and March 1919. Britain reported 200,000 influenza deaths. President Woodrow Wilson was seriously ill with influenza while in Paris with American officials negotiating the Versailles Treaty.

The 1918 influenza virus exhibited a remarkable pathogenicity, which greatly multiplied its destructive force. The mass movement of troops over an extended period, populations in Europe weakened by malnutrition, and a complete lack of medical care in many areas in the world combined

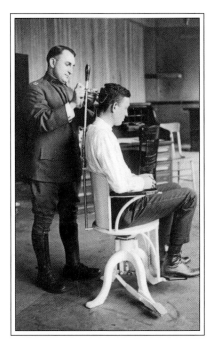

LEFT: *Captain Louis Levy tests the vestibular function of an air force recruit after being spun in a bárány chair. The procedure is used to test spatial disorientation in pilots and astronauts.* BELOW: *The U.S. prepares for the influenza pandemic. The photo shows a demonstration at the Red Cross Emergency Station for handling the sick, Washington, D.C., 1918.* OPPOSITE PAGE: *The influenza ward at Walter Reed General Hospital, Washington, D.C., about 1918. Masks were not only worn by medical personnel for protection, but by public service workers, office workers, barbers, and by any group which was exposed to the public.*

to exact a devastating death toll. In the United States, there were 675,000 reported deaths with a civilian illness rate of twenty-eight percent. Cities in the Northeast suffered greatly. Philadelphia sustained some 7,000 deaths in such a short time that people were buried in mass graves, and deaths in New York City exceeded 30,000.

Enormous research has gone into identifying the 1918 influenza virus, believed to be an H1N1 avian virus, and determining why this pandemic was so devastating. Although researchers have not reached a conclusive determination as to where this virus first appeared, scientists tracked what is believed to be the first outbreak in this country to Fort Riley, Kansas. The fort bred its own poultry and swine. Those who later studied frozen tissue specimens concluded that the H1N1 virus was an especially virulent strain new to its human population. It had features which enhanced transmission and caused an increased incidence of influenza pneumonia, making it unusually deadly. Since 1918, influenza pandemics have been caused by viruses which have some of the same genes existing in the 1918 virus. It is safe to say that all of the A viruses since that time are a descendant of this virus. The current H1N1 virus has some of the same genes but also incorporates genes from an influenza virus found in swine.

Especially disturbing was the fact that this flu attacked the young and healthy. In some cases, people collapsed suddenly and died within 24 to 48 hours. Across the United States, schools, churches, and entertainment facilities were closed, and travel greatly restricted. All non-essential services and meetings were cancelled, and some industries temporarily shut down. Hospitals were full and overflowing. Doctors and nurses as well as the general public contracted influenza.

Influenza Reaches Memphis—September 1918

Influenza cases were first reported in Memphis in late September 1918. By October 4, there were 243 cases. The progress of the disease was so swift that within three days all schools, churches, theaters, and some businesses were closed, as hundreds of new cases were reported. The October 13, 1918, *Commercial Appeal* headline, "Flu is still raging," was confirmed by columns of death notices. Because local hospitals were full, the Red Cross organized Central High School as a temporary emergency hospital for influenza cases.

In Memphis there were 6,531 reported cases of flu and 172 deaths, although some cases may not have been reported. Many deaths came from the secondary infections of bacterial pneumonia or septicemia (blood poisoning from the toxic overload in the body). In the period between the two World Wars, Memphis had three major flu epidemics—1928, (15,000 cases reported during the fall and winter); 1936; and 1940, which the Health Department declared equal to the 1918 pandemic.

By the spring of 1919, most American military hospitals in Europe had closed and the remaining wounded transferred to hospitals in the United States. Memphis doctors returned from the war having experienced trauma surgery and disease on an unprecedented scale. Great progress had been made in surgical techniques, supporting technology, and sanitation. Dr. Semmes joined other neurosurgeons in urging rehabilitation therapy for spinal cord injuries. New burn treatments came from the terrible injuries caused by mustard gas and explosives. Although research that identified disease-causing viruses and companion bacterial infections was still in the future, the great laboratory of war accelerated the development of medical specialties in the years between the two world wars.

CHAPTER EIGHT

1920s PROSPERITY TO 1930s GREAT DEPRESSION

1920–1939

The twenties were prosperous years for Memphis. With a population of 162,351, the Bluff City was the business and medical center of the Mid-South. In downtown Memphis, new office buildings, hotels, and ornate movie theatres were constructed and the housing market expanded eastward. The twenties were also a time of growth for medical specialties supported by the addition of new hospitals, clinics, and sanatoria. But, the twenties and thirties also held challenges that taxed the city's economic and human resources. In 1927 and 1937, Memphis was stretched to provide assistance for thousands of refugees when two devastating floods ravaged the lower Mississippi River. With the nation, Memphis would endure the Great Depression of the 1930s.

In February 1921, shortly after the Willis C. Campbell Clinic opened at 869 Madison, the *Commercial Appeal* reported that Memphis doctors were moving east to the medical center for the convenience of their patients, a process that continues to the present, expanding east to Collierville and south into DeSoto County, Mississippi. Madison Avenue became the medical office corridor for specialty practice groups. The newspaper noted in 1922 that Drs. Otis Warr, R. L. Sanders & Associates, and Drs. Edward C. Mitchell and Eugene Rosamond had moved their offices to the hospital area. The Henry G. Hill Orthopedic Clinic opened at 847 Madison, and Dr. Max Goltman announced that his son Alfred would join him in the Goltman Clinic at 995 Madison.

In 1926, Dr. Elizabeth Kane moved her Woman's Clinic to 1099 Madison, the Shea Clinic located at 1018 Madison, and the Memphis Eye, Ear, Nose and Throat Hospital opened at 1060 Madison. The Physicians and Surgeons Building, constructed in 1928 as part of the Baptist Memorial Hospital complex, was designed so that each floor of the nine-story building was connected to the hospital. In addition to medical offices, the building included a hotel floor for families of patients, a pharmacy, and restaurant. However, not all doctors moved out of the downtown area. In 1924, the new Medical Arts Building on Madison at Fourth was designed expressly for physicians, and the Exchange Building on Second at Madison remained a popular medical office location until well after World War II.

Formation of Medical Societies

Reflecting the growth in specialization, new medical societies were formed during the twenties. The Memphis Pediatric Society, organized in 1921, elected Edward C. Mitchell as its first president. A professor of pediatrics at

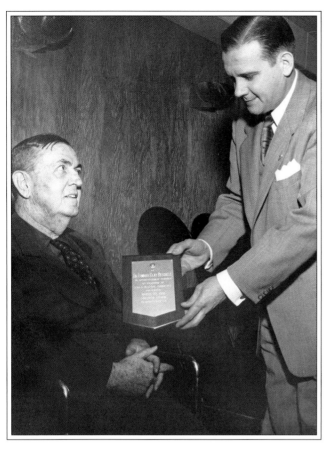

the University of Tennessee and pediatrician-in-chief of the Memphis General Hospital, Mitchell was made a charter member of the American Academy of Pediatrics (AAP) at its formation in 1930 and elected president in 1941. A recognized leader in the field of pediatrics, Mitchell's first partner was Percy Toombs, followed by Eugene Rosamond, and then Arthur Quinn. Rosamond served as president of the Memphis Pediatric Society (1924) and later as president of the Tennessee Pediatric Society.

The importance of x-ray technology in World War I prompted organization of new radiology organizations. One of them, the American College of Radiology (ACR), was formed in 1923. Walter S. Lawrence, dean of Memphis radiologists, was a charter member. In 1926, Lawrence and Charles Heacock organized the Memphis Roentgen Society.

LEFT: *Dr. Edward Clay Mitchell, left, receives an award from the American Legion for his services as chairman of the Child Welfare Committee, 1951. It is presented by Judge George T. Lewis.* BELOW: *Interior of the "First Aid and Clinic," Memphis Fairgrounds Refugee Camp, Flood of 1937. Note the armed soldier and the physician handing medicine to a patient. The clinic was in the Smith-Hughes Vocational Agriculture building.*

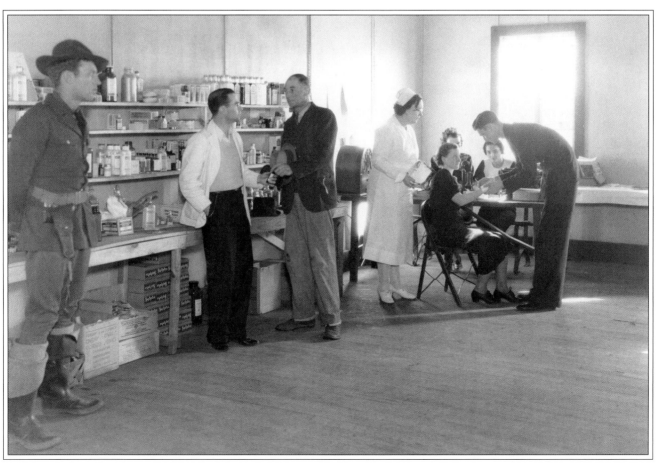

THE CAMPBELL CLINIC AND WILLIS CAHOON CAMPBELL

In the decades known as the era of surgical giants, Memphis was graced to have men of extraordinary abilities whose accomplishments garnered national and international acclaim. One of them, Willis Cahoon Campbell, established a clinic and hospital recognized worldwide for orthopedic excellence. Born on December 18, 1880, in Jackson, Mississippi, Campbell was a 1904 Phi Beta Kappa graduate of the University of Virginia School of Medicine. He served internships at the Norfolk Virginia Protestant Hospital and at the City of New York Infant Asylum.

When Campbell came to Memphis in mid-1906, he opened an office in the Randolph Building and worked for a time in general practice and pediatrics, with some work in anesthesia. His interests led in other directions and, in 1908, the 27-year-old doctor determined to enter the field of orthopedic surgery, a decision of far-reaching consequence. Campbell went to London in 1909 for postgraduate work at the Royal National Orthopaedic Hospital followed by study at the University of Vienna with Dr. Adolph Lorenz, then the world's leading orthopedic surgeon. He did additional work in New York and Boston, the orthopedic centers of this country.

By December 1909, Campbell was back in Memphis. He occupied several downtown office locations before building the famous Campbell Clinic on Madison Avenue, opened in December 1920. The *Commercial Appeal* reported, on February 27, 1921, that Dr. Willis C. Campbell had erected a brick office building at 869 Madison Avenue in the hospital center. A four-story, 80-bed, hospital building was added in 1922, which was enlarged several times. Orthopedic surgery was performed at this location until 1967 when surgery and inpatient care was moved to Baptist Memorial Hospital.

After the University of Tennessee moved to Memphis in 1911, Campbell organized the Department of Orthopaedic Surgery and was professor of orthopedic surgery until his untimely death in 1941. Dr. Speed followed Campbell as department head (1941-1959). In 1924, the Campbell Clinic established a postgraduate training program in orthopedic surgery, a residency program now conducted in partnership with the University of Tennessee Department of Orthopaedics.

ABOVE: *The Campbell Clinic at 869 Madison Avenue, 1930s.* OPPOSITE PAGE: *Dr. Willis Cahoon Campbell celebrates his 60th birthday, December 18, 1940.*

DR. SPENCER SPEED COMES TO MEMPHIS

Dr. James Spencer Speed received his M.D. degree from The Johns Hopkins University Medical School in 1916, followed by two years of internship. With other American doctors, he gained extensive surgical experience during World War I. Following a postwar residency at the Hospital for Women in Baltimore, Maryland, Speed came to Memphis in 1920 and through a chain of fortuitous circumstances was persuaded to join Campbell, who promised to train him in orthopedic surgery. Thus was formed a unique partnership. Although different in personality—Campbell a mass of dynamic energy; Speed, orderly and thoughtful—and different in physical appearance; Campbell tall, broad,

230 pounds; Speed slight of stature, 135 pounds—they combined complementary surgical skills and leadership abilities that created an internationally acclaimed hospital and teaching institution.

Hospitals for Crippled Children and Crippled Adults

Recognized by his peers as a renowned pioneer in orthopedic surgery, Willis Campbell, in the public mind, is indelibly associated with the founding of The Crippled Children's Hospital and School, opened at 2009 Lamar Avenue in 1919. Before antibiotics, polio vaccine, and other chemotherapy, hundreds of children in the Mid-South suffered from tuberculosis, osteomyelitis, infective arthritis, and poliomyelitis, as well as congenital deformities. Families of these children were unable to afford multiple surgeries and long-term medical treatment. During the existence of this hospital, the Campbell Clinic staff provided all orthopedic care without charge. Doctors in other specialties, as well as dentists, generously provided care for the children. Surgical facilities were provided by Baptist Memorial Hospital. Support for operating the hospital came from charitable organizations, private donors, the children of Memphis, and an annual fund-raising gala. In the depths of the Depression, the public was generous in supporting work that made young bodies whole. Campbell was greatly beloved for his work with crippled children.

In addition to care for children, Campbell sought support for a hospital for crippled adults because the children's hospital did not accept patients older than age 14. In 1923, the Hospital for Crippled Adults opened in the former Presbyterian Home Hospital on Alabama. Surgery was performed at St. Joseph Hospital. With the aid of generous donors and the support of area Rotary clubs, in 1928 a new Hospital for Crippled Adults was constructed on La Paloma Circle adjacent to the Crippled Children's Hospital.

With post-world War II advances in orthopedic procedures, children needing corrective surgery no longer required long-term hospitalization and, in 1982, the hospital facility was sold to the Methodist Health Care System. Proceeds from the sale of the hospital and its assets established the Crippled Children's Foundation (now the Children's Foundation of Memphis), which provides support for the Children's Foundation Research Center of Memphis established by the University of Tennessee Health Science Center at Le Bonheur Children's Medical Center in 1995. The Hospital for Crippled Adults closed in 1970, the result of changes in government requirements for healthcare delivery.

A group of children enjoy a tea party on the lawn of the Crippled Children's Hospital and School, 1920. All medical services were provided without charge. In 1955, a physical and occupational therapy wing and a new section for African-American patients were added.

Campbell, Ryerson, and the American Academy of Orthopedic Surgeons

Edwin Ryerson, a prominent orthopedic surgeon at Northwestern University Medical School, and Willis Campbell were two of the seven founding members of the American Academy of Orthopedic Surgeons (AAOS) organized in 1932 in Chicago. Ryerson was named founding president and Campbell followed in 1933 as the first elected president. Campbell also held membership in the American Orthopedic Association (serving as president in 1931), the International Society of Orthopedic Surgery, the American College of Surgeons (board of governors), the American Medical Association (chair of the orthopedic section), the Mid-South Post Graduate Medical Assembly, the Tennessee Medical Association, and the Memphis and Shelby County Medical Society.

Campbell's contributions to orthopedic literature include the world-famous *Campbell's Operative Orthopaedics*, (now in its eleventh edition), cowritten with Hugh M. Smith in 1939, *Orthopaedics of Childhood*, and *Orthopaedic Surgery*. Dr. Speed, also a prolific writer, was the author of some 40 published articles, many classics in orthopedic procedures. He was coeditor with Hugh M. Smith, of the second edition of *Campbell's Operative Orthopaedics* (1949) and editor, with Robert A. Knight as associate editor, of the third edition of this classic text (1956).

Willis C. Campbell died on May 4, 1941, at the age of 60. Extensive coverage of his honors and accomplishments affirmed, "No other man did more to make Memphis known as a great medical and hospital center." Acclaimed as "one of the outstanding orthopedic surgeons of the world," Campbell's legacy was his vision for excellence in orthopedic surgery and the establishment of a residency program that has successfully trained generations of doctors in the orthopedic surgical techniques he pioneered.

After Campbell's death, Dr. Speed became chief of staff at Campbell Clinic and the hospitals for crippled children and crippled adults. A nationally recognized orthopedic surgeon, Speed was active in leading medical societies and served as president of the Clinical Orthopedic Society (1944), which Campbell had organized in 1930. He was president of the American Orthopedic Association (1950-1951), a founder of the American Board of Orthopedic Surgery, and a member of the American Academy of Orthopedic Surgeons.

Named Tennessee Physician of the Year in 1965 by the Tennessee Medical Association, the award honored Speed whose "incomparable contributions to the crippled children

Top: *The Campbell Clinic staff, 1935. Left to right: John Lovejoy, Hugh Smith, Joe Frank Hamilton, Willis C. Campbell, Joseph Mitchell, Harold Boyd, and J. Spencer Speed, who succeeded Dr. Campbell as chief of staff.* Above: *The second Hospital for Crippled Adults, 1933.* Right: *While Mrs. Richard Wooten holds an oversize card from the women's auxiliary to the Memphis and Shelby County Medical Society, saluting Dr. Rocco Calandruccio, Mrs. Hoyt Crenshaw hands him his coffee, 1966.*

and crippled adults of Tennessee and of the nation have been boundless." A surgeon of internationally recognized ability, Speed's professional accomplishments embraced a warm humanity and a gentle manner in working with patients. Respected by the Campbell staff for his superior judgment and surgical skills, Speed's wise counsel was sought by many. He retired as chief of staff in 1962, closing a long career of service and achievement in orthopedics. James Spencer Speed died on April 30, 1970.

After World War II, Alvin J. Ingram and Marcus Stewart completed their residencies at the clinic and joined the staff, Ingram specializing in pediatric orthopedics and Stewart in sports-related orthopedic surgery. Ingram followed Harold Boyd as chief of staff in 1967. Andrew Hoyt Crenshaw joined the Campbell staff in 1951. A skilled surgeon, Crenshaw made a major contribution as editor of five editions of *Campbell's Operative Orthopaedics*. Rocco Calandruccio and Lee W. Milford were added to the staff in 1954. Calandruccio served as chief of staff (1976-1983) and was followed by Milford (1983-1987), a pioneer in hand surgery. As surgical programs expanded, Robert W. Tooms, an expert in orthopedic prosthetics, served as an adviser to companies manufacturing prosthetic devices.

When Dr. Campbell died, his partners, Drs. Speed, Hamilton, Boyd, and Smith, purchased the clinic from Campbell's heirs. Dr. Speed planned to donate his interest in the clinic to the surviving partners; however, they determined that the tax liability would be prohibitive. In 1946, the partners established a non-profit educational foundation, and the clinic and its assets were transferred to the Campbell Clinic Foundation, which oversees and supports the residency program, advanced research, and ongoing revisions of *Campbell's Operative Orthopaedics*.

The Memphis Urological Society was organized on December 9, 1924, at Baptist Memorial Hospital and George R. Livermore was elected president. Members included Oswald S. McCown; Jerome L. Morgan; Hubert K. Turley; J. H. Smith; O. P. Walker; John Ragsdale; Carroll H. Morgan; and I. G. Duncan. Meetings of the society rotated among various hospitals; however, in the late 1930s, members decided to hold monthly dinner meetings at the University Club. In 1933, Dr. Livermore served as president of the American Urological Association.

Although obstetrics may rightly be considered the oldest specialty in Memphis, it was not until 1928 that an Obstetrical and Gynecological Society was organized in this city. Dr. William T. Black was elected president at the organizational meeting on January 13, 1928. Among the founding members were Percy Toombs; Jesse Cullings; William T. Pride; Wilson Searight; J. C. Ayers; W. L. Williamson; Percy H. Wood; John Metcalf Maury; James L. Andrews; and George Gartly. Associate members were William H. Brandon; James O. Gordon; James Reinberger; Phil Schreier; Walker L. Williamson; and S. B. Williamson. Also in 1928, Dr. Black organized and was the first chairman of the Southern Medical Association Section on Gynecology. Drs. Black, Toombs, and Pride were given original memberships on the American

ABOVE: *Born in England in 1868, Dr. Walter Sibley Lawrence's interest in radiology began in 1896. He is seated, conducting an experiment with an associate, unknown date.* RIGHT: *Dr. James Bassett McElroy, c. 1920. Born in Columbus, Mississippi, Dr. McElroy opened his practice in Stovall, Mississippi. In the 1890s, he studied malaria in his small, but well-equipped laboratory. He opened his practice in Memphis in 1904 and became the first chair of pathology at the Memphis Hospital Medical College before joining the faculty at the University of Tennessee College of Medicine.*

Founding members included: William Bethea; Guy Campbell; Steven Coley; Horace Gray; Cash King; Robert Pain; H. N. Pulliam; and Walter Robinson. Dr. Robinson, the author of numerous scientific monographs, was instrumental in developing spot film radiography. At the initial meeting of the society, Drs. Lawrence and Edward G. Campbell, an internist, presented, "Radiographic Findings in Certain Common Bone Diseases." In 1933, representatives from the five radiologic societies joined together to create The American Board of Radiology (ABR), incorporated that year. The oldest radiologic accrediting organization, the ABR, in 1934 offered the first examinations for certification in three locations: Cleveland, Pittsburgh, and Memphis.

Board of Obstetrics and Gynecology, formed in 1927. Dr. James Reinberger was the first Memphis doctor to take the certification exam (1932). For many years, obstetrics and gynecology were two separate departments at the University of Tennessee. Dr. John M. Maury headed the Department of Gynecology, followed by Dr. William T. Black, and Dr. Percy Toombs headed the Department of Obstetrics, followed by Dr. William T. Pride. The two departments were combined in 1945 when Dr. Frank E. Whitacre was made professor of obstetrics and gynecology.

Practitioners of internal medicine did not organize into a formal group until 1946, but their leadership in Memphis medicine was well established by the 1920s. Dr. James Bassett McElroy, acknowledged by his peers as the most influential member of that field, was professor of medicine and chairman of the Department of Medicine at the University of Tennessee from 1920 to 1939. An astute diagnostician, pathologist, teacher, writer, and scholar, McElroy is credited with laying the foundation that led the Department of Medicine to greatness. In appreciation of his leadership, alumni and faculty presented his portrait to the university in 1934. Stories about "big Jim" McElroy were legion, but all affirmed his commitment to the highest principles of medicine. Recognized for his brilliant research in malaria

ABOVE: *Dr. Percy Walthall Toombs, c. 1920. Born in Greenville, Mississippi, Dr. Toombs came to Memphis in 1907. He became professor of physiology at the College of Physicians and Surgeons, then professor of obstetrics at the University of Tennessee College of Medicine, and chief obstetrician at Baptist Memorial Hospital and at Memphis General Hospital.* LEFT: *Dr. William Thomas Black, the first president of the Memphis Obstetrical and Gynecological Society, c. 1923.*

and kidney disease, McElroy was a "regular guy," who enjoyed baseball, fishing, and his prize-winning azalea garden. University of Tennessee colleagues who shared the teaching of internal medicine included Louis Leroy; Frank A. Jones; Reuben S. Toombs; Otis S. Warr; and Conley H. Sanford, who followed McElroy as chairman of the Department of Medicine (1939-1953).

Dr. Harry Christian Schmeisser came to Memphis in 1921 as professor and chair of the department of pathology (1921-1944). In 1922, he began the school for medical technologists in the Department of Pathology, one of the first postgraduate programs for technologists. He also served as chairman of the Section on Pathology of the Southern Medical Association. During the 1920s and 1930s, Drs. W. W. Robinson, Thomas C. Moss, and T. C.

Gladding operated private pathology laboratories. Dr. Douglas H. Sprunt, who succeeded Schmeisser as chairman of the department of pathology (1944-1968), organized the Memphis Society of Pathologists in 1946. Among the first members were Alfred Golden; Thomas C. Moss; I. N. Durbin; I. D. Michelson; A. C. Goss; E. D. Murphey; H. D. Chipps; and R. T. Shields.

Without question, Dr. Lemuel W. Diggs was the leading hematologist in Memphis. He taught clinical pathology at the University of Tennessee from 1929 until 1968, except for 1944 to 1947, when he worked at the Cleveland Clinic Foundation. Diggs organized the city's first blood bank at John Gaston Hospital (see page 109). His textbooks and journal contributions on blood coagulation, blood banking, and sickle cell disease were extensive. His first published paper, "The Blood Picture in Sickle Cell Anemia," appeared in the June 1932 issue of *The Southern Medical Journal*. Diggs said when he came to the university in 1929 there were fewer than 20 scientific papers on sickle cell disease. "Practical Points in Blood Examination" was his first article for the *Memphis Medical Journal*. His major work on hematology, *The Morphology of Human Blood Cells*, was written with Ann Bell and illustrated by Memphis artist Dorothy Sturm. In 1969, Diggs was named professor emeritus in medicine and hematology.

New Hospitals—Methodist Hospital and Memphis Eye, Ear, Nose and Throat Hospital

Methodist Hospital, the third major private hospital organized in Memphis, opened on November 2, 1921, at Lamar and Dudley. Funded by regional Methodist conferences, construction of the hospital was delayed by World War I. Dr. Battle Malone, named chief of staff, led Methodist Hospital from its opening in 1921 until his death on September 4, 1939. His distinguished service during World War I is noted in Chapter 7. A member of numerous medical organizations, he was a life fellow of the American College of Surgeons and an officer in state and local societies. Malone was ranked with Eugene J. Johnson and James A. Crisler as one of the city's preeminent surgeons.

Six months after Methodist Hospital opened, it was sold to the Veterans Bureau, then seeking a hospital in the Mid-South to provide care for service veterans. Identified as Veterans Hospital 88, one floor was reserved for black veterans, and ten beds were set apart for female ex-service nurses. The Methodist staff moved back to the Lucy Brinkley Hospital (acquired in 1918), while a second hospital was under construction at 1265 Union Avenue.

Of classic Georgian design, Methodist Hospital opened on September 16, 1924, and the School of Nursing was

ABOVE: *Built on the site of the Overton Home at 1265 Union Avenue, the new Methodist Hospital opened on September 16, 1924.* OPPOSITE PAGE: *Rear view of Methodist Hospital, Lamar at Dudley, 1921. The nurses' home is on the far right. To its left is the kitchen, laundry, and sterilization department. Six months after its opening, it became a Veterans Administration hospital.*

transferred from the Lucy Brinkley location to the new hospital. Three years later, an addition provided space for an x-ray department and a pathology laboratory. Dr. Thomas C. Moss was named director of the Methodist Hospital laboratories in 1933. Additions to the hospital and to the nurses' dormitory were made in 1938, and a doctors' building, with office facilities for staff physicians, a large pharmacy, and popular soda fountain, was completed in 1940. A church-based institution, Methodist Hospital provided charity care for twenty percent of its patient admissions between 1921 and 1936.

An important smaller institution, the Memphis Eye, Ear, Nose and Throat Hospital, opened to receive patients on August 2, 1926. Dr. Louis Levy was the driving force behind the organization and construction of this specialty hospital located at 1060 Madison Avenue. Dr. Levy moved his office to the hospital and for years was its unpaid administrator. From the opening, accommodations were provided for African-American patients, making this facility the city's first integrated private hospital. A free daily clinic brought many patients from across the Mid-South. Staff physicians in private practice donated their services to the outpatient clinic.

The Memphis Eye, Ear, Nose and Throat Hospital was the first specialty hospital between St. Louis and New Orleans to have an accredited residency in ophthalmology and otolaryngology. In 1928, 20 Memphis doctors were board certified by the American Board of Otolaryngology. At that time, Memphis had a higher percentage of specialists who were board certified in otolaryngology than any city in the country. In 1942, the Eye, Ear, Nose and Throat Hospital was sold to Methodist Hospital, and the residency program was later limited to ophthalmology and lengthened to three years.

The Marcus Haase Nurses Home at Memphis General Hospital (formerly City of Memphis Hospital) was dedicated in January 1927 as a memorial to Dr. Haase, honoring his years of service as unpaid hospital administrator (1906-1924). He organized the systematic maintenance of hospital records, establishing the first records department in any Memphis hospital. Memphis General Hospital was accredited by the American College of Surgeons as the first Class A hospital in Memphis. Dr. Haase died in August 1924 from a streptococcus infection.

The Great Flood of 1927

Spring 1927 brought overwhelming disaster to the Mississippi River valley. Heavy rains during the winter, which were more than ten times the yearly average, turned the river into a raging force. April rains were an unending torrent. On April 15, the *Commercial Appeal* warned: "fear is felt over the prospects of the greatest flood in history." Levees along the Mississippi River gave way. When the levee at Mounds Landing near Greenville, Mississippi, failed, water with the force of Niagara Falls surged over the rich Delta farmland engulfing some 13 million acres. A year's worth of crops was lost. The human tragedy was overwhelming but would have been even greater without assistance from Memphis and other cities that provided food, shelter, and medical care for thousands of refugees. Throughout the Mississippi valley, some 95,000 people received typhoid shots, 85,000 were vaccinated against smallpox, and many thousands more were given quinine to protect against malaria. Ten years later, a more devastating flood would bring greater havoc and thousands of refugees to Memphis.

Expansion of Medical Specialties

Psychiatry was represented in Memphis medical practice beginning in the late nineteenth century when Dr. Benjamin F. Turner was appointed professor of neuropsychiatry at the

Dr. Benjamin Franklin Turner in the anatomy lab, Columbia University, 1890. Turner served as professor of neurology at the University of Tenneseee College of Medicine and chief of staff at St. Joseph Hospital.

JOHN RICHARD BRINKLEY: A SURGICAL CHARLATAN

John Richard Brinkley (1885-1942) was the most famous medical charlatan of his day. On a trip to Memphis in 1913, Brinkley married Minerva (Minnie) Telitha Jones, daughter of Memphis physician Tiberius Gracchus Jones. In 1918, Brinkley established a medical practice in Milford, Kansas. He gained notoriety from surgery that claimed to restore sexual virility by transplanting goat gonads into the patient's scrotum, earning the nickname "Goat Balls Brinkley."

From Kansas, Brinkley moved to the border town of Del Rio, Texas, in 1932. He purchased a powerful radio station in Mexico, over which he broadcast popular entertainment features and promoted his patent medicines, prostate treatments, and medical advice. A master marketer, Brinkley amassed a fortune from his sham medical operations and sale of patent remedies, which were little more than flavored water.

Finally driven out of business by numerous malpractice lawsuits and the unrelenting efforts of the American Medical Association, Brinkley declared bankruptcy in January 1941. He died on May 26, 1942, in San Antonio, Texas, following multiple heart attacks and the amputation of a leg. He is buried in the Jones family plot at Forest Hill Cemetery in Memphis, his grave marked by a large winged victory statue purchased in Naples, Italy. His amazing medical career is part of the lore of the Great Depression. An anonymous voice in the crowd was heard to say, "I knowed he was bilking me, but I liked him anyway."

Dr. John Richard Brinkley's grave marker, with its winged victory statue, Forest Hill Cemetery, Memphis. Brinkley was so popular that he ran three times for governor of Kansas, winning thirty percent of the vote in 1930.

Memphis Hospital Medical College. He became the first professor of neuropsychiatry at the University of Tennessee and was head of the school's Department of Neurology and Psychiatry. Also prominent in the field of psychiatry, Dr. Walter Wallace was an instructor in diseases of the nervous system at the university. A partner in the Pettey and Wallace Sanitarium, he later formed a partnership with Dr. William G. Somerville. In 1921, Somerville succeeded Dr. Turner as professor of neurology and psychiatry.

Dr. John Hayes joined the University of Tennessee faculty about 1926 as assistant professor of psychiatry and also opened a sanatorium in the county on Raleigh LaGrange Road. The facility was purchased in 1937 by Dr. Carroll Turner, son of Dr. Benjamin F. Turner, and Dr. Nicholas Gotten, Sr. and operated as the Turner-Gotten Sanatorium. In his memoirs, Dr. Gotten wrote that neither he nor Dr. Turner had the money to purchase the facility; however, a loan made to the doctors by the husband of a patient at the sanatorium enabled them to make the necessary downpayment. When Dr. Gotten enlisted in the military service in 1942, he sold his interest in the sanatorium to Dr. Turner, who operated it until 1967. After the war, Dr. Gotten elected to limit his practice to neurosurgery and founded the Neurosurgery Group at Methodist Hospital. He was professor of neurology at the University of Tennessee and taught there for more than 30 years.

1929 Stock Market Crash and the 1930s Depression

In spite of earlier prosperity, warning signs marked the last years of the twenties. Agriculture, in a crisis situation from overproduction begun during World War I, was on the brink of collapse. In the South, cotton prices fell sharply. Memphis was a cotton town, and cotton went from 20.2 cents per pound in 1927 to 5.7 cents per pound by 1931. The all-time low of 4.4 cents per pound came in 1932.

As grim as was the crisis in agriculture, the Stock Market crash on "Black Thursday," October 24, 1929, dealt a disastrous blow to the American economy. Unregulated market speculation and easy credit drove stock prices to unrealistic highs. The market began to fall apart in October and efforts to shore it up failed. The New York Stock Exchange collapsed on October 29, 1929, losing $26 billion almost overnight (closer to one trillion in today's dollars). The failure of the American banking system wiped out the deposits of millions of Americans before the Federal Deposit Insurance Corporation and other agencies were created to provide protection for depositors. Although hundreds of banks across the South failed, no bank in Memphis defaulted.

ABOVE: *Dr. Nicholas Gotten, Sr., 1953. After eight years of training in neurosurgery in Philadelphia and at the New York Neurological Institute, Dr. Gotten returned to Memphis in 1936 to begin his practice of neurological surgery which lasted until 1972. During those years, he was also in the Department of Neurology at the University of Tennessee Medical School and on the staff of Methodist and St. Joseph Hospitals.* RIGHT: *Psychiatrist Dr. Dick Cauthen McCool, 1939. Originally from Canton, Mississippi, Dr. McCool received his medical degree from the University of Illinois in 1931. His lifelong love was music. He led "The Mississipians," a dance band at the University of Mississippi and played clarinet with the Memphis Symphony Orchestra.*

Dr. Carroll Turner was a founding member of both the Southern and Memphis Psychiatric Associations (1937). In 1935, Drs. Dick Cauthen McCool and Justin Adler entered psychiatric practice in Memphis. Dr. McCool was on the staff of Gartly-Ramsay, which began the use of electric shock therapy in 1941. Dr. Adler introduced insulin shock therapy, one of the first such programs in the United States. Gailor Memorial Hospital, opened in 1942 as part of the John Gaston Hospital complex, was the first public psychiatric facility in Memphis.

Advertisements for Grove's Chill Tonic, Nash's chill-liver tonic, and 666, three patent medicines for malaria, on a garage in Natchez, Mississippi, 1935. Grove's Chill Tonic, manufactured by E. W. Grove, Paris, Tennessee, was one of the most popular. Grove added a sugary syrup to his product which became an overnight sensation in the 1880s. In 1896, he manufactured Grove's Laxative Bromo Quinine, the first tablet remedy for colds.

The full impact of the market crash was not felt until 1931 when unemployment had penetrated all sectors of the economy. Thousands of homeless sharecroppers and jobless transients came to Memphis seeking work. In the winter of 1930 to 1931, breadlines formed outside hospital kitchens, as the hungry begged for surplus food. Poverty caused an increase in disease and malnutrition was widespread. The sharp rise in the number of suicides testified to the impact of a shattered economy. Infant mortality, always high in Memphis, rose as a direct result of poverty. Dr. L. M. Graves, director of the Health Department, organized 18 city-supported clinics in poor neighborhoods to provide care for mothers and babies.

Business bankruptcies and home mortgage foreclosures added to the financial misery. Millions of hard-working Americans were on relief or existed as best they could. Starving veterans of World War I marched on Washington in 1932. The medical profession was not spared as patients were unable to pay their bills and many hospitals fell on hard times, with some closing.

Former mayor E. H. Crump, elected to the U.S. House of Representatives in 1930, worked with Tennessee Senator Kenneth D. McKellar to obtain relief programs and construction projects for Memphis and Shelby County. When President Franklin D. Roosevelt took office on March 4, 1933, his bold legislation addressed the failure of the banking system, unemployment then nearing thirty percent of the work force, and the crisis in agriculture. His ability to convey courageous optimism brought hope to the American people.

Drs. Semmes and Murphey—1934

Even in the worst of times, good things happen, and that proved to be the case in the association formed by Dr. R. Eustace Semmes and Dr. Francis Murphey. Dr. Paul Bucy of Chicago, a friend of Dr. Semmes, initiated the contact with Francis Murphey, then an intern at Northwestern University Hospital in Chicago. In 1934, Murphey came to Memphis and became Semmes's first neurosurgical resident. Their partnership led to the later formation of the Semmes-

DR. EUSTACE SEMMES AND DR. FRANCIS MURPHEY, NEUROSURGICAL PIONEERS

Raphael Eustace Semmes

Among the men of medicine whose extraordinary abilities shaped Memphis as a medical and research center, Raphael Eustace Semmes established one of the earliest neurosurgical practices in the United States. Born in Memphis on August 15, 1885, young Semmes received his secondary education at Christian Brothers College (both a high school and college), where he ran track and was football team captain.

Semmes received a B.A. degree from the University of Missouri in 1907. While there, he formed a close friendship with Walter E. Dandy who would become professor of neurosurgery at The Johns Hopkins University, celebrated for his research and classic monographs on neurology. Semmes and Dandy enrolled at Johns Hopkins, where they had the benefit of training with William Halsted and Harvey Cushing. Semmes received his M.D. degree (Phi Beta Kappa) in 1910 and interned at Johns Hopkins Hospital (1910-1911) followed by a surgical residency at the Women's Hospital in New York (1911-1912).

On his return to Memphis in mid-1912, Semmes opened an office in the just-completed Baptist Memorial Hospital, later moving his practice to the hospital's new Physicians and Surgeons Building in 1928. Semmes was appointed assistant professor of surgery in charge of neurosurgery at the University of Tennessee College of Medicine in 1913 and in 1932 was named professor of the newly organized Department of Neurosurgery. When Semmes retired in 1956, his partner Francis Murphey followed in that position (1956-1972).

Semmes's medical service during World War I is covered in Chapter 7. In the two decades after World War I, Semmes taught students and general practitioners how to recognize symptoms of brain tumors and other neurological lesions.

OPPOSITE PAGE: *Guests of Dr. Cushing, renowned neurosurgeon, attend the first meeting of the Harvey Cushing Society, Boston, 1932. Dr. Raphael Eustace Semmes stands second from left.* ABOVE: *Dr. Semmes, right, greets Mrs. Richard Ranson at an event to honor Dr. Francis Murphey, left, at the Memphis Country Club, 1972. To the left of Dr. Semmes is Dr. Guy Odom, Duke University, and Richard Ranson. Many colleagues and pupils from the United States, South America, and Canada came for the celebration.*

At a meeting of The Memphis Medical Society in June 1925, Semmes read a paper, "Diagnosis and Treatment of Tumors of the Brain," asserting that "tumors of the brain are of common occurrence" and success in treating the tumor depends on the examining doctor's acumen and promptness of action. In presentations to other physicians, Semmes not only demonstrated knowledge of acute neurological disease but also the ability to communicate critical information to his colleagues.

SEMMES'S MAJOR CONTRIBUTION TO NEUROLOGICAL SURGERY

One of Semmes's major contributions to neurological procedures was the use of local anesthesia during surgery, which was much safer for the patient, especially when postoperative effects of general ether anesthesia, such as vomiting and coughing, posed a serious hazard for delicate neurological surgery. Regarding another of Semmes's significant contributions, Dr. D. J. Canale writes: "Dr. Semmes made a major contribution to the diagnosis and surgical treatment of herniated lumbar discs and his method would become the standard operative approach for lateral lumbar disc herniations."

During the twenties and thirties, Memphis doctors engaged in specialty practice became members of national organizations formed to support these fields of medicine. (See profiles of Drs. Ellett and Campbell.) Disappointed that he was not elected to membership in the Society of Neurological Surgeons (formed in 1920), Semmes and three other neurosurgeons met in Washington, D.C. in October 1931 and organized the Harvey Cushing Society, which received Cushing's blessing. The first meeting of the 23 charter members took place on May 6, 1932, in Cushing's Boston clinic. This group subsequently became the American Association of Neurological Surgeons (AANS), the largest neurological society in America.

Semmes served as AANS president from 1939 to 1940, and Francis Murphey held the presidency from 1965 to 1966. Other members of the Semmes-Murphey group who have led the AANS include Richard DeSaussure (1975-1976), James T. Robertson (1991-1992), and his brother Jon H. Robertson (2007-2008). A fellow of the American College of Surgeons, Semmes took an active role in the formation of leading professional organizations, including the American Academy of Neurological Surgery and the Southern Neurological Society (1949), serving as founding president of the latter in 1950. When the American Board of Neurosurgery was formed in 1940, Semmes was certified without examination, one of 50 original board members. In 1959, he was organizing president of the Memphis Neurological Society, as well as a fellow of the American College of Surgeons.

In 1976, Semmes received the Distinguished Service Award from the Society of Neurological Surgeons. (The society had belatedly voted him a member in 1933.) Two years later, at the 1978 meeting of the Society, Semmes, then 93, read a paper "Walter Dandy, M.D., His Relationship to the Society of Neurological Surgeons." Although one of the country's leading neurosurgeons, friends and colleagues knew Semmes as warm and sensitive with a ready wit. Totally uncompromising in his surgical principles, an acclaimed pioneer in neurosurgery, a gifted teacher and mentor, Dr. Raphael Eustace Semmes died on March 2, 1982, at the age of 96. He had lived to see amazing accomplishments in all fields of medicine, most notably the development of neurological subspecialties.

Francis Murphey Comes to Memphis

When Francis Murphey accepted Semmes's invitation to come to Memphis and train in neurosurgery, it marked the beginning of a partnership that lasted through the profes-

Dr. Francis Murphey, 1953. For years he was associated with Dr. Eustace Semmes in the Semmes-Murphey Clinic.

sional careers of both men. Their extraordinary personal and working relationship was a close parallel to orthopedic partners Willis Campbell and Spencer Speed described elsewhere. A native of Mississippi, Murphey received his B.A. degree from Vanderbilt University and his M.D. degree from Harvard Medical School in 1933. He was one of the last students to train with Harvey Cushing, who had gone to Harvard University in September 1912 to be chief of surgery at the Peter Bent Brigham Hospital, which opened in 1913. Cushing retired in 1932.

Murphey enlisted in the U.S. Army Medical Corps in 1942 and was appointed chief of the neurological service at O'Reilly General Hospital, built in 1941 as a temporary military hospital at Springfield, Missouri. (Kennedy General Hospital constructed in Memphis during World War II was the same type of military facility.) After the war, Elmer C. Schultz (1948) and Richard L. DeSaussure (1950) joined Semmes and Murphey. The group operated as a partnership until 1961, when the practice was incorporated as the Semmes-Murphey Clinic. The Clinic had long tenure at Baptist Memorial Hospital, where Murphey was chief of the hospital neurological service.

During his lifetime, Murphey received numerous professional honors, including the Harvey Cushing Medal (1985), awarded by the American Association of Neurological Surgeons for the original contributions he and Semmes made in the surgical management of ruptured cervical discs. Murphey was also recognized internationally for developing spinal disc surgery and for his research monographs on that subject. In 1965, he was named Neurosurgeon of the Year by the American Academy of Neurosurgery. When Murphey retired in 1972, medical colleagues and friends honored him at a special celebration in Memphis. Former students, who were leading neurosurgeons, paid tribute to an outstanding teacher. Dr. Paul Bucy, who was instrumental in bringing Murphey to Memphis, was a special guest. Dr. Francis Murphey died in Naples, Florida, on June 3, 1994.

Acknowledgment is made to Dr. D. J. Canale et. al., for information adapted from "The History of Neurosurgery in Memphis: the Semmes-Murphey Clinic and the Department of Neurosurgery at the University of Tennessee College of Medicine," Journal of Neurosurgery, June 12, 2009.

LEFT: *John Gaston Hospital, 1939. Completed in June, 1936, the hospital's construction was financed by a bequest from the widow of French-born Memphis restauranteur John Gaston, supplemented by local and federal funds.* BELOW: *Dilapidated shanties, such as these, with outhouses and no indoor running water, dotted the Memphis landscape for years. In 1938, they were replaced by Lauderdale Courts and Dixie Homes, housing built by the WPA. They were both located near hospitals.*

Murphey Clinic. In 1932, Semmes had been named first professor of neurosurgery at the University of Tennessee, a position he held until his retirement in 1956. Murphey held the professorship from 1956 to 1972. Traditionally, the professorship in that department has come from the Semmes-Murphey group.

John Gaston—The New City Hospital

Dedicated on June 27, 1936, the new John Gaston Hospital, a handsome brick and stone building on Madison Avenue, replaced the outdated and over-crowded Memphis General Hospital. A bequest of $350,000 from the estate of Theresa Gaston Mann, in memory of her first husband John Gaston, was augmented by appropriations from the city and county and a grant from the Public Works Administration, providing a total of $800,000 for the new hospital. The new city hospital continued its role as the primary clinical facility for the University of Tennessee.

A new marine hospital, officially known as the U.S. Public Health Hospital, opened at the old Fort Pickering location in June 1937. The surgeon general invited members of the Memphis Medical Society to tour the modern brick hospital and meet the commanding officer, Dr. E. H. Carnes, a 1921 graduate of the University of Tennessee. The U.S. Public Health Service Hospital closed in 1965, as smaller government hospitals where phased out. The Memphis

Metal Museum now occupies part of the site in a structure originally built as a nurses' dormitory.

Federal funds were approved for two important slum clearance projects in Memphis. Completed in 1938, Lauderdale Courts for white residents replaced hundreds of decaying shacks west of St. Joseph Hospital, and Dixie Homes for black residents cleared slums north of Poplar between Ayers and Decatur near the medical center. Between 1937 and 1940, 3,000 public housing units replaced disease-ridden slums. Dixie Homes, leveled in 2008, is being rebuilt as Legends Park, a mixed-use development that is part of the revitalization of the medical center area.

was deeply eroded, and coal chutes and other industrial eyesores gave the city's front door a dirty, unsightly appearance. In a major engineering project begun in 1934, the bluff was cut back, evenly graded, and planted with grass, trees, and flowering shrubs. A roadbed was laid, and landfill used to create the space that became Tom Lee Park. When Riverside Drive was dedicated on March 28, 1935, the reconfiguration and cleanup of the entire area gave Memphis one of the most beautiful riverfronts on the Mississippi River.

1937 Flood Devastation

The fearsome power of the Mississippi River and its primary tributaries, the Ohio and the Missouri rivers, was shown in the flood of 1937, "the most destructive flood in American history." Work that had been done to repair levees after the great flood of 1927 was not equal to the enormous volume of water that rose in some places to 55 feet above flood stage. At Memphis, the Mississippi crested at 48.7 feet on February 10, 1937. In the areas affected by the flood, every means of transportation was crippled, roads and bridges were washed out, railroad beds failed,

TOP: *The overhead coal conveyer of the West Kentucky Coal Company still crossed Riverside Drive in July 1949. The road was built to eliminate the public dumping of garbage and other waste and to control serious erosion of the bluffs.* ABOVE: *Refugees from the Flood of 1937 in the women's and children's barracks, Memphis Fairgrounds. Even the large cattle barns were converted to living quarters with beds lined shoulder to shoulder.*

In other improvements, the Work Projects Administration (WPA) financed new schools, community centers, Crump Stadium, sewer and drainage projects for malaria control, dormitories at the University of Tennessee Medical School, and the TVA electric power distribution system. A major improvement, certainly the most aesthetic, was the construction of Riverside Drive and cleanup of the riverfront. For years, garbage and trash had been dumped over the south bluff directly into the Mississippi River. The bluff

utilities were destroyed, water supplies contaminated, public services paralyzed, and disease was rampant. More than one million people were made homeless throughout an enormous region.

Recalling the disaster of 1927, the American Red Cross organized massive rescue and relief efforts. Led by Dr. Otis S. Warr, Sr., the Memphis and Shelby County Medical Society organized physician teams to provide care for children and adults. Some 80,000 refugees were brought to Memphis, with 60,000 arriving in a single week. Refugees were processed at the receiving station organized in the Ellis Auditorium (located at Main at Poplar), where an emergency hospital was set up on one side of the building. Each individual was inoculated for typhoid and those not previously vaccinated against smallpox were vaccinated. About twenty percent of those arriving in Memphis had pneumonia or influenza as a result of exposure in the cold January weather.

All centers housing refugees were checked daily for communicable diseases. A second isolation hospital under construction on the John Gaston Hospital campus was rushed to completion and received cases of spinal meningitis. Those suffering from general illness were distributed among Memphis hospitals, and refugees with tuberculosis were sent to Oakville Memorial Hospital. The Red Cross organized a base hospital in Fairview School at Central and East Parkway adjacent to the large barracks refugee center at the Fairgrounds. The new juvenile court building was used as a children's hospital.

When the crisis had subsided, Dr. Otis Warr left for Charleston, South Carolina. While stopping at Gatlinburg, he was stricken with pneumonia and died on March 22, 1937, at St. Mary's Hospital in Knoxville. As clinical professor of medicine at the University of Tennessee, Dr. Warr was recognized for his outstanding diagnostic abilities. Dr. O. W. Hyman of the University said: "Otis Warr was the most important and the most beloved man on our staff." He was the first of four generations of Memphis physicians.

Dr. Diggs and the Memphis Blood Bank

The last pre-World War II medical development for which Memphis received national recognition was the Memphis Blood Bank established at John Gaston Hospital in 1938, the fourth blood bank in the country and the first in the South. Dr. Lemuel W. Diggs directed the research that solved problems associated with direct patient-to-patient blood transfusions. Direct transfusion was time-consuming and involved finding a donor and typing the blood, a delay that often resulted in the patient's death.

Working with a team of technicians, interns, nurses, and medical students, procedures were developed for the sterile collection, processing, and refrigeration of blood for later use. To prevent clotting of the blood, medical student Colin F. Vorder Bruegge (later Brigadier General, U.S. Marine Corps) created the first mechanical shaker, and intern Dr. Edward Atkinson invented a slide shaker that hastened cross matching of blood. Diggs later recalled that private citizens donated money to purchase the refrigerators used in 1938.

In 1940, medical technologist Jeanette Spann was named director of the blood bank at John Gaston Hospital, one of the first in the country to process plasma, a critical compo-

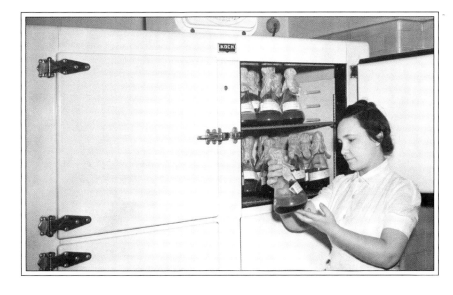

Jeanette Spann, director of the Blood Bank program at John Gaston, examines a sample of blood in an Erlenmeyer flask, 1941. That year, she was also named "Miss Blood Bank." Alice Jean Keith was the first technician to work in the facility.

nent in saving the lives of soldiers wounded in World War II by preventing death from shock. Dr. Russell Patterson was employed by the American Red Cross to assist in the blood bank. The Red Cross utilized technology developed at John Gaston Hospital to collect and process blood for America's massive war effort. After the war, blood banks were established in all Memphis hospitals, a life-saving benefit pioneered by Dr. Diggs and his research team at John Gaston Hospital. In 1973, Dr. Diggs received the Tennessee Association of Blood Banks first Lemuel W. Diggs Meritorious Service Award for his pioneering contributions to blood banking. (His internationally acclaimed research in sickle cell disease is covered in Chapter 12.)

By the end of 1939, new industries such as Firestone Tire and Rubber Company and Ford Motor Company provided jobs for several thousand workers in Memphis. The federal Flood Control Act passed in August 1937 put hundreds of men to work on levee construction. Under the well-designed Bartholomew Plan, the stage was set for future city growth. As the medical center of the Mid-South, Memphis was recognized for its leadership in general surgery, orthopedics, ophthalmology, neurology, radiology, and hematology. Europe was already at war, and the great conflagration of World War II was only two years away. The massive scale of America's involvement in the war would bring full employment for the workforce, increased opportunities for women, especially in medicine and nursing, and dramatic advances in all branches of medicine.

CHAPTER NINE

THE WORLD AT WAR AND THE BATTLE AGAINST INFECTION

1940-1949

WORLD WAR II BROUGHT THE UNITED States together as no other event in history. Able-bodied men were mobilized for fighting or for industrial production. Rationing of food, tires, and gasoline began, and women joined the workforce, doing jobs men left behind. We were fighting for survival, and President Roosevelt proclaimed America "the arsenal of democracy."

Physicians in War

In September 1940, with war raging in Europe and looming in the United States, Congress authorized the first peacetime draft in American history, requiring men aged 21 to 35 (later expanded to men 18 to 45) to register with local draft boards. After Pearl Harbor, machine guns guarded the Harahan Bridge, the airport, and Memphis Light, Gas and Water facilities.

Selective service boards handled the selection process, with boards advised by physicians on local advisory committees for procurement and assignment. Physicians from a single city were not designated "hospital groups" as in World War I, but were assigned by specialty.

In October 1942, an article in the *Memphis Press-Scimitar* bemoaned the scarcity of physicians, with "87 of 346 practicing physicians called" and "other doctors waiting to be called." However, the Tennessee Medical Association reported that almost 600 Tennessee physicians had entered military service without causing distress in local communities.

Injuries in World War II were often blast injuries, burns, and shock. The effects of blast injuries were little known before the war. Neurological damage from blasts and lung damage from burns might not be immediately evident. Use of plasma on the battlefield was a major advance in

OPPOSITE PAGE: James McLemore salutes Kennedy General Hospital in Memphis, 1946. By the end of the war, it was the second-largest of the Army's general hospitals, specializing in neurology, neurosurgery, psychiatry, orthopedics, and thoracic surgery.

110 MEMPHIS MEDICINE: A HISTORY *of* SCIENCE AND SERVICE

ABOVE: *Dr. Giles Coors with members of Memphis Draft Board No. 9, October 1940. Left to right: William Loeb, Robert Galloway, John Vorder Bruegge, Dr. Coors, Swayne Latham. President Roosevelt had approved 12 draft boards for Memphis and Shelby County. Dr. Coors was a medical reservist during World War I.*

RIGHT: *Dr. Harwell Wilson received the Legion of Merit as a member of the United States Army Medical Corps during World War II. He was professor of surgery and chairman of the department at the University of Tennessee from 1948 to 1974.*

effectiveness. Some units experienced one hundred percent rates of malaria and other tropical diseases. Oil was spread on mosquito breeding grounds, and DDT was sprayed widely. Medical personnel dispensed quinine and later atabrine for malaria. Researchers at the University of Tennessee—Drs. Arthur Richardson, Lloyd Seager, Reginald Hewitt, and James Etteldorf—searched for an alternative to Japanese cinchona to treat malaria and recommended the synthetic drug atabrine, which is still used.

From the first attention by an Army medic or a Navy corpsman, the wounded were moved to a battalion aid station and then to an evacuation hospital, either fixed with 750 beds and major medical facilities, or "semimobile" with 400 beds and capable of moving ten hours after patient evacuation and becoming operational six hours after moving. Known as Mobile Army Surgical Hospitals (MASH) in Korea, these units were immortalized in the popular media, though portrayed as smaller than in reality. Severely wounded soldiers were sent to field hospitals or to general hospitals of 1,000 beds. Particularly in the Pacific, hospital ships provided initial care and sometimes surgery. Air evacuation developed late in the war.

combating shock and, by 1945, whole blood was available near the battlefield.

Combat surgeons brought American ingenuity to the treatment of injuries. Debridement and delayed closing of wounds reduced infection. Orthopedic surgeons reestablished blood flow in limbs. Memphian Dr. Marcus Stewart and his crew in England rehabilitated the wounded. New drugs—first sulfa and then, in 1944 with America's mass production and use at D-Day, penicillin—saved countless lives.

Soldiers were vaccinated for typhoid, tetanus, typhus, and yellow fever, diseases endemic to some battle areas. The Pacific theater presented special problems, including tropical diseases and malaria, which could prevent combat

Psychological damage—"shell-shock" in World War I—was labeled "combat fatigue." Removal from combat, rest, regular meals, and clean surroundings brought recovery to most patients. Later, in Vietnam and both Gulf Wars, the condition was known as post-traumatic stress disorder (PTSD).

In 1946, papers from a Symposium on War Medicine and Surgery, sponsored by the Memphis and Shelby County Medical Society, were published in the *Memphis Medical Journal*. Dr. Harwell Wilson discussed sympathectomy for claudication in wounds, use of penicillin and sulfa in managing soft tissue wounds, and advanced techniques

DR. C. BARTON ETTER AND PATTON'S SLAPPING INCIDENT

Memphis pediatrician Dr. C. Barton Etter (pictured below) was receiving officer when General George Patton, desperate for men in the fierce battle for Sicily and having heard of soldiers complaining of battle fatigue, toured the 93rd Evacuation Hospital. In an infamous incident, General Patton interrogated, slapped, and threatened a soldier in his twenties, who was later revealed to have malaria. Major Etter submitted this report to Colonel Richard T. Arnest, his superior officer:

"On Monday afternoon, August 13, 1943, at approximately 1330, General Patton entered the Receiving Ward of the 93rd Evacuation Hospital and started interviewing and visiting the patients who were there. There were some ten or fifteen casualties in the tent at the time. The first five or six that he talked to were battle casualties. He asked each what his trouble was, commended them for their excellent fighting, told them they were doing a good job, and wished them a speedy recovery.

He came to one patient who, upon inquiry, stated that he was sick with high fever. The general dismissed him without comment. The next patient was sitting huddled up and shivering. When asked what his trouble was, the man replied, 'It's my nerves,' and began to sob. The General then screamed at him, 'What did you say?' He replied, 'It's my nerves. I can't stand the shelling any more.'

He was still sobbing. The General then yelled at him, 'Your nerves Hell, you are just a Goddamn coward…. He then slapped the man and said, 'Shut up that Goddamn crying….' He then struck at the man again, knocking his helmet liner off and into the next tent. He then turned to the Receiving Officer and yelled, 'Don't you admit this yellow bastard, there's nothing the matter with him. I won't have the hospitals cluttered up with these sons of bitches who haven't the guts to fight.

He turned to the man again, who was managing to 'sit at attention' though shaking all over, and said, 'You're going back to the front lines and you may get shot and killed, but you're going to fight. If you don't, I'll stand you up against a wall and have a firing squad kill you on purpose. 'In fact,' he said, reaching for his pistol, 'I ought to shoot you myself….' As he went out of the ward he was still yelling back at the Receiving Officer to send that yellow son of a bitch to the front lines." (Quoted by Ladislas Farago in *The Oxford Book of Military Tales*)

Dr. Etter returned to Memphis after the war and became a prominent pediatrician and a driving force in establishing Le Bonheur Children's Hospital. Through his wife's career, the name survives today in Coleman-Etter Real Estate.

ABOVE: *Major James G. Hughes, right, awaited the arrival of his new executive officer at Fort Benning, Georgia, in 1943. It turned out to be his twin brother, Captain John D. Hughes.* RIGHT: *In 1942, Dr. Henry B. Gotten first joined the Naval Reserves in Charleston, South Carolina, as a lieutenant commander. Subsequently, he served at Palermo and at other Italian bases before being promoted to commander. This photo was taken before the U.S. entry into the war when he worked at the Gartly-Ramsay Hospital.*

in thoracic and vascular surgery. Dr. William M. T. Satterfield, who had worked in the Iranian desert, detailed his experience with leishmaniasis. Dr. Edward D. Mitchell, Jr., discussed evacuation of the injured to hospitals, noting that a large percentage of injuries were caused by mortar shells and hand grenades, with fewer wounds from rifle fire. In Okinawa, physicians saw many burns, and gas gangrene was particularly troublesome. In all theaters of war, the need for blood was widespread and urgent. Dr. John D. Hughes discussed the neuropsychiatric aspects of war medicine, noting that peptic ulcers (often related to stress) were common, although most had existed before military service. Dr. Thurman Crawford discussed aviation medicine. Symposium presenters also considered problems of returning doctors, many of whom had developed specialized knowledge in wartime but returned to less specialized problems, a rising cost of living, scarce living and office space, and concerns about socialized medicine.

More than 120 physicians from the Memphis Medical Society served in World War II. At least eight alumni of the University of Tennessee College of Medicine died in combat or in military accidents: Everett B Archer, Army major, died serving as an Army physician in New Guinea, 1945; Alton C. Bookout, Navy lieutenant, was killed in Manila Bay, December 7, 1941, when a Japanese plane sank his destroyer; Newton A. Cannon, Army 1st Lieutenant, was killed in action, Luzon, Philippines; Claude R. Huffman, Navy lieutenant, died of chest wounds from shrapnel in an attack on his destroyer in the South Pacific; Lewis C. Ramsey was killed in an airplane accident; Robert H. Robbins, Army captain, died of a fractured skull in North Africa; Wendell F. Swanson, Army major, a prisoner of war after Bataan and Corregidor, died in the sinking of a Japanese prison ship; and John C. Walker, Army Air Force major and flight surgeon, was killed in a plane crash in England.

Many physicians won medals and awards. Others served and returned home to build successful careers. Among Memphis physicians serving were the following:

★ Shed Caffey—Navy, Pacific and Atlantic
★ Lawrence E. Cohen—Army, 1943-1944

★ Francis H. Cole—U.S. Marines in New Orleans; distinguished chest surgeon
★ George A. Coors—Navy surgeon and prominent general surgeon
★ Joseph Crisler, Jr.—First Army in Europe; cared for frontline casualties
★ McCarthy DeMere—plastic surgeon; Croix de Guerre from Luxembourg for service in the Battle of the Bulge
★ Richard L. DeSaussure, Jr.—Army, 1944-1945, 2nd Lieutenant, Cavalry, Battalion surgeon, 3rd Armored Division; Bronze Star; Combat Medic Badge
★ Thomas G. Dorrity—99th Infantry, Battle of the Bulge; chief of general surgery at Methodist Hospital, 1970-1979; president, The Memphis Medical Society, 1977
★ Francis D. Gibson—Captain in Army Medical Corps; Bronze star; worked at VA Hospital after war for 30 years; assistant chief of spinal cord injury
★ Henry Gotten—Lt. Commander, Navy; California. Well-known Memphis internist
★ Nicholas Gotten, Sr.,—Lt. Commander, Navy; Hawaii and Oklahoma; founder of Neurosurgical Group in Memphis
★ Arthur E. Horne—Army; received Medical Society's "50-Year Physician" award
★ James G. Hughes, twin of Dr. John D. Hughes—awarded battle stars and Legion of Merit as commanding officer of 225th Station Hospital, which gained the Meritorious Unit Award; became commanding officer of 330th General Hospital, U.S. Army Medical Reserve; appointed to Surgeon General's Advisory Council for Reserve Affairs; retired as Brigadier General after 22 years in Reserve; prominent Memphis pediatrician
★ John D. Hughes, twin of Dr. James G. Hughes—prominent internist in Memphis

ABOVE: *World War II veterans Drs. Thomas G. Dorrity, left, and McCarthy DeMere at a meeting of the Tennessee Medical Association, April 1976. At the time, Dr. Dorrity was chief surgeon at Methodist Hospital and a past president of the American Association of Physicians and Surgeons. Dr. DeMere helped prepare the first brain death law for Tennessee which was adopted nationally.*
LEFT: *Dr. J. Pervis Milnor, Jr., 1960. A surgeon in a mortar company in New Guinea, he also served in the Philippines campaign during World War II.*

★ Alvin H. Ingram—Army Medical Corps, four years; discharged as major
★ J. Cash King—served aboard U.S.S. *Monterrey* in Gilbert and Marshall Islands, Saipan, and Guam; eminent radiologist at Methodist Hospital
★ William P. Maury, Jr.—Army Medical Corps, four years; discharged as major
★ Pervis Milnor—left Yale for the University of Tennessee's accelerated quarter program in 1939 to avoid draft

ABOVE: *Lieutenant Commander Moore Moore and his wife, December 1942. Dr. Moore became a rear admiral in the Navy in November 1964. He also became chief of orthopedic surgery for Methodist Hospital.*

RIGHT: *Radiologist Dr. J. Cash King, left, served on the U.S.S.* Monterrey. *He presents a portrait of Dr. Charles Hunter Heacock to Dr. Homer F. Marsh, vice-president of the University of Tennessee, 1963. Dr. Heacock was chairman of the University of Tennessee Department of Radiology from 1938 to 1958.*

1944-1945; recalled during Korean Conflict, 1954-1955
★ Marcus J. Stewart—orthopedic surgeon and rehabilitation specialist, Army, Europe and England; received Legion of Merit; colonel in reserves
★ Webster Riggs, Sr.—5th Army Dispensary, St. Louis, MO
★ Henry Stratton, Jr.—Naval air lieutenant; fighter pilot with 285 carrier landings; two Distinguished Flying Crosses; 12 stars of merit; noted orthopedic surgeon
★ Hubert K. Turley—Captain in Medical Corps
★ Harwell Wilson—Surgeon in war; later longtime head of surgery at University of Tennessee
★ John Wilson—Army; North Africa and Italy; prominent radiologist

THE HOME FRONT

Memphis faced wartime realities. The wounded arrived almost daily at the U.S. Naval Hospital in Millington, built in 1942 to serve Navy personnel, and at the Army's Kennedy General Hospital. When it was built in Memphis in 1943, Kennedy Hospital—named for Brigadier General James N. Kennedy, a World War I surgeon whose last assignment was as commanding general in the Army Medical Center in Washington, where Dr. Walter Reed also worked—was originally planned for 1,500 patients, but the Army, realizing it was too small, added another 1,500-bed hospital beside it. By 1945, it was the second-largest of the Army's general hospitals. Kennedy Hospital specialized in neurology, neurosurgery, psychiatry, orthopedics, and thoracic surgery.

interruption of medical education; graduated 1942 and interned at Boston City Hospital; volunteered for the Air Force; served at aid station, New Guinea, and mortar company support, Philippines
★ Moore Moore—Admiral in Navy; at one time served on the yacht (sold to Navy in bankruptcy) of charlatan Dr. John Richard "Goat Balls" Brinkley; prominent Memphis orthopedic surgeon
★ Donald Pinkel—first head of St. Jude Hospital; served

The VA facility at 1025 East Crump Boulevard—VA No. 88, built in 1922—served veterans with a 225-bed capacity until the war ended. When the number of veterans became too great for the facility, it moved to the Kennedy facility

DR. MARCUS STEWART

Dr. Marcus J. Stewart joined the Army Medical Corps in 1941 and later commanded several field hospitals. For establishing Army rehabilitation programs, he received the Legion of Merit. He remained active in the Army Reserve, retiring as a colonel.

During World War II, Eleanor Roosevelt, Marlene Dietrich, and Bob Hope visited the hospital. Stewart worked with Britain's Dr. Howard Florey, who was developing penicillin for mass use. Florey sent small amounts of the experimental drug, which Stewart injected daily into the infected knee of a patient, curing the infection.

A key figure in developing rehabilitation methods further expanded in VA hospitals after the war, Stewart toured 26 hospitals east of Paris, where he was called to address problems with amputees during the Battle of the Bulge. He was assigned a sergeant who had lost a leg in the Pacific and a corporal who had lost his right arm in North Africa. The two rehabilitated men performed a routine in the amputee wards, first incensing patients with tales of high living in New York and then demonstrating their agility with prostheses. After their visits, even depressed soldiers who had refused to eat began chatting, ready for rehabilitation.

As an orthopedic surgeon in Memphis, Stewart developed a strong relationship with orthopedic device manufacturer Richards Medical. With fellow orthopedic surgeons Drs. Hugh Smith, Alvin Ingram, and Lee Milford (a prominent hand surgeon), he developed devices for surgery and reconstruction.

A consultant to the surgeon general of the U.S. Army for many years, Stewart inspected American orthopedic services and worked with rehabilitation. He served on the President's Committee for Employment of the Physically Handicapped and chaired Tennessee's equivalent organization under two governors.

Dr. Stewart taught at the University of Tennessee and helped found orthopedic groups. He was president of both the American Orthopedic Society and the Clinical Orthopedic Society. He was team physician for the University of

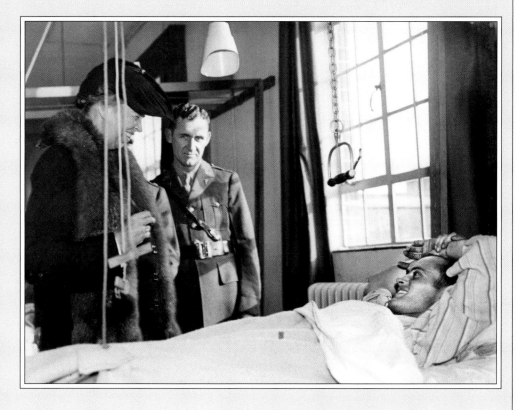

Dr. Marcus J. Stewart, center, watches as Eleanor Roosevelt visits with a patient at Second General Hospital, Oxford, England, 1943.

Mississippi's football program from 1955 to 1975, and he published widely, including an edited collection, with Dr. William T. Black, titled *History of Medicine in Memphis* (1970). He helped organize the American Orthopaedic Society for Sports Medicine and was its sixth president (1977-1978).

Dr. Stewart never forgot his primary concern. In his autobiography he wrote, "The greatest reward and joy for a doctor comes from serving his patients."

ABOVE: *The Veterans Administration Hospital, 1025 East Crump Boulevard, was initially the Methodist Hospital before being purchased by the Veterans Bureau in 1922.* RIGHT: *Named for General James N. Kennedy, Medical Corps, the Kennedy General Hospital became the Veterans Administration Medical Teaching Group Hospital on July 1, 1946. Located at Park Avenue and Getwell Street, few of the buildings remain. Note the narrow links connecting the buildings.*

vacated by the Army. VA No. 88 was converted in 1946 to a 300-bed tuberculosis hospital until the West Tennessee Tuberculosis Hospital was built in 1948 on Dunlap between Jefferson and Adams.

The Kennedy Hospital was given to the Veterans Administration and, in 1946, was renamed the Veterans Administration Medical Teaching Group Hospital. The facility kept its "Kennedy" association in the public mind, however. The facility comprised 128 buildings distributed over 146 acres off Shotwell Street. When it opened as a VA Hospital, the name "Shotwell" was changed to "Getwell" south of Park Avenue. When the VA's plans to collaborate with the University of Tennessee did not materialize, Dr. Paul R. Magnuson, chief medical director of the VA in Washington, organized a professional staff, and the facility became one of perhaps two VA hospitals in the United States with a teaching program not affiliated with a medical school. That status would change with completion of a new VA Hospital in 1967. The hospital provided expert patient care and teaching and is remembered particularly for the surgical training it offered under Dr. Joseph McCaughan, a vascular surgeon who performed some of the longest composite aortic grafts in the world, and Dr. Ralph Downs, a surgeon from Walter Reed Hospital.

Members of civic organizations provided gifts and entertainment for the wounded veterans. The staff—often

ABOVE: *Dr. John Lucius McGehee, Jr., 1948. Near the end of World War I, Dr. McGehee was placed in charge of maxillofacial surgery at American Base Camp No. 15 at Chaumont, General Pershing's headquarters. He became president of the Memphis and Shelby County Medical Society in 1920, and was professor and chairman of the Department of Surgery, University of Tennessee, from 1932 to 1948.* RIGHT: *Continuing a long tradition of Red Cross volunteerism, Mrs. E. L. Goelz, seated, holds her award as Mrs. W. H. Bade, chair of Memphis Volunteers, gives her approval, 1970. Mrs. Goelz received the "Gold Award," the highest honor, for her 15,000 volunteer hours.*

confronting new and frustrating challenges in repairing damaged bodies—developed an active program of research and publication.

At the University of Tennessee, older professors held the Medical School together, and older practitioners took on additional duties while younger physicians went to war. Dr. Thomas H. West, after a surgical residency at Boston's Lahey Clinic and an invitation to remain on staff, received a message from Dr. Lucius McGehee, then chair of the University's Department of Surgery: "Come back and help me train doctors for the war." Dr. West returned to Memphis and led the Department of Surgery from 1941 to 1951. Dr. Russell Patterson, Jr., prominent general surgeon at Baptist, was also part of this generation of surgeons.

Medical school enrollments grew during the war, leveled off, and then grew again later. The unique admissions policy at the University, initiated in 1932, meant accepting a class of students each quarter, enabling students to complete medical school in three years. The federal government instituted a program to ensure a supply of doctors with the Armed Specialized Training Program (ASTP) and the Navy V-12 program. Eligible medical students received commissions but were given "detached service" in medical school, receiving military instruction and participating in weekly drills while in school. Later in the war, the 9-9-9 program was instituted. Instead of attending medical school for nine months and working in the summers, students were given $132 per month, plus uniforms, and summer vacations were eliminated so that students could graduate more quickly.

VOLUNTEER EFFORTS

The Red Cross in Memphis trained 8,000 volunteers. Rationing coupons were issued to each man, woman, and child, and travel required saving gasoline coupons. The

Coast Guard Temporary Reserve relieved regular members. Older cars were kept running as metal was used for the war. Patients were urged to go to the doctor's office to save a house call. Buying war bonds was considered a patriotic duty, and significant purchases brought naming rights for aircraft. The *Memphis Belle* was the first bomber to complete the 25 missions required to leave combat. Named for the pilot's girlfriend, the plane was filmed by William Wyler on its last mission. The African-American community in Memphis bought $300,000 in war bonds for a B-24 Liberator bomber named *The Spirit of Beale Street*.

Prosperity at Home

War industries brought jobs and prosperity to the area. Men and women not in military service flooded into Memphis, doing office work or serving the wounded. By 1942 in Millington, the Naval Air Technical Training Center trained aircraft mechanics, the Naval Air Station provided a large base, and the Naval Hospital treated casualties. The Navy

ABOVE: *The* Memphis Belle, The Spirit of Beale Street, *and* City of Memphis *were three World War II bombers tied to Memphis. When this photograph was taken in 1945, the* City of Memphis *had flown 16 missions against industrial and military targets in Japan. The only member of the crew from Memphis was Lieutenant John J. Handwerker, airplane commander, standing center.*

OPPOSITE PAGE: *At the Jones & Laughlin Steel Corporation plant on the Wolf River in Memphis, the Pigeon-Thomas Iron Works launched its first LCT (Landing Craft Tanks) in July 1943. By October 1944, 185 had been built.*

rented 300 acres from Bragg Brothers Farm in northern Shelby County for pilot training. The Army Depot and the Mallory Air Force Depot were built in 1943. The Army Air Force had a supply depot on Jackson Avenue. Private industries in Memphis supported the war effort by producing smokeless powder, Navy assault vehicles, ordnance, precision gears, precision-machined parts for aircraft engines, aircraft gears, and frozen food equipment.

In 1940, anticipating conflict, the War Department divided the country into four areas, with Memphis as headquarters for the Second Army, whose tasks were to bring National Guard units up to regular Army standards and to train selectees. In 1942, an air support base at the Memphis airport began ferrying supplies to the war. Even German prisoners housed at the Defense Depot voluntarily fought levee breaches. A detachment of the Army Air Force's College Training Program received housing and instruction at Memphis State Teachers College.

Public Health

Although war occupied the minds of Memphians, disease took no holiday. In late February 1940, Memphis hospitals were filled with influenza patients. Public Health Director Dr. L. M. Graves observed that there was as much flu in 1940 as in 1918, though the 1940 strain was less virulent. When flu returned at the end of the year, the Rockefeller Institute sent 80 doses of a new vaccine to Memphis, and prominent Memphians volunteered to test the vaccine. A month after Pearl Harbor, 50,000 Memphians were reported stricken with flu.

In 1941, when new outbreaks of polio began, the Health Department organized the Crippled Children's Service, which received funds from the U.S. government under the Social Security Act to aid needy children. Dr. Robert Knight of Campbell Clinic worked with Ms. Betty Speltz, a physical therapist trained in the Kenny method, during the war years, and pediatrician Dr. Gilbert Levy was in charge of the Isolation Hospital behind John Gaston Hospital (see Chapter 10).

Tuberculosis, the leading cause of death in the early twentieth century, remained widespread in Memphis, which had a high incidence rate because of poverty, crowding, and lack of sanitation in some neighborhoods. No cure was yet available for the "white death," whose victims withered away as white scar tissue formed in the lungs. The Medical Society held special clinics to educate people and to isolate TB cases. In 1941, the Health Department established a tuberculosis division where "TB nurses" cared for patients.

In a much-needed move of cooperation between Memphis and Shelby County governments, city and county health departments merged in 1942. The new department opened at 879 Madison Avenue. It would move in 1959 to new quarters at 814 Jefferson Avenue, designated the L. M. Graves Health Building in honor of Dr. Graves, who headed the department from 1928 until his death in 1964. A tall, imposing figure, Graves spoke to the populace via radio on health-related topics, emphasizing disease prevention. With the Medical Society's support, he helped Memphis win awards for its health initiatives.

MEMPHIS PRACTICES

In 1948, when management of hypertension with medication was years away, Dr. Francis Murphey reported in the *Memphis Medical Journal* a surgical procedure to relieve hypertension. Requiring removal of the twelfth rib, an incision through the diaphragm, and resection of part of the sympathetic ganglion and the splanchic nerves, the procedure was complex. Today, blood pressure is controlled with medications, such as clonidine, that inhibit the sympathetic nerves.

Neurosurgery. Contemporary with the Semmes-Murphey Clinic (still a partnership at this time, not taking the name Semmes-Murphey until 1961) at Baptist, the Neurosurgical Group at Methodist was founded by Dr. Nicholas Gotten, Sr., on his return from war service. The group grew to include ten neurosurgeons, including Drs. A. Roy Tyrer; William Ogle, Jr.; Joseph Miller; David Cunningham; and C. Douglas Hawkes. Dr. Gotten, Sr., taught neurology at the University of Tennessee for 30 years and established the first private electroencephalography laboratory in Memphis. Pediatrician Dr. Evelyn Ogle joined him, running the laboratory and establishing a strong reputation as a pioneer in electroencephalography in Memphis.

Dr. Douglas Hawkes was instrumental in establishing a residency in neurosurgery at Methodist, and Dr. Roy Tyrer established the Foundation for International Education in Neurological Surgery, which strengthened neurosurgical training worldwide.

Dr. Tyrer, who continued to see patients into his nineties, noted in 2009 that neurosurgery came into its own during World War II. Dr. Tyrer saw advances, such as myelograms, that developed during the war while he was at Letterman Hospital in California, where the injured from the Pacific theater came. He recalled the development of vascular surgery in the 1970s and of spine surgery, which began with the organization of a Spine and Peripheral Nerve section by the Congress of Neurological Surgeons in 1979 and became a leading subspecialty in the 2000s. Dr. Tyrer was active at the national level of the American Medical Association.

Obstetrics and gynecology. During the war years, obstetrician and gynecologist Dr. Webster Riggs, Sr., who was past draft age, taught at the University of Tennessee and had one of the largest practices in Memphis. When caudal anesthesia became available, he was the first obstetrician in Memphis to use it. Dr. Walter Ruch, Sr., was another prominent obstetrician and gynecologist. His sons, Walter, Jr., and Robert, would later join him in practice. At Methodist, Drs. Carey Bringle, Curtis Ogle, and Peter Ballenger were stalwarts.

Ophthalmology. Since the days of Dr. Edward Coleman Ellett, ophthalmology remained prominent in Memphis. Dr. Ellett was joined by Drs. Ralph Rychener, Roland Meyers, Richard Miller, and William Murrah. Dr. Rychener first brought photocoagulation to Memphis with the Zenon Arc Coagulator, a major innovation that was the fourth of its kind in the country. The Mid-South Eye Bank for Sight Restoration was organized in 1940, providing eye transplant tissue for patients blinded by corneal disease.

Dr. Philip Meriwether Lewis succeeded his brother, Dr. A. C. Lewis, as clinical professor and chair of ophthalmology at the College of Medicine in 1945 and served in that position until 1969. His original clinical practice was in the Exchange Building near Court Square in downtown Memphis, the most prestigious physician office address at

Tuberculosis was still a public health threat well into the twentieth century. The placard advertises a teaching clinic, sponsored by the Shelby County Tuberculosis Society, for African-American physicians, 1934.

the time and the same building where Dr. Willis Campbell had practiced for a time. Tall, with the bearing of a Southern gentleman and known for his fast driving, Dr. Lewis practiced for 60 years, from 1924 until 1984, the first 55 years in the same office in the Exchange Building. He was president of the Memphis and Shelby County Medical Society in 1955 and of the American Ophthalmological Society from 1966 to 1967. The P. M. Lewis Eye Clinic in the Gailor Outpatient Clinic building was dedicated in 1967 in his honor.

From the late 1930s until the 1960s, most private eye surgery was done at the EENT Hospital, which was owned by Methodist Hospital from 1942 until it was sold in 1966 to Dr. John Shea, Jr. Until his death in 1952, Dr. Louis Levy practiced EENT and used the facility. When the facility was sold, the equipment, personnel, and training program were transferred to Methodist Hospital, with eye patients hospitalized on the third floor of the Thomas Wing, and ENT on the third East wing.

Memphis had two eye residency training programs, one at the EENT Hospital—a combined ophthalmology and ENT residency until 1952—and the other at John Gaston Hospital—begun in 1945 as an ophthalmology residency. The two ophthalmological residencies merged in 1956 under the University of Tennessee's direction, and the ophthalmology training program at the University gained full departmental status in 1967.

Orthopedics. Dr. Willis C. Campbell remained the dominant figure in orthopedics and orthopedic surgery until his death in 1941. Campbell wrote the first edition of his *Operative Orthopaedics*, published in 1939. Thorough and well-conceived, the book became the standard reference for orthopedic surgery and remains so in 2010. Regular updates edited by Memphis orthopedic surgeons have kept the book current.

The first edition of *Campbell's Operative Orthopaedics* was one volume with 1,126 pages. The eleventh edition, the largest, had four volumes with a total of 4,899 pages and was accompanied by surgical techniques on DVDs and for the first time offered on-line access and an iPhone application.

Pathology. Early key figures in pathology included Dr. Thomas C. Moss, who worked at Methodist Hospital in the 1930s and at St. Joseph Hospital in the 1940s, and later developed an independent pathology laboratory. Dr. N. E. Leake worked in pathology at Baptist, and Dr. John A. McIntosh worked at St. Joseph.

Pediatrics. Dr. Gilbert Levy (1893-1975), a captain in World War I, was one of the few pediatricians in Memphis during World War II. For 30 years he was an attending pediatrician at Memphis General and later at Gaston Hospital, where he a treated a variety of contagious diseases. A member of The Memphis Medical Society for 50 years, he also was president of the Memphis Pediatric Society and a member of the board of governors of St. Jude. In 1937, young Dr. Levy, walking by the emergency room, saw an anoxic four-year-old with diphtheria. Grabbing a scalpel from his medical bag, he performed an emergency tracheotomy. Despite blood everywhere, the boy began breathing before being taken to surgery for insertion of a tube. Thirty years later, Dr. Levy recognized the patient by the scar on his neck.

CAMPBELL'S OPERATIVE ORTHOPAEDICS, EDITIONS

Editor	Date	Edition
Dr. Willis C. Campbell, author	1939	1
Drs. James S. Speed, Hugh Smith	1949	2
Drs. James S. Speed, Robert Knight	1956	3
Dr. A. Hoyt Crenshaw	1963, 1971	4, 5
Drs. A. Hoyt Crenshaw, Allen S. Edmondson	1980	6
Dr. A. Hoyt Crenshaw	1987, 1992	7, 8
Dr. S. Terry Canale	1998, 2003	9, 10
Drs. S. Terry Canale, James H. Beaty	2008	11

ABOVE: *The Thomas F. Gailor Psychiatric Hospital and Diagnostic Clinic on Dunlap between Madison and Jefferson, October 1942.* RIGHT: *Dr. George R. Livermore, Sr., urologist, 1939. From 1932 to 1933, he was vice-president of the American Urological Association and, during the 1920s and 1930s, chairman of the Department of Urology, University of Tennessee.*

Psychiatry. As early as 1941, five months after its introduction, electroshock therapy was used at Gartly-Ramsay Hospital. The Gailor Psychiatric Hospital opened in 1942. Named for Episcopal Bishop Thomas F. Gailor, the hospital was converted into the Gailor Clinics for outpatients when the West Tennessee Psychiatric Hospital was built in the 1960s. Dr. D. C. McCool, who had come to Memphis in 1935, was active during the war years and later was called "the finest Christian psychiatrist anywhere" by Dr. Frank Lathram, author of a typescript history of psychiatry given to the Medical Society. In 1942, there were five practicing psychiatrists in the city. Later the following arrived: Dr. Charles Miller (1946); Dr. Elizabeth Gehorsam (1947), and Dr. Sam Paster, chief of psychiatry at Kennedy Army Hospital (then at Kennedy VA) until 1948, when he began private practice. Dr. Frank Lathram—psychiatrist, historian, and avid fisherman—came to Memphis in 1948.

When the Kennedy Army Hospital was turned over to the VA in 1946, its physicians continued to diagnose and treat large numbers of psychiatric patients, as they did later after the new facility was built in the Medical Center in 1967. The VA Hospital also had a 200-patient drug rehabilitation unit.

Urology. Dr. George R. Livermore remained chair of urology at the University of Tennessee until 1944. The next year, Dr. Thomas D. Moore, trained at Mayo, became chair, and organized an accredited residency. For several years, trainees at Mayo spent six to 12 months at the University. Under Dr. Moore's leadership, the program trained 13 residents. Dr. Hubert Turley was a prominent urological surgeon at Baptist during this period, serving as president of the American Urological Association in 1947.

POSTWAR MEMPHIS MEDICINE

In 1945, the Tennessee Medical Association established a Postwar Planning Committee parallel to one from the American Medical Association. This group compiled information to help returning doctors find practice opportunities and information about educational opportunities through the G.I. Bill of Rights and also offered refresher courses for returning veterans.

After the war, when the Veterans Hospital and the Marine Hospital needed blood but found donors scarce, paid donors were sought, and some blood banks opened without hospital connections. In 1947, the American Association

to the 2000s. He graduated from the University of Tennessee with an M.D. degree in 1943, interned in Chicago, and then entered the Marine Corps during World War II. He returned to Memphis to complete his residency at Methodist Hospital, and worked with chest surgeons Drs. F. H. Alley and Francis H. Cole before being recalled during the Korean War. He was assigned to the chest surgery service at St. Alban's Hospital on Long Island, New York, where Naval and Marine veterans were returning from Korea. After his service, George joined his father, Dr. Giles Augustus Coors (son of Dr. George Augustus Coors), a general surgeon with 30 years' experience. In the early 1950s, Coors observed Dr. Denton Cooley operate in Houston and became interested in cardiovascular and peripheral vein surgery. He performed the first successful aortic graft in Memphis.

Dr. Coors went on to become Methodist chief of staff, a Federal Aviation Administration examiner for 50 years, a FedEx flight surgeon for 21 years, and an instrument-rated pilot.

In an interview, Dr. Coors recalled prominent Memphis surgeons of this era: Drs. Gus Crisler, Joseph J. McCaughan, R. L. Sanders, and Ferrell Varner. Coors remembers friendships with other prominent Methodist Hospital physicians, including Dr. Sam Sanders, Drs. William Milton and his

ABOVE: *Dr. George A. Coors with his wife, Jeanne Parham, 1944. He and his father, Dr. Giles Coors, operated the Coors Clinic at 1304 Union. Dr. Coors was a surgeon at Methodist Hospital beginning in 1949 and became chief of staff in 1976.* RIGHT: *Major C. Ferrell Varner, 1945. He left Memphis as an army physician in July 1941. Assigned in the Pacific to a portable surgical hospital, he returned to Memphis in November 1945*

of Blood Banks began; Dr. Merlin Trumbull and Dr. E. Eric Muirhead of Memphis served as national presidents.

Most of Memphis's hospitals and clinics benefited from the 1946 Hill-Burton Act, which provided grants and guaranteed loans to build or improve the nation's hospital system with a goal of 4.5 beds per 1,000 people.

A major surgeon at this time was Dr. George Alcorn Coors, whose recollections bring back this era before sophisticated medical treatments developed, an era when surgery was particularly important. A veteran of World War II and Korea, Coors practiced in Memphis from the 1940s

WILLIAM MILTON ADAMS, M.D.

Dr. William Milton Adams, a third-generation physician from northeast Mississippi and a cousin of William Faulkner, graduated from Tulane Medical School in 1928. After a year in general practice, he took specialty surgical training in New York with Dr. H. Sheehan Eastman, a maxillofacial specialist who had learned reconstructive techniques in World War I and helped lay the foundation for modern plastic surgery. Back in Memphis, Dr. Adams became widely known for both his cosmetic talents and his dedication to reconstruction.

During World War II, he was assigned to teach courses in plastic surgery at Tulane University, after which he enlisted in the Navy and was assigned to the Naval Hospital in Oakland, California. As a lieutenant commander, he honed his skills in reconstructing injuries, developing innovations for facial fractures that became the national standard. So intense were his work habits that he was called the "Memphis Fireball," a name that reflected what Dr. Charles Steiss of New York called "his dynamic spirit and zest for work" (Eulogy, *Journal of the American Medical Association*, 1957).

Returning to Memphis, he became chief of plastic surgery at all the major hospitals as well as chair of the Division of Plastic Surgery at the University of Tennessee. Eager to expand the field, he established a training program in plastic surgery and trained a number of graduates, including his brother Lorenzo, who joined him in practice in 1950. Known for his high standards, he considered plastic surgery an artistic exercise and taught his students "to always treat the tissues with respect." One evening, reconstructing the face of an injured patient, he stepped back and said, "Damn! That's not right." While his resident groaned, thinking of early rounds the next morning, Dr. Adams began again. When he finished, he said, "Now that's right."

Representing the specialty of plastic surgery, he became internationally known for his work. He was a founding member and chairman of the National Board of Plastic Surgery and its president in 1954. In 1947, Adams hosted a scientific program in Memphis that Dr. Steiss called "a display of southern hospitality which will probably never be surpassed."

Dr. Adams had a crippling heart attack at the age of 44. Later, he underwent the first cardiovascular surgery

brother Lorenzo Adams; Phillip Meriwether Lewis; Battle Malone and his uncle ("old Dr. Battle Malone"); Brewster Harrington; Joseph Francis, uncle of Dr. Hugh Francis; and Cash King.

Coors recalls Drs. Pervis Milnor, Ed Garrett, Sr., Ed Cocke, and R. L. Sanders at Baptist.

At the time, Baptist and Methodist hospitals, led by administrators Frank Groner and James Crews, respectively, maintained a close relationship. Coors had great respect for Drs. Harwell Wilson, chief of surgery at the University of Tennessee, M. K. Callison, longtime dean of medicine

Dr. William Battle Malone II prepares for an operation at the 101st United States Army General Hospital, England, April 1945.

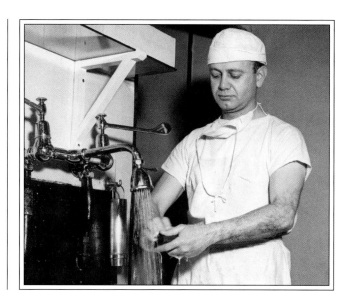

in Memphis. Dr. Robert Glover of Philadelphia, who flew to Memphis "for Milton Adams," used an internal mammary artery to improve blood flow to the heart at Methodist Hospital. He died in 1957 at age 51. Concluding in the *Journal of the American Medical Association*, Dr. Steiss called Adams a man of "complete professional integrity; a restless man forever seeking perfection; a man of unlimited compassion for those in need."

Dr. Adams's son, Dr. R Franklin Adams, says of his father's work: "Although my dad became renowned for cosmetic surgery (including several Hollywood stars), he believed that plastic surgery was not just for the privileged, and it should always be balanced with reconstructive surgery. He originated the multispecialty charity clinic for cleft palate deformities and actually did as much charity work as he did private."

Dr. Milton Adams's legacy in Memphis extends to the present day. His brother, Lorenzo Adams, whom he trained, proved to be highly talented, and he continued the tradition of excellence in Memphis for many years. Dr. Milton's son, Dr. William M. Adams, Jr., and Dr. Lorenzo's son, Dr. R. Louis Adams, are prominent plastic surgeons. Dr. Milton Adams also trained several other notable Memphis plastic surgeons including Drs. Leigh K. Haynes, Robert Reeder, and James Hendrix, the latter two becoming presidents of the National Society of Plastic and Reconstructive Surgery.

OPPOSITE PAGE: *Dr. William Milton Adams served at the Navy Hospital on Oaknall Navy Base, Oakland, California, during World War II.* ABOVE: *Dr. William Milton Adams, holds a child with cleft lip and palate, c. 1950. Watching, rear row, left to right, are: Dr. Jimmy Hendricks (plastic surgeon), Gertrude O'Kelly (surgical nurse), Dr. Fausten Weber (orthodonist), Dr. Lorenzo A. Adams (plastic surgeon), Dr. L.M. Graves (Department of Public Health), and Dr. Alan Vinicoff (trainee, plastic surgery). Front row, left to right, are: Ann Hamilton Shoat, unknown, unknown, unknown (trainee, plastic surgery), Dr. Dick McCool (psychiatrist). The scene was a multispecialty, charity clinic held in Dr. Adams's office on Madison.*

at the University, and Jim Gibb Johnson, respected teacher and administrator. Dr. Coors regrets the loss of friendly interaction among physicians that came with increased specialization and widely separated hospitals. Back then, he recalls, "If a colleague had a problem in surgery with a patient, and he asked for your help, you went, regardless of the time of day or night."

From the 1940s through the 1960s, when surgery was king, Memphis was home to some of the country's most prominent surgeons. Others affiliated with Baptist included Drs. William F. (Chubby) Andrews; Joseph Brock; Fenwick Chappell (first full-time physician at Baptist in charge of the emergency room); David Dunavant (whose son and grandson are now surgeons at Baptist); George Livermore; Robert Miles (who invented the Miles clip to prevent blood clots from moving from a vein to a vital organ, later used on President Richard Nixon); Eugene Nobles; Robert McBurney; John Nash; and Ernest Kelly.

Major surgeons at the University of Tennessee included Drs. Thomas West, Clarence Gillespie, Harwell Wilson, and Russell Patterson, Jr. Prominent surgeons at Methodist included Drs. Malcolm Aste; Joe Laugheed; Malcolm and Edward Stevenson (twins); Gus Crisler; Battle Malone; Hugh Francis; James B. Green; and Francis Cole, Sr.

Even amid the tragedy of war, the 1940s became a watershed for Memphis medicine, with new drugs, new techniques, and nuclear medicine. From wartime experience, Memphis doctors learned to save the lives of blast and burn victims, to reconstruct the injured, and to minimize infection.

CHAPTER TEN

MEMPHIS: A CITY OF THE 1950s

1950-1959

THE FIFTIES WERE THE BEST OF TIMES FOR Memphis, which abounded with optimism. The first Holiday Inn opened on Summer Avenue in 1952. The city's musical heritage—blues, country, gospel, and pop—blended in the form of "rock and roll" at Sun Studio on Union Avenue and virtually exploded into world consciousness. Memphis medicine came onto the world stage. So great was Memphis's impact that *The Times* of London named three Memphians among the prime "Makers of the 20th Century": Kemmons Wilson, Elvis Presley, and Dr. John Shea, Jr. The first Miss America from Memphis, Barbara Jo Walker (1947), married Memphian Dr. John Hummel; a second Miss America from the Mid-South, Lynda Lee Mead (1960), would later marry Dr. John Shea, Jr. In 1949, the Nobel Prize for Literature went to William Faulkner, whose fictional Yoknapatawpha County drew on characters and mores of the Mid-South. So exciting was that time that, in 1986, the Memphis Brooks Museum of Art held an exhibition titled *Memphis 1948-1958*. Commenting on the era, British historian Sir Peter Hall called Memphis "the place to be" in the 1950s because of cultural changes that began here.

THE AGE OF ANTIBIOTICS

In medicine, the 1950s became the Age of Antibiotics. The "miracle drugs" (sulfa and the first antibiotic, penicillin), highly effective with injuries late in World War II, brought seeming "miracles" to the larger public in the 1950s, followed by newer antibiotics such as chloramphenicol (1949), erythromycin (1952), vancomycin (1955), and methicillin (1960), with more than 130 antibiotics by 2009. Streptomycin (first used in 1944), was the first drug effective against TB. Isoniazid (1952) was dramatically effective against TB, and improved sanitation further reduced incidence of the disease.

The Memphis medical complex in October 1955, excluding Methodist and St. Joseph Hospitals. New facilities are the addition to Baptist Memorial Hospital with its wings at an oblique angle, Le Bonheur, and E. H. Crump Memorial Hospital. In the 1960s, Russwood Park, seen at the right, became the site for the James K. Dobbs Medical Research Institute, William F. Bowld Hospital, and Baptist Hospital's ten-story Doctor's Building. During that decade, the Memphis Mental Health Institute was added north of Le Bonheur in the upper right of the photo. Baptist also built its new south unit, extending toward Union and the University of Tennessee, seen at the left.

Le Bonheur Children's Hospital

Established to sew garments for orphans of the flu epidemic of 1919, the Le Bonheur Club was named by Elizabeth Jordan Gilliland, the name meaning 'happiness' or, literally, 'the good hour.' Later the group visited the Crippled Children's Hospital and bought clothes and necessities for the children.

The club sponsored the Children's Bureau, operated a boarding home, and in 1940 established the Le Bonheur Charity Horse Show, which drew celebrities including Dick Powell and Bob Hope. With funds from the shows and time donated by private pediatricians, the club started a clinic for poor children at Methodist Hospital, serving more than 1,000 children in 1948.

Asked for suggestions for future projects, pediatrician Dr. C. Barton Etter challenged the group to build a hospital, which was needed in the early 1950s for polio patients. Working with officials from the University of Tennessee, the club sought Hill-Burton funds, incorporated the hospital as a non-profit, and raised $50,000 more than anticipated.

ABOVE: *Le Bonheur Children's Hospital, June 1952. Located at 848 Adams, The four-story wing had huge, double-thickness windows to provide plenty of natural lighting for the children.* RIGHT: *Dr. F. Tom Mitchell and family with his portrait that was commissioned by fellow physicians to commemorate his long service to Le Bonheur Children's Hospital, 1953. He was chief of staff at Le Bonheur and head of pediatrics at the University of Tennessee.*

At the 1952 dedication of the hospital, club president Elise Pritchard released a cluster of balloons with the hospital's keys attached to signify that Le Bonheur was always open. Four physicians were the "founding fathers" of Le Bonheur Children's Hospital: Dr. Frank Thomas Mitchell, chief of the Division of Pediatrics at the University of Tennessee (1940-1960), who became the hospital's first chief of staff; Dr. James Etteldorf, the first full-time pediatrics professor at the University, who later established pediatric specialty training at Le Bonheur; Dr. C. Barton Etter; and Dr. James Hughes, who chaired the Department of Pediatrics from 1960 to 1975 and wrote textbooks on pediatrics. Dr. Emmett Bell, Jr., first resident at Le Bonheur, became a prominent pediatrician.

Dr. Raphael Paul was the first pediatric cardiologist in the Mid-South. He came to Memphis in 1948 after training with Drs. Blalock and Taussig of the famous blue-baby shunt operation at Johns Hopkins Hospital. In 1952, he opened

his private practice at Le Bonheur Children's Hospital with one of the first non-segregated physicians' waiting rooms in Memphis.

In pediatric surgery, Le Bonheur and the University of Tennessee brought two physicians from the country's top program at Boston Children's Hospital, one of whom, Dr. Robert Allen, performed the first open-heart surgery in Memphis on a child in 1972. Tragically, Dr. Allen died in an automobile accident in 1981. Dr. Earle Wrenn became a prominent pediatric surgeon, as did Dr. Robert Hollabaugh,

LEFT: *Dr. Robert Allen, left, and Jack Schmollinger of the Memphis Heart Association with a heart-lung machine, Le Bonheur Children's Hospital, February 1962. The equipment was part of the Heart Association's annual research tour highlighting the most recent advances in the field.* BELOW: *J. Everett Pidgeon, second from left, talks with, left to right, Dr. O. W. Hyman, dean of the University of Tennessee Medical School, Allen Morgan, and Dr. James Etteldorf, retiring president of the Pediatric Society, December 1949. Mr. Pidgeon was president of the board of the hospital and Allen Morgan, chairman of the campaign fund. Dr. Etteldorf had worked on the synthetic drug atabrine during World War II and was one of Le Bonheur's founding fathers.*

MEMPHIS: A CITY OF THE 1950S

The "Bunny Room" at Le Bonheur, established when the new hospital opened in 1952. A toy or doll was chosen by the young patient before going into surgery.

who trained at Philadelphia General Hospital followed by a pediatric surgery residency at the University of Tennessee. These physicians helped establish Le Bonheur as one of the nation's outstanding children's hospitals.

Opened in the same year as the first Holiday Inn, Le Bonheur marked a time of growing prosperity for Memphis. It was designated "Hospital of the Year" by *Modern Hospital* (1954). Children remembered the "Bunny Room," where they would choose a toy before surgery and find it with them when they awoke. As a teaching hospital, Le Bonheur trained pediatricians from around the world, many of whom remained in the Mid-South. The first intensive care (IC) unit in Memphis was established at Le Bonheur in 1957, and other hospitals followed with IC units.

In the 1960s, Mr. and Mrs. J. Everett Pidgeon donated funds for a cardiovascular treatment program and for developing a child-sized artificial kidney machine. When St. Jude opened in 1962, the two hospitals cooperated. In 1968, Le Bonheur consolidated the city's pediatric orthopedic service. Later in the decade, Raymond C. Firestone made a donation in memory of his wife, the former Laura An Lisk, and Le Bonheur opened the Firestone Children's Outpatient Center for patients with muscular dystrophy, rheumatoid arthritis, hemophilia, and cystic fibrosis.

In 1974, Le Bonheur established a board of directors and pledged equal pediatric care to all children. Dr. John F. Griffith, the new director in 1975, closed the pediatric unit of the John Gaston Hospital (Tobey Wing) and brought all pediatric patients to Le Bonheur. In 1975, administrator Eugene K. Cashman, Jr., spearheaded an expansion, and in 1985, Le Bonheur opened a satellite facility in East Memphis, followed by other offices in Bartlett and DeSoto County, Mississippi, plus a one-day surgery center in East Memphis.

Technologically, Le Bonheur remained in the forefront, beginning liver transplants in 1983, followed by kidney, allergy immunology, and diabetes centers in the late 1980s, when it also joined the Children's Miracle Network telethon and launched a campaign to promote awareness of its services. Drs. Lloyd Crawford and Fred Grogan were important allergy specialists at Le Bonheur.

With Dr. Russell Chesney as vice president of academic affairs at Le Bonheur in 1988, the patient tower on Adams Avenue was demolished, and a new, seven-story tower was built on North Dunlap. In 2001, Dr. Stephanie Storgion became the hospital's first woman medical director.

When TennCare began in 1994, Le Bonheur's independent status was threatened by lower utilization and reimbursement. Guided by Chief Financial Officer Robert Trimm, Le Bonheur merged with Methodist Healthcare Systems in 1995. The Sam Walton Children's CT/MRI Center opened in 1990 and, in 1993, the Crippled Children's Foundation established the Children's Foundation Research Center, centralizing research on pediatric asthma, arthritis, diabetes, kidney disease, and cystic fibrosis. In 1998, Les Passees merged with Le Bonheur, forming Le Bonheur Early Intervention and Development Program. In 2001, Le Bonheur was designated the only Comprehensive Regional Pediatric Center in West Tennessee.

Le Bonheur has the largest pediatric brain tumor program in the United States, with a strong affiliation with St. Jude. Dr. Robert A Sanford built this program with three other pediatric neurosurgeons, Drs. Michael Muhlbauer, Frederick A. Boop, and Stephanie Einhaus.

On the site of the former Tennessee Psychiatric Hospital at 865 Poplar (demolished in 2007), Le Bonheur built a new research and treatment center (2010) befitting its goal of becoming a premier children's hospital.

New Facilities

In 1956, the 128-bed E. H. Crump Hospital was built at 852 Jefferson for African Americans in the customarily segregated system. In 1983, this facility became part of The MED.

Under director Dr. W. S. Martin, Collins Chapel Hospital (1909) continued to serve African-American physicians and patients, although some Caucasian physicians worked there as well. In 1955, a new, 48-bed facility opened at 409 Ayers. Dr. Martin guided it until his death in 1958. Collins Chapel was home to many of Memphis's best-known African-American physicians, including Drs. Wheelock A. Bisson, Edward Reed, Clara Brawner, and C. O. Daugherty.

Dr. Ralph Braund, Memphis's first surgical oncologist, began a cancer clinic at Gaston that became the Van Vleet Cancer Center in 1951. Trained at Memorial Sloan-Kettering Institute in New York, he brought new ideas and procedures to Memphis. With a special laboratory for research with isotopes from the Oak Ridge National

Mayor Edward Orgill spoke at the dedication of the E. H. Crump Memorial Hospital, March 19, 1956. Located at the southeast corner of Jefferson and Dunlap, the facility was a teaching hospital for African-American physicians and nurses. The three African-American physicians at the left, in the back row, are, left to right, unknown, W. Oscar Speight, Sr., and Leland L. Atkins.

Laboratory, radiation was studied as a treatment. In 1956, the clinic was incorporated as the West Tennessee Cancer Clinic on Dunlap and, in 1959, it purchased the first cobalt radiation unit in Memphis. Supported by prominent Memphians, including the Belz family, the facility developed a training program for oncologic head and neck surgeons at the University of Tennessee. Under Dr. Braund, the department achieved national recognition and helped form the National Head and Neck Surgical Society.

After Dr. Ralph Braund's pioneering work in surgical oncology, a second generation of surgeons followed with

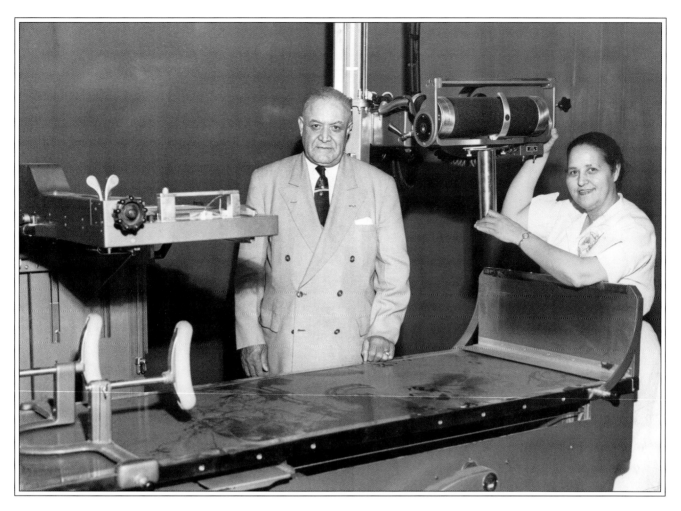

ABOVE: *Dr. William S. Martin stands with Nurse Eva Cartman in one of the operating rooms in the new Collins Chapel Hospital, 1955. Nurse Cartman was made superintendent, while Dr. Martin, because of his long association with the hospital, was made president of the staff.* RIGHT: *Dr. Ralph Braund, right, with Memphis mayor Walter Chandler, December 1964. The first surgical oncologist in Memphis, Dr. Braund was a pioneering researcher into the use of radiation as a treatment for cancer.*

training in equally prestigious centers. Drs. Irvin Fleming, Jack Piggott, and brothers Alfred, Roy, and Gene Page were among the most prominent leaders in cancer care for the next four decades. They were dedicated physicians, greatly admired for their clinical skills and professionalism. Dr. Fleming, who has 130 publications in surgical oncology, was president of the American Cancer Society (1993-1994) and in 2001 received the Society's Distinguished Service Award. He was instrumental in establishing Hope Lodge in Memphis to provide temporary lodging for out-of-town patients.

In 1954, Methodist planned a building fund drive. With an architectural drawing of Methodist as a backdrop, left to right are Dr. F. H. Alley, Dr. S. Fred Strain, Leslie M. Stratton Jr., and Dr. Joseph H. Francis.

Polio

Although there had been serious polio epidemics before, particularly in 1916 with 27,000 cases in the United States, the most serious were in the late 1940s and early 1950s. Across the country, polio cases averaged about 20,000 per year from 1945 to 1949, but during the country's worst outbreak in 1952, nationally there were almost 58,000 cases amid "polio hysteria," and in 1953 there were 35,000 cases.

The polio scares of the 1950s struck fear in the hearts of Memphis parents. Although the disease attacked adults as well as children (Franklin D. Roosevelt's struggle with polio was well known), children seemed especially vulnerable. In Memphis, orthopedist Dr. Alvin Ingram supervised polio patients at Campbell Clinic. Newly diagnosed patients were sent to the Isolation Hospital on Jefferson behind the John Gaston Hospital (demolished in 1990 for the expanding MED) where pediatrician Dr. Gilbert Levy was in charge. There were 75 to 100 beds for anyone, regardless of ability to pay.

In an interview with Anita Shugart of Memphis State University's Oral History Program in 1989, Dr. Ingram said that Dr. Levy probably did not charge the poor at the Isolation Hospital or Crippled Children's Clinic, nor did Dr. Ingram, except for private patients at the Campbell Clinic. Dr Ingram recalled, "We had nothing that we could do…. The biggest problem was bulbar polio. We had to use a respirator—the old iron lung." In the Isolation Hospital, the severest cases—those confined to 7-foot, 750-pound "iron lungs"—lay in rows while the machines delivered pressure to force air into and out of their lungs. Patients who exercised during the acute phase suffered increased paralysis. After the acute phase, indigent patients were sent to the Crippled Children's Hospital or to crippled children's clinics in Tennessee, Mississippi, and Arkansas.

At the time, the Kenny treatment, developed by an Australian nurse called "Sister Kenny" though not a nun, applied hot packs to relieve pain and used passive exercise to help prevent withering of paralyzed limbs. The hot packs

ABOVE: *Dr. Alvin Ingram, orthopedic surgeon, c.1990, oversaw the management of polio patients at Campbell Clinic.* RIGHT: *Dr. Louis P. Britt of the Campbell Clinic, standing right, at a conference of the Muscular Dystrophy Association of America, Memphis, June 1957. Others are, left to right, Mrs. William B. Smith, Dr. L. M. Graves, Arthur A. Gallway (signing), and Dr. F. L. Roberts. Dr. Britt was moderator of a panel discussion.*

hospitals overflowing, each major hospital took 20 patients at a time. Memphis newspapers printed regular reports of the number of new polio cases. Ingram sensed panic in Memphis. Many physicians worked seven days a week, and Ingram's wife asked him to change his clothes on the back porch when he returned from the hospital. As author of the chapter on polio in *Campbell's Operative Orthopedics*, Dr. Ingram shared front-line experience with the world.

In Memphis, several physicians who worked with polio patients tell of approximately 1,000 cases, though the number of cases reported to the Health Department varied from that number. Dr. Ingram reported that he was consulted on more than 600 of almost 1,000 new polio patients in Memphis between mid-June and mid-September 1952, and pediatrician Dr. George Lovejoy confirmed Dr. Ingram's estimate of about 1,000 patients. Dr. Hugh Smith reported in *History of Medicine in Memphis* almost 1,000 cases of polio in Memphis. Totals from the Health Department, however, suggest lower numbers, and at least one physician wrote about "polio hysteria" for the *Memphis Medical Journal*. Numbers of cases reported in that journal are reflected in the table opposite.

Dramatic tales remain from the era of polio. Dr. William King Payne tells of a woman eight months pregnant who suddenly grew worse and was dying in the Isolation Hospital. The senior resident in surgery dashed across the

would sometimes burn, and in Memphis a "modified" Kenny treatment was adopted. Before Sister Kenny, patients had been immobilized in casts, and their limbs withered. Dr. Ingram went to Warm Springs, Georgia, in 1951 to observe the Kenny treatment; afterwards, he hired Aline Belcher as head physical therapist at Campbell Clinic. Pediatrician Dr. Louis P. Britt, a specialist in physical medicine and rehabilitation (no relation to transplant surgeon Dr. Louis G. Britt), came to Memphis in 1951 and worked at Campbell Clinic and Les Passees for 15 years until his death.

In 1952, swimming pools and movie theaters were closed, and tonsillectomies, thought to increase vulnerability to polio, were postponed. At the Isolation Hospital, even parents could visit patients only through a window in the room. With

MEMPHIS AND SHELBY COUNTY POLIO CASES, 1949-1956*

1949	1950	1951	1952	1953	1954	1955	1956
138	116	159	109	122	174	55	58

*Polio case numbers were highest in late summer from June through October.

Source: Memphis and Shelby County Health Department statistics published in *Memphis Medical Journal*, Vols. 24-31 (1949-1956)

parking lot to meet the gurney headed for the operating room. Realizing that time had run out for the mother, he performed an emergency caesarian on the parking lot; the infant survived. Although no accurate count is available of polio patients surviving in iron lungs into the twenty-first century, Dianne Odell lived near Jackson, Tennessee, for 56 years in an iron lung—from acute polio at age three in 1952 until May 28, 2008, when the backup generator at her home failed during a power outage.

At least three famous Tennesseans were polio survivors. Dinah Shore contracted polio when she was 18 months old.

LEFT: *Dr. Jonas Salk who developed the Salk vaccine using a killed virus, March 1968.* ABOVE: *Dr. L. M. Graves, director of the Memphis & Shelby County Health Department, reads about the introduction of the Salk vaccine to Shelby County.*

Treated with the Kenny method, she recovered to become an athlete in high school, a Vanderbilt graduate, and a star of radio and television. Wilma Rudolph contracted polio at age four, with paralysis of her left leg. Five years of exercises and massage enabled her to walk without a brace and eventually to outrun the neighborhood boys. At the age of 16, she won a bronze medal in the 1956 Summer Olympics. She won three gold medals in the 1960 Summer Olympics. Steve Cohen of Memphis, son of a psychiatrist, contracted polio at age five and recovered; he went on to serve in the Tennessee Senate and is currently a member of the United States House of Representatives.

Reviewing twentieth-century orthopedics, Dr. George Omer, Jr., noted the cumulative effects of polio in the United States, citing federal statistics of 185,000 handicapped children and 219,000 handicapped adults in 1947. In that year, the American Board of Physical Medicine and Reha-

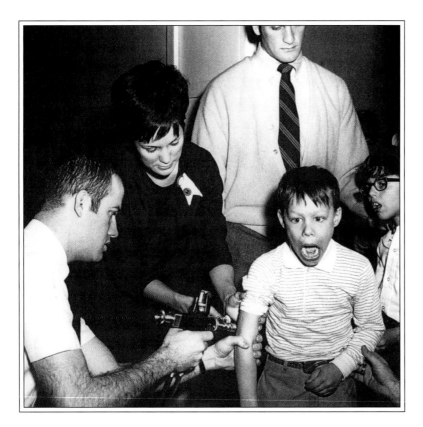

Russell Riggs, son of Dr. Webster Riggs, makes a face during the administration of polio vaccine by public health personnel, Presbyterian Day School, 1969.

bilitation was established and, in 1953, the National Polio Foundation designated Memphis's Baptist Hospital as the area's rehabilitation center, with a new unit allocated for physical medicine and rehabilitation of polio survivors.

Under political and social pressure from the frightened public, Dr. Jonas Salk and Dr. Albert Sabin raced to develop vaccines. Salk used a killed virus in a unique tissue culture, while Sabin used a live but attenuated virus in an oral vaccine. Early versions of the Salk vaccine were tested in 1952, with field trials in 1954 that included Memphis, sponsored by the National Foundation for Infantile Paralysis.

When the Salk killed-virus vaccine was reported safe in 1955, a nationwide vaccination program began. In Memphis, the Health Department and private physicians vaccinated younger children, targeting first the highly vulnerable eight-year-olds. From 174 cases and eight deaths from polio in1954, the number of cases dropped dramatically to 55 cases and two deaths in 1955. In 1957, the Health Department nurses gave 25,518 polio injections through the Health Department, primarily in the schools, aided by Medical Society volunteers.

Dr. John Nash, an intern in the early 1950s, remembers that the Isolation Hospital had two floors of polio cases early in the decade. After the Salk vaccine, the floors were empty. The facility was renamed the Jefferson Pavilion and served as a children's hospital until the Tobey Children's Hospital was established in 1957. Polio soon became a disease of the past. Retired Memphis internist Dr. John Barron called the polio vaccine the most important development in his 40-year medical career.

The new vaccine was not perfect, however. An improperly processed batch from Cutter Laboratory in California contained some live viruses and caused 200 cases of polio in the United States, including two deaths and several paralytic cases. Some physicians and parents feared the vaccine, and heartbreaking stories came from parents who did not have their children vaccinated and later regretted the decision. In 1962, the Sabin oral (attenuated) vaccine became available and was widely administered.

Cardiology

In the 1950s, cardiology developed as a subspecialty of internal medicine. Before his death in 2007, Dr. Thomas Stern wrote a history of cardiology in Memphis, drawing on his and his father's experiences.

Begun in 1948, the Memphis Heart Association brought attention to the need to understand and treat heart disease. Founded by laymen and physicians, the association emphasized home visits and reduction of death and disability. The organization raised funds for x-ray machines, supplied an infant heart-lung machine for Le Bonheur in 1961, and sponsored community efforts to prevent, diagnose, and treat heart disease. From 1967 to 1969, the association provided Health Department scanning of schoolchildren for heart problems. Noted physicians who served as presidents of the association included Drs. Neuton S. Stern; Otis S. Warr; James D. Hughes; Charles J. Deere; Robert G. Allen; Walter K. Hoffman; and James W. Pate. The American College of Cardiology was chartered in 1949 in Manhattan.

By 2009, the Memphis Yellow Pages listed 52 cardiologists. In addition to the Stern Cardiovascular Group, other major groups included the Sutherland Group (founded by Dr. Art Sutherland), the University of Tennessee Medical Group—Cardiology, Cardiovascular Physicians of Memphis, Cardiology Associates of Memphis, Cardiovascular Physicians of Memphis, Heart Center of Memphis, Heart and Vascular Institute, and Memphis Heart Clinic.

MEMPHIS CARDIOLOGY IN RETROSPECT

CONDENSED FROM AN ARTICLE BY DR. THOMAS N. STERN

Cardiology as a subspecialty was just beginning to be recognized in 1920 when my father, Neuton Stern, returned to Memphis after studying electrocardiography with Sir Thomas Lewis in England. Although there were no cardiologists as such, some internists and university medical centers had a special interest in cardiology. After Dr. Stern gave a paper on electrocardiology for the Medical Society, a senior physician announced that any doctor who needed such a machine should not call himself a doctor.

In the 1940s, young internists joined Dr. Stern, including Dr. Dan Brody, later a professor and a leading EKG expert. After Dr. Neuton Stern's first myocardial infarction in 1949, I returned to Memphis. Our long, wonderful joint practice was interrupted only by my two years in the Army (primarily in Korea) from 1953 to 1954, and by his death in 1969.

In the 1960s, I did cardiac catheterizations in our office in a seven-foot-wide x-ray room with space only for the patient, the physician, and a few instruments. I converted an old EKG machine into a recorder stationed in the adjoining darkroom, creating a crude but effective setup. When Baptist Hospital started its laboratory, Dr. J. Leo Wright from the Mayo Clinic headed the Cardiac Unit, and Dr. Frank Kroetz came from Iowa to do both left heart and right heart catheterizations.

In the 1950s, our therapeutic armamentarium was limited. We had digitalis and quinidine. Procainamide had appeared, and other antiarrhythmics trickled into use. Nitroglycerin was available, but the only diuretics were mercurials. Then hydrochlorothiazide, both a weak diuretic and an antihypertensive, arrived, and eventually furosemide.

The treatment for hypertension was phenobarbital, also given as supportive treatment to patients with acute myocardial infarction. In the trunk of my car, I carried a tank of oxygen and an EKG machine. When a patient called with chest pain, I sped to the home, took an EKG, started oxygen, and gave intravenous morphine, usually with aminophylline. In the hospital, the patient was placed in an oxygen tent. Prolonged bed rest was universally prescribed for myocardial infarction in the 1940s. Some patients received heparin, and coumadin was given in the 1950s. Gradually the time to ambulation was shortened.

Beta blockers came in the 1970s and ACE inhibitors early in the 1980s. Defibrillation became routine when needed in the early 1970s, and right-sided pacemakers, first temporary and then permanent, were implanted. The panoply of tools

Dr. Thomas Stern listens to Dr. Nancy Flowers describe her research project. Dr. Flowers was the recipient of a grant from the Memphis Heart Association.

of the electrophysiologist came into use, but there was no acute intervention until the 1980s, when thrombolytics, then angioplasty, became available.

We kept up with technology. In 1952, following instructions in a heart journal, I devised a simple telemetry unit using a couple of capacitors and a twenty-five-cent piece to transmit the EKG from the walking patient to the recorder. Later, we used radionuclide stress testing, echocardiographic stress testing, and then the CT scanner.

The changes in medical practice have been amazing. From a small office, we grew to our well-equipped building. Few people are fortunate enough to enjoy their work as much as I have. Even more wonderful has been my splendid group of colleagues. Few are so privileged.

phis, Dr. Thomas Stern, Dr. Louis Britt, and Dr. George Coors served. Dr. William L. Wallace (University of Tennessee Medicine, 1932), a medical missionary, died in a Chinese Communist prison camp in 1951. His classmates dedicated the Wallace Memorial Book Collection at the University library to him.

Medical personnel in Korea battled hepatitis, TB, enteric viruses, neurotropic viral diseases, Far-Eastern-type hemorrhagic fever, and yellow fever. They encountered fewer injuries from shells, rockets, and bombs than in World War II, but more from small arms, grenades, and land mines. Air evacuation was crucial, and mobile army surgical

LEFT: *The original Les Passees Treatment Center for children with cerebral palsy, 822 Court, March 1954. On May 5, 1957, a new Les Passees Treatment Center at the corner of Court and Dunlap was dedicated.* BELOW: *Mrs. Thomas Dye, staff nurse at the Mid-South Defense Blood Center, draws blood from Billie Ockleberry, assistant editor of the* Memphis World, *October 1951. All blood was to go to the armed forces fighting in Korea.*

LES PASSEES REHABILITATION CENTER

In 1949, some young Memphis women organized Les Passees, a treatment center for children with cerebral palsy, in a house across the street from Gaston. The center utilized medical staff from the University, advice and aid from the Society for Crippled Children and Adults, and further counsel from the Health Department. Unique in Memphis, it had a new building on North Dunlap in 1957, funded by Les Passees and matching Hill-Burton funds.

After physician referrals, children with cerebral palsy and other neuromuscular conditions were evaluated by specialists, given a collaboratively devised treatment plan, and assigned appropriate learning activities, supervised by a city teacher and by Les Passees volunteers. Success stories from the program became evident from the start. In 1959, with $200,000 from the state, matched by Hill-Burton funds, a second floor was added for an adult division.

WAR IN KOREA

From 1950 to 1953, young men were sent to fight in Korea. New military personnel were recruited, and World War II warriors were recalled to duty. Among others from Mem-

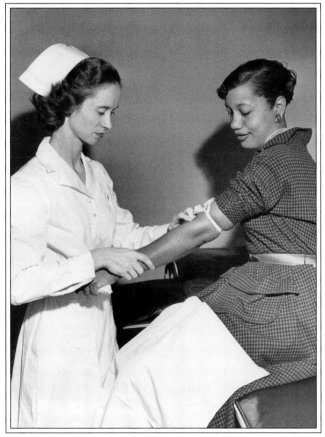

Children from Porter-Leath Home, 850 Manassas, enjoy the ear examination of a newcomer by Dr. George S. Lovejoy, 1965. Dr. Lovejoy worked with the Downtown Exchange Club to give check-ups to the children.

hospitals (MASH units) were designed not only to perform life-saving surgery but to move the entire unit quickly if necessary.

Memphis Practices

Memphian Dr. George S. Lovejoy (born in 1917) learned about medical practice in his father's home office. After a broken leg at age nine, he developed a deformity that Dr. Willis C. Campbell corrected. He graduated from the University of Tennessee in 1943, interned at Grady Memorial in Atlanta, and then entered the Army, serving in California.

After the war, Lovejoy took his residency in pediatrics with Dr. Thomas Mitchell at the University of Tennessee and then opened his office on Highland Street. Reflecting on his life in medicine, Dr. Lovejoy says the greatest advance during his career was penicillin. When he was an intern in Atlanta from 1943 to 1944, penicillin was available in limited quantities, and physicians used it only for osteomyelitis (bone infection). Burn patients at the hospital were subject to terrible infections, and doctors, knowing that penicillin was excreted through the urine (which was sterile), collected urine from those treated with penicillin and used it to make wet packs for burn patients. Penicillin cured "almost anything bacterial," says Dr. Lovejoy.

The other great advance, Dr. Lovejoy says, was the vaccine for polio. Like Dr. Ingram, he notes that physicians could only consign their patients to the isolation ward and treat them with the Kenny method.

In the 1940s, Dr. Lovejoy heard a prominent physician comment that medicine could hardly be improved. "He said this before polio was conquered, before tuberculosis was conquered," said Dr. Lovejoy. In a hospital on the edge of the city, tuberculosis patients wore shorts in winter because fresh air and sunshine were prescribed.

In 1969, Lovejoy became director of public health for Memphis and Shelby County (plus Fayette County). The Health Department could do nothing for mental patients, who were sent to the state hospital at Bolivar, where Dr. Edwin Cocke, Sr., was in charge.

Of medicine as a career, Dr. Lovejoy says, "I miss it. Retirement is the worst job I ever had. I loved the practice of medicine."

Internal Medicine. During the 1950s, specialists in internal medicine became increasingly prominent in Memphis. Established in 1957, the Tennessee Society for Internal Medicine had its largest local chapter here. Drs. J. D. Young, I. Frank Tullis, John Hughes, and Henry G. Rudner, Sr., were prominent. Dr. Henry B. Gotten practiced and taught medicine. Dr. William C. Chaney, discoverer of the Achilles tendon response test for hypothyroidism, spent 40 years practicing, teaching, and serving professionally. Other prominent internists included Drs. Walter K. Hoffman, Jr., N. Edward Rossett, Bernard M. Zussman, Bernard M. Kraus, Otis S. Warr, Sr., and Otis S. Warr, Jr.

Dr. John M. Barron, who entered practice in 1952, similarly recalls the impact of antibiotics and polio vaccines, plus advances in diagnosing and treating cancer. He recalls Dr. Neuton Stern as the only cardiologist in Memphis; treatments for heart attacks were bed rest and sometimes nitroglycerin. Then came early treatments for hypertension, followed by thiazide diuretics. Dr. Barron names Drs. Warren Kyle, M. K. Callison, Hall Tacket, and Frederic Strain as major practitioners and "students of medicine." From 1952 until 2009, he saw subspecialties develop from internal medicine and he adds, "Tobacco causes more disease and death than any single substance."

During the 1940s through the 1960s, Dr. Barron says, Memphis became "a great medical center," particularly in orthopedics, neurosurgery, and otolaryngology, and he

ABOVE: *Dr. Henry G. Rudner, Sr., standing, with Frank Magoffin, administrator of Oakville Memorial Hospital, a tuberculosis sanatorium, 1954. Dr. Rudner was on the staff.* RIGHT: *At a Memphis Heart Association dinner, Dr. Hamel B. Eason, president, looks at the shadowbox of mementoes given to Dr. Otis S. Warr, Jr., right, in 1974 to honor his 25 years of service to the association. Mrs. Joseph Goodman, chair of the annual meeting, looks on.*

JOHN J. SHEA, JR., M.D., AND THE TEFLON MIRACLE

SON OF NOTED OTOLARYNGOLOGIST JOHN J. SHEA, Sr., a distinguished World War I veteran and founder of the Shea Ear Clinic (1926), John J. Shea, Jr., received his medical degree from Harvard University, interned at Bellevue in New York, and returned to Harvard to study further before completing his residency in otolaryngology at the Massachusetts Eye and Ear Infirmary in Boston. Called to active duty, he served in the Marine Corps during the Korean conflict, after which he spent a winter in Vienna dissecting ears in a hospital morgue and seeking to understand otosclerosis, the formation of extra bone on a tiny bone (stapes) in the ear that leads to hearing loss. Wearing an overcoat against the cold, he persevered until he realized the need to replace the tiny stapes bone. He experimented with materials, starting with pieces of bone, but they developed the same calcium deposits as the original stapes.

Back in Memphis, Dr. Shea asked Harry Treace of Richards Medical to create an artificial stapes of Teflon. At Baptist Hospital in May 1956, Shea performed the first operation, removing the hardened stapes and replacing it with the delicate sliver that Harry Treace had prepared. Although the results were not ideal, he spoke to the American Otological Association that year about the operation, encountering skepticism. Still, he persevered, and two years later when he spoke to the same group, he reported performing the operation on 88 patients, of whom fifty percent reported fully restored hearing, with the others reporting varied but significant improvement.

Dr. Shea's story is the stuff of legend. He has performed thousands of stapedectomies, and though the procedure is delicate and the technique difficult, he has given new hearing to untold numbers. Patients have come from around the world for the operation that has become the gold standard for restoring hearing to those with calcified stapes. Shea was featured in *Life* in the issues of January 1, 1962, and July 1, 1962. Later he developed a technique utilizing external perfusion for inner ear disorders.

Dr. Shea was named an honorary fellow of the Royal College of Surgeons of England in 1992; the same honor was granted in Australia. In 1957, the Shea Ear Clinic Founda-

Dr. John J. Shea, Jr., boarding a Quantas jet for Sydney, Australia, to deliver a paper at a meeting, 1960.

tion was established to support research as well as training of doctors.

Shea bought the Memphis Eye and Ear Building on Madison from Methodist Hospital in 1966 and practiced there until the Shea Clinic moved to Poplar at Ridgeway in 1985, with an outpatient surgery center. Dr. John Emmett, who trained in otology at the University of North Carolina, joined Dr. Shea in 1976, cofounding the clinic's Cochlear Implant Program. Dr. Shea has two sons in otology—Dr. John Shea, III, and Dr. Paul Shea— the latter practicing at the Shea Clinic. Dr. J. Gregory Staffel completes the clinic staff.

remembers the camaraderie among doctors in spite of competition among hospitals.

Neurosurgery. Dr. Bland Cannon established a neurosurgical clinic at Baptist Hospital. Practicing in the mid-1950s through the 1980s, Dr. Cannon was highly respected. He served as president of the national Congress of Neurosurgery and of the Tennessee Medical Association. The Bland W. Cannon Scholarship at the University of Tennessee was established in 1989.

Neurology. Neurology, dealing with complex diseases of the central nervous system, including multiple sclerosis, Parkinson's disease, and seizure disorders, has a distinguished history in Memphis. Dating to its separation from neuropsychiatry in the 1930s, through the postwar years when Dr. Nicholas Gotten, Sr., was chair, the Department of Neurology at the University of Tennessee was led by Dr. Gene M. Lassiter from 1957 to 1959 and by Dr. Robert Utterback from 1959 to 1974. Dr. Utterbeck stressed neurology as a scientific discipline and led the department to independent status in 1967. In 1991, Dr. William Pulsinelli was recruited from Cornell Medical Center to chair the department. Later, Dr. Tulio Bertorini developed a prominent private referral practice.

Pathology. Dr. Thomas C. Moss, a Memphis pioneer in pathology, established the independent Moss Pathology Laboratory in 1951. It was the predecessor of the Memphis Pathology Laboratory, which eventually became part of Baptist Hospital.

After establishing the first blood bank in Memphis in 1938, Dr. Lemuel Diggs became a clinical pathologist at the Cleveland Clinic in 1945, returning to Memphis in 1947 as professor of medicine and chairman of the Division of Hematology and Laboratory Medicine at the University of Tennessee. He never lost his fascination with sickle cell disease (SCD), which he had encountered within a week of arriving in Memphis in the 1930s. Of his studies he said, "It was just a virgin field. The more you got into it the more you didn't know." Diggs studied this "important disease" throughout his long life.

For years, Dr. Diggs dreamed of a special clinic for treating the little-known disease, which, although occurring frequently in African Americans, is not limited to any race or ethnicity. In the 1950s, Diggs's work in sickle cell pathology bore fruit. Dr. Alfred Kraus joined the University's Division of Hematology soon after Dr. Diggs's return to Memphis in 1947, and his wife, Dr. Lorraine Kraus, joined in 1956. In 1958, the Sickle Cell Center, affiliated with the College of Medicine, became the only recognized center for the disease in the United States. The National Heart and Lung Institute chose the University as one of ten Comprehensive

ABOVE: *Dr. Lemuel W. Diggs in his office, known as the Sickle Cell Information Center, 1968.* OPPOSITE PAGE, ABOVE: *Dorothy Sturm at the drawing board in her basement office in the Medical and Dental Building, 915 Madison, 1936.* OPPOSITE PAGE, BELOW: *A Dorothy Sturm illustration of an eosinophil.*

Centers for Research and Service of Sickle Cell Disease in 1972. Its purpose was to use a multidisciplinary approach to patients with the disease, a practice universally accepted today. From 1979 to 1989, the Cooperative Study for the Natural History of Sickle Cell Disease was established in Memphis, with the focus varying with the patient's age. Children went to Le Bonheur and adults to The MED. In 1994, the Diggs-Kraus Sickle Cell Center was rededicated to honor Drs. Diggs, Alfred Kraus, and Lorraine Kraus, who had spent their lives studying and treating this disease. The goal remains to conquer the disease.

Dr. Diggs also wrote the standard laboratory reference for blood cells, *Morphology of Human Blood Cells*. He commissioned well-known Memphis artist Dorothy Sturm, who taught at the Memphis Academy of Art, to make painstakingly accurate drawings of blood cells as educational tools. Using watercolors and pen and ink, she produced detailed illustrations of the morphology of blood cells in healthy and diseased states. Dr. Diggs, Dorothy Sturm, and medical technologist Ann Bell, M.S., wrote *The Morphology of*

Human Blood Cells, published in 1954, 1956, and 1970 by Abbott Laboratories. At the time, photomicrography could not produce the needed quality of images, and Dorothy Sturm's work was crucial.

Ann Bell, who taught medical technology at the University for many years and was awarded an honorary doctor of science degree upon retirement, helped produce two series of color transparencies showing blood cells. These were distributed by the American Society of Clinical Pathology and by the Armed Forces Institute of Pathology. She presented workshops for technologists, pathologists, residents, and others. For her special contributions, she was named a "Specialist in Hematology." She remains active in retirement.

In the late 1950s, Dr. Diggs visited a number of U.S. pediatric centers to help Danny Thomas conceptualize St. Jude Hospital, urging that the hospital focus not only on treating children with leukemia and other diseases, but on research. Dr. Diggs received awards including the Park-Mason Award as Virginian of the Year, from UVA Press Association; the Martin Luther King, Jr., Award from the Southern Christian Leadership Council for sickle cell disease research; and awards from the Tennessee Medical Association, the American Society of Clinical Pathology, the American Lebanese Syrian Associated Charities (ALSAC), and the University of Tennessee. He remained active in medicine and research until his death in 1995 at age 95.

Before 1956, autopsies for medical legal cases in Shelby County were performed gratis at the University of Tennessee. The Quarterly County Court contracted with the University in 1956, which shortly thereafter selected a medical examiner, whose salary was minimal at first but gradually increased. Dr. Jerry T. Francisco served as medical examiner until retirement in 1999. In 2004, Shelby County contracted with Forensic Medical, which now staffs the Shelby County Medical Examiner Office and the West Tennessee Regional Forensic Center. In 2010, Dr. Karen Chancellor was chief medical examiner, and Dr. Marco Ross was deputy chief.

Psychiatry. Gartly-Ramsay Hospital, located downtown on Jackson Avenue, which had been a private general hospital since 1909, became a private psychiatric hospital in 1950. R. G. Ramsey, Sr., and later his son, R. G. Ramsey, Jr., operated the highly regarded private psychiatric institution.

Private facilities, or sanataria, for drug and alcohol rehabilitation served many patients. When Dr. John Barron moonlighted as a resident at the Wright Sanatarium, a small private facility near Holly Springs, Mississippi, owned by Dr. Leonard Wright, Sr., literary giant William Faulkner would come for treatment. Dr. Wright's son, Leonard, Jr., later a physician in Memphis, sometimes rode in the car with Faulkner. A binge drinker, Faulkner went to Gartly-Ramsay in Memphis and to Westhill Sanatarium in the Bronx, but usually to Wright Sanatarium. In 1962, he attempted to taper off his bourbon use with beer and seconal at home. Earlier diagnosed with depression and anxiety, Faulkner had experienced frequent thoughts of death in the previous year, and he asked to go to Wright, though he usually resisted hospitalization. On July 5, 1962, Dr. Wright, Sr., admitted him and gave him vitamins and an antihistamine. Within eight hours, despite quick attention and 45 minutes of external heart massage, Faulkner died, apparently of a heart attack.

Treatment for Cancer. In the 1950s and beyond, cancer was treated mainly with surgery and radiation therapy. Dr. David Carroll was an outstanding diagnostic and therapeutic radiologist during that era.

Urology. In the 1950s, two Memphis urologists held major offices in organized medicine. Dr. Thomas D. Moore was president of the American Urological Association in 1951, and Dr. Samuel L. Raines was president of the Southeastern Section of the Association in 1955. Dr. Henry K. Turley was a highly respected Memphis urologist.

TESTING THE PAP SMEAR AT THE UNIVERSITY OF TENNESSEE HEALTH SCIENCE CENTER

IN 1952, UNDER THE AUSPICES OF THE UNIVERSITY OF Tennessee, one of the largest, most important studies of the century occurred in Memphis. Dr. Douglas Sprunt, chair of pathology, helped plan the study that tested the effectiveness of a cervical smear devised by Dr. George Papanicolaou. The Public Health Service's Cytology Laboratory, which had recently moved to the University's Cancer Center, was in charge, and Dr. Sprunt was instrumental in obtaining funding from the U.S. Public Health Service for large-scale testing of the smear—the first use of the Pap technique in mass screening on a large scale. The program was designed to detect cervical dysplasia and cervical cancer at the most curable stage, before signs and symptoms occurred.

ABOVE: *Dr. Cyrus C. Erickson instructs his cytotechnology students, July 1957.* OPPOSITE PAGE: *Pathologist Dr. Douglas Sprunt, center, in his laboratory with Commissioner David C. Mitchell, left, and Attorney General Phil M. Canale, right, February 1960.*

After three and a half years, more than half the women in Shelby County, a total of 108,000, had been tested. Of this number, cancer was detected in 773. Chairing the committee was Dr. Cyrus C. Erickson, who in 1954 published an article in *CA: A Cancer Journal for Clinicians*, which came to three major conclusions: 1) that the exfoliative cytology technique via mass screening was practicable; 2) that detection must be followed by tissue biopsy; and 3) that early detection provided the best hope for increasing the survival rate for cervical cancer.

Other physicians participating in the study were Drs. Henry Gotten, Frank Whitacre, Phil Schreier, and L. M. Graves, professor of medicine and director of the Memphis Health Department. Dr. Erickson was associate director of the University of Tennessee Institute of Pathology (the first named institute at the University) from 1950 through 1965 and director of the Memphis-Shelby County Population Cancer Survey from 1952 through 1958. When the U.S. Postal Service issued a special stamp honoring Dr. Papanicolaou, a special ceremony was held in Memphis honoring Dr. Erickson's part in the study. In 1965, he received the Papanicolaou Award in New York City.

After the success of the University screening program, similar screenings took place in Washington, D.C., and other cities. Soon the procedure's popularity highlighted the need for technologists to do the screening because too few pathologists were available. In Memphis, the University trained 15 "girls" as laboratory technicians, and three other "girls" were trained at Massachusetts General. A Memphis newspaper photograph of the "girls" was captioned, "Cancer Hunters—The success of the test hinges on highly trained cyto-technologists like these girls who painstakingly study slides under the microscope to track down cells suspicious of cancer."

In 1952, the University of Tennessee formalized its Cytotechnology Program, later incorporated into the College of Allied Health Sciences in 1985, and developed one of the few master's programs in cytopathology in the United States, a Master of Cytopathology Practice, begun in 2006.

CHAPTER ELEVEN

ST. JUDE HOSPITAL, MEDICARE, AND SOCIAL CHANGE

1960-1969

The social and political turmoil of the 1960s brought profound changes to Memphis. Medical practices and private hospitals—once divided into black and white—rather smoothly accepted integration. Drug treatment programs were developed to supplement alcohol treatment programs. The path would not always be smooth, but Memphis medicine would make strides from mass use of the Sabin polio vaccine, to new vaccines for common diseases, to new techniques for heart surgery, to founding a world-class hospital for catastrophic diseases of children, to studying risk factors for chronic diseases.

Milestones

A number of important milestones occurred in this decade, one of the most important being the opening of St. Jude Children's Research Hospital, one of Memphis's great medical accomplishments. So poignant are its success stories, so touching are its patients, that St. Jude raises support from around the world. When talk of St. Jude's moving to St. Louis began in the 1980s, the Memphis community mobilized. St. Jude would remain in Memphis, launching a massive new building program and compiling an ongoing list of successes in research and chemotherapy.

The decade also saw the opening of two significant medical facilities, the first being the state-owned Tennessee Psychiatric Hospital and Institute (TPHI) for research and short-term treatment (later the Memphis Mental Health Institute) which opened in 1962 at 865 Poplar Avenue. Conceived by Dr. David H. Knott and James D. Beard, Ph.D., the Alcohol Rehabilitation Research Center was another first in the state. Drs. Knott and Beard were nationally recognized in 1963 for their rehabilitation work. A Children's Unit opened at TPHI in 1967. Emphasizing training, research, and treatment, this facility coordinated efforts of residents, psychologists, social workers, pharmacists, and nurses from local institutions.

The first superintendent of TPHI was Dr. Bruce Walls. Dr. J. A. Wallace was superintendent from 1964 to 1974, after which Dr. Bill Webb became acting superintendent and also chair of psychiatry at the University of Tennessee.

The second of these facilities was the William F. Bowld Hospital, which the University opened in 1965 at 951 Court. It was a teaching and research hospital for the faculty, named for the philanthropist and former vice president

OPPOSITE PAGE: *Danny Thomas, founder of St. Jude Children's Research Hospital, unveils the statue of St. Jude Thaddeus, February 4, 1962.*

148   MEMPHIS MEDICINE: A HISTORY *of* SCIENCE AND SERVICE

ST. JUDE HOSPITAL, MEDICARE, AND SOCIAL CHANGE

ST. JUDE CHILDREN'S RESEARCH HOSPITAL

Founded in 1962 by entertainer Danny Thomas, who proclaimed, "No child should die in the dawn of life," St. Jude Children's Research Hospital exists because of a promise. As a struggling comedian in the 1940s, Thomas had only seven dollars, and few jobs were available for comedians. In a Detroit church, kneeling before a statue of St. Jude Thaddeus, patron saint of the lost and the hopeless, he made a vow: "Show me my way in life, and I will build you a shrine where the poor and the helpless and the hopeless may come for comfort and aid." Not long afterward, Thomas broke into show business in Chicago. Later the star of the long-running 1950s television sitcom *Make Room for Daddy* and the producer of other successful shows, he never forgot his promise to St. Jude.

Thomas consulted his mentor, Chicago's Samuel Cardinal Stritch, a Nashville native whose first parish had been in Memphis. Suggesting Memphis as the hospital's site, Stritch referred Thomas to the manager of the Memphis Chicks baseball team, lawyer Edward Barry, for help with fundraising. Thomas and Barry began raising money in the early 1950s. In 1957, Thomas established the American Lebanese-Syrian Associated Charities, the fundraising arm of his great project.

Thomas first proposed a children's hospital, but Memphis already had Le Bonheur Children's Medical Center. With

Ground-breaking for St. Jude Children's Research Hospital. Left to right: Rabbi James A. Wax, Rt. Rev. Msgr. J. Harold Shea, the Rev. Dr. Donald Henning, and Danny Thomas.

ABOVE: *Danny Thomas, seated, with members of the St. Jude medical advisory board, June 1957. Standing, left to right: Drs. William F. Mackey, Gilbert Levy (president of the board), Lemuel W. Diggs, Robert Raskind, Charles G. Allen Jr., and Ralph O. Rychener.* RIGHT: *Dr. Donald Pinkel in his laboratory, 1962. He was chief of pediatrics at the Roswell Memorial Institute, Buffalo, New York, before becoming medical director of St. Jude.*

help from Dr. Lemuel Diggs and Mayor Frank Tobey, he envisioned a hospital for children with catastrophic illnesses, children without hope. At the time, children diagnosed with leukemia only lived a year or two as no effective treatment existed. Thomas's proposed hospital would establish research programs and treat children with leukemia and other catastrophic conditions such as sickle cell disease and inherited immune disorders. All would receive equal treatment regardless of race, religion, or financial circumstances. Those who could not pay would not be denied care. Meanwhile, research would attempt to control or conquer their diseases.

With the help of his wife, Rose Marie, and Dr. Diggs, and with Mayor Tobey's support, Thomas repeatedly crossed the country by car, raising funds. In 1962, St. Jude Children's Research Hospital opened in Memphis and began the long struggle to conquer diseases that, by the 1980s, included childhood HIV/AIDS.

When St. Jude opened, fewer than four percent of children diagnosed with acute lymphocytic leukemia (ALL), the most common pediatric cancer, survived as long as five years. As the first director of St. Jude Children's Research Hospital in 1962, Dr. Donald Pinkel quickly began laboratory work, trying various protocols. Over the years, the survival rate increased, but Pinkel was still not satisfied when the cure rate reached fifty percent. He devised a combined regimen of chemotherapy and radiation that dramatically increased the survival rate. Dr. Pinkel received the Albert Lasker Prize for his work. In the 2000s, the survival rate grew to ninety-four percent.

In the 1990s, St. Jude began bone marrow transplants and gene therapy. The success with leukemia led its researchers to a broadening spectrum of diseases. St. Jude willingly exported its expertise through its International Outreach Program, sharing knowledge with hospitals around the world and remaining the only children's cancer center in the United States where families were never asked to pay for treatments not covered by insurance. In addition, St. Jude assisted with travel, lodging, and meals. Three facilities for patient and family housing were built: Grizzly House, Ronald McDonald House, and Target House. In 2009, the hospital spent more than $1.5 million each day to run the hospital.

St. Jude has 29 International Outreach Partner sites in 15 countries around the world. The hospital also brings the latest advances in treating pediatric cancer to developing countries through Web-based initiatives, educating local healthcare providers, and establishing pediatric cancer clinics in areas with limited resources.

In 2002, an affiliate of St. Jude, the Children's Cancer Center of Lebanon (associated with American University of Beirut Medical Center) opened in Beirut, Lebanon. Through its Domestic Affiliates Program, St. Jude offers partnerships with other pediatric programs in a network of hospitals, hematology clinics, and universities. Clinics are located in Johnson City, Tennessee; Peoria, Illinois; Shreveport and Baton Rouge, Louisiana; Huntsville, Alabama; and Springfield, Missouri.

Dr. Donald Pinkel was the hospital's first director (1962-1973), followed by Dr. Alvin Maur (1973-1983); Dr. Joseph Simone (1983-1992); Dr. Arthur Nienhuis (1993-2004); and Dr. William S. Evans (2004-present). In 2008, St. Jude became the first and only NCI-designated Comprehensive Cancer Center devoted solely to children.

ABOVE: *Dr. David H. Knott peers at a flask held by Dr. James Beard, 1970. The machines at the left constitute the Atomic Absorption Spectrophotometer used to determine concentrations of various metals in the body.*
RIGHT: *Dr. David Shields Carroll, 1966, when he was a member of the Heacock-King Radiological Group. Among his many achievements, he was also president of the American Cancer Society.*

of Buckeye Cellulose Division of Procter and Gamble in Memphis.

Also opened at this time was the Dobbs Research Institute, named for businessman and philanthropist James K. Dobbs (1894-1966), which was established through the Dobbs Foundation and housed in the Chandler Building at 865 Jefferson. It included basic research facilities for the University, plus the Department of Radiology.

Major milestones were also reached in the field of medical techniques and procedures. In a world-changing event, Dr. H. Edward Garrett, Sr., performed the world's first coronary artery bypass graft in 1964, while working with Dr. Michael DeBakey in Houston. The patient did well but returned to South America and disappeared from follow-

TOP: *The James K. Dobbs Medical Research Institute, 1965. It was built on the site of Russwood Park.* ABOVE: *President Lyndon B. Johnson signs the Medicare Bill at the Harry S. Truman Library, Independence, Missouri, July 30, 1965. To the left is Lady Bird Johnson. At the right is President Harry Truman, seated, and Vice-President Hubert Humphrey and Bess Truman, standing.*

where they injected a radio-opaque dye that enabled visualization of blockages by cholesterol plaques. Coronary angiography was performed by cardiologists specifically trained in cardiac catheterization. At Baptist Hospital, Drs. Thomas Stern and Pervis Milnor pioneered the procedure; at Methodist, Drs. Dwight Clark and Arthur Sutherland were pioneers.

Coronary surgeons of the day were led by Drs. Edward Garrett, Sr., Brewster Harrington, Glenn Crosby, and Robert Richardson. Comparable benefits came to patients who suffered with blockages of peripheral arteries. General surgeons who became experts in performing graft procedures on the aorta and on femoral artery occlusions included Dr. George A. Coors at Methodist, who performed the first surgery in Memphis for an abdominal aortic aneurysm, and Dr. Randolph Turner, who trained with Dr. Denton Cooley in Houston.

Advances were also made in the field of radiology, which was divided into diagnostic and therapeutic disciplines, each having its own board examination. Subspecialties would continue to develop through this decade and the next. A turning point in Memphis and nationally was the determination that radiologists, highly trained medical specialists, should not be considered hospital employees and reimbursed by the hospital. Instead, radiologists would bill patients independently as medical specialty consultants. Memphian Dr. David Carroll, later president of the American College of Radiology, spoke in Washington, D.C., in 1962 on behalf of his colleagues. Dr. Webster Riggs, Jr., was active in implementing separate billing and was the first in Memphis to bill independently for radiological services, at Le Bonheur in 1969.

up for several years; when he returned for reevaluation, the graft was still good. Shortly after performing this procedure, Dr. Garrett moved to Memphis to join Dr. Hector Howard in the practice of cardiovascular surgery.

The breakthrough that allowed for these coronary artery bypass grafts was the development of coronary angiography in 1964, in which physicians passed a catheter into the groin, up the aorta, and into the arteries of the heart,

MEDICARE, LANDMARK LEGISLATION OF THE TWENTIETH CENTURY

In 1965, Congress passed the Social Security Act Amendments, known as Medicare, with Medicaid added, as a program of hospital insurance for those aged 65 and older and for younger disabled individuals. The program was funded by a tax on employees' earnings, matched by employers' contributions. Consideration of a national health program had begun with President Harry S. Truman, who proposed a national health insurance plan. Healthcare costs continued to rise, and private insurers considered those over age 65 poor risks. The Medicare legislation was approved through the efforts of Congressman Wilbur Mills and champion arm-twister President Lyndon Johnson, who invited a reluctant Senator Harry Byrd to his office and immediately led Byrd into the packed pressroom, taking Byrd's elbow and asking, "Harry, is there any reason why you can't move the Medicare legislation?"

Medicare established its own rates of reimbursement for medical and hospital services, and private insurers based their own rates on Medicare rates. The effect on hospital and physician reimbursements was dramatic and, as Medicare continued to refine its plan and rates, the effects would continue.

The Medicare stipulation that changed Memphis medicine was that reimbursement would not be made to any hospital that denied access to individuals on the basis of race, creed, or color. This requirement, in effect, ended racial segregation in Memphis health care.

The Worst of Times: The Assassination of Dr. Martin Luther King, Jr.

Dr. Martin Luther King, Jr.—civil rights leader, political activist, and opponent of the Vietnam War—came to Memphis in the spring of 1968 to support striking sanitation workers. These men, nearly all African Americans, worked long hours for pay below the poverty level. They had no health benefits. Strongly influenced by Mahatma Ghandi, King believed in nonviolent protest, in "passive resistance" that, by show of numbers and sheer moral force, would draw attention to the protesters' plight.

On April 3, Dr. King spoke to followers at Mason Temple (headquarters of the Church of God in Christ) in Memphis, recalling the bomb threat that had delayed his flight and foreseeing difficult times ahead. With seeming prescience, he spoke of wishing to live a long life but promising "to do God's will," adding, "... I've been to the mountaintop....And I've seen the promised land. I may not get there with you, but I want you to know tonight that we as a people will get to the promised land."

OPPOSITE PAGE: *A strike by the sanitation workers in February 1968 was the impetus which first brought Dr. Martin Luther King to Memphis on March 18. Here, strikers walk west on Beale across Second Street.*
ABOVE: *Martin Luther King and Jesse Jackson at Mason Temple, April 3, 1968. Dr. King was horrified over the March 28 riot, and planned to return to Memphis and a peaceful march on April 5. However, the city had asked for a restraining order. A telegram from the American Civil Liberties Union was sent to Lucius Burch, asking him to defend Dr. King in order to lift the injunction. On April 4 he succeeded, but Dr. King was assassinated that evening.*

The next day at 6:01 p.m., from the second-floor balcony outside his room in the Lorraine Motel, Dr. King spoke his last words to Ben Branch, a musician scheduled to play that night: "Ben, make sure you play 'Take My Hand, Precious Lord' in the meeting tonight. Play it real pretty." A single shot from a window across the street struck Dr. King in the jaw and traveled to the back of his head, severing the spinal cord and moving downward, not leaving his body. Minutes later, at nearby St. Joseph Hospital, a young doctor, finding no response, opened the chest and directly massaged his heart. But the man who had changed America could not be resuscitated; he was pronounced dead at 7:05 p.m.

As word spread that Dr. King was dead, riots broke out in Memphis and more than 100 other cities. It was the worst of times for Memphis, smugly labeled by *Time* "a backwater river town." Supporters of Dr. King, black and white, grieved for him and for their city. Schools and colleges were closed. City services, already affected by the strike, shut down. In the coming weeks, the strike in Memphis was settled in favor of the sanitation workers, and the city began a long and painful struggle to rebuild.

The face of Memphis would change in the next 40 years. The assassination of Dr. King in Memphis and *Time*'s label of a "backwater" forced Memphians to rethink their values. Volunteer agencies such as the Metropolitan Inter-Faith Association (1968) were organized. Gradually, "diversity" would signal recognition of years of contributions by African Americans, and the city would be restored with appreciation of its many racial and ethnic influences. Medicine in Memphis would become fully integrated at all levels, including African Americans as hospital board members and chiefs of staff at Memphis hospitals.

African-American Physicians

The dominant event in Memphis medicine's cultural history in the 1960s was integration of previously segregated hospitals. A brief review of key dates in the movement away from segregation helps explain the conditions under which African-American physicians labored. In 1945, at the end of World War II, 26 of the 78 accredited medical schools in the United States were closed to African Americans, all 26 in Southern or border states. In 1947, 27 of the 127 VA hospitals maintained segregated wards, and 19 of these 27 refused to admit African Americans except for emergency treatment. In 1948, President Truman ended segregation in the military by executive order. In 1955, the VA desegregated. In 1960, 12 of 26 medical schools in the South were still closed to African Americans. In 1964, the U.S. Fourth Circuit Court of Appeals ruled against race discrimination in federally funded hospitals. Also in 1964, the American Medical Association passed a resolution saying that it was

PERSPECTIVES ON INTEGRATION: MR. MAURICE ELLIOTT, DR. EDWARD REED, DR. LAWRENCE WRUBLE

In a paper presented to the Memphis Medical Society in 2006, Maurice Elliott, retired administrator at Baptist and at Methodist, recalled the 1960s, when integration came to Memphis health care. At the time, the Medical Center was the hub of Memphis medicine and no satellites existed. The private hospitals were more closely affiliated with churches than some are today. Black and white communities were separate—restaurants, schools, hospitals, doctors' offices. John Gaston Hospital served both races, but with separate wards. The Civil Rights Act of 1964 mandated that hospitals receiving federal funds hold policies of nondiscrimination on the basis of race, nationality, ethnic group, religion, or gender. Hospital administrators had to sign forms ensuring compliance.

In Memphis's large private hospitals, nearly all black employees held low-level jobs and had separate locker rooms and cafeterias. The limited number of African-American physicians practiced at John Gaston (with separate facilities for whites and blacks), E. H. Crump Hospital (built by the city in 1958 for African Americans), and the private Collins Chapel Hospital. White physicians sent their African-American patients to Crump, and a few white physicians worked at Crump.

In March 1965, the Baptist Board met to discuss signing a letter signifying compliance with the 1964 act. Dr. R. Paul Caudill of First Baptist Church urged the group to sign, saying it was the right thing to do. The board agreed. At Methodist, the board hesitated at first, but a medical staff representative advised signing. In its editorials, the *Commercial Appeal* took a negative view. For years, private hospitals had subtly discouraged African-American physicians by requiring two years of postgraduate training for staff membership, while most African-American physicians at the time had taken a single year of internship. St. Joseph integrated in 1966. Rumor held that an African-American employee once was brought into the hospital, but no one confirmed the incident. Harry Mobley, assistant administrator at Methodist from 1967 to 1975, encountered a "plantation mentality" when he came to Memphis from California.

Dr. Lawrence Wruble had just opened his gastroenterology practice in 1964 when an African-American patient, a school principal, came to see him. When Dr. Wruble advised hospitalization, the man said, "I have insurance, and I want to go to Baptist." Because Baptist had no black patients at the time, Dr. Wruble called the hospital. Mechanically, the voice said, "We refuse admission to no one." For Wruble, who admitted the patient, the time was difficult; he received threatening phone calls, and the medical community's support was mixed. When Medicare came a year later, Frank Groner,

to Memphis in 1962, Dr. Reed encountered not only overt segregation, but also "a tone and a philosophy" echoed by newspaper editorials about health care and segregation. Dr. Reed could practice only at Crump and Collins Chapel.

In 1966, Dr. Reed received privileges at Baptist, but although he was board certified in surgery, he was assigned a "preceptor." A member of the executive committee said, "If you know any other qualified Negro doctors, please tell us." He was also invited to join the faculty at the University, but he was not admitted into the Memphis Surgical Society until later. Eventually he was on the staffs of Bowld, St. Joseph, and Le Bonheur, in addition to Baptist. At St. Joseph and Le Bonheur, he was the first African-American physician. Ten years later, nominated by 13 agencies, Dr. Reed was awarded the prestigious L. M. Graves award.

In later years, Reed did much cancer surgery and introduced surgical staples to the Mid-South, after training in New York with the technology. He was on the hospital authority before The MED received its current name, having started in 1966. At The MED, he worked for both access to and quality of health care. In 1989, he presented a paper in Las Vegas about people who were too poor to get sick. He chaired the Medicaid Advisory Committee from 1979 to 1990.

OPPOSITE PAGE, ABOVE: *Maurice Elliott, assistant administrator at Baptist Memorial Hospital, 1968.*
OPPOSITE PAGE, BELOW: *Dr. D. Lawrence Wruble, gastroenterologist.*
ABOVE: *Dr. Edward W. Reed receives the L. M. Graves Memorial Health Award from Irvin Bogatin, a trustee of the Haspel Foundation which sponsors the award, 1976.* RIGHT: *Dr. Neal S. Beckford, 2003. He was a past president of both the Bluff City and The Memphis Medical Society.*

the administrator at Baptist, thanked Dr. Wruble, who added, "They just needed someone to break the barrier. Besides, it was a pain in the neck to separate patients in the office." At Methodist, neurosurgeon Dr. Billy Ogle admitted one of the first African-American patients.

The personal remembrance of a highly respected African-American physician and surgeon, Dr. Edward Reed, provides another perspective. A graduate of Meharry Medical College in 1955, he remained there for internship, surgery residency, and teaching. Coming

Dr. Reed's story of expertise, determination, concern, and professionalism is one of many. Today, Memphis's medical community is enriched by the many contributions of devoted and knowledgeable African-American doctors, nurses, and technicians. African-American physicians serve in key positions at all private hospitals and at the University. This number includes former chancellor Dr. William F. (Bill) Owen, Jr. (2005-2006), originally from Memphis. Dr. Neal Beckford has served as president of the Baptist medical staff and was elected the first African-American president of The Memphis Medical Society in 2004.

In the late 1970s, the Frank Tobey Wing of Gaston closed, and all pediatric patients went to Le Bonheur. Earlier, the University of Tennessee, Le Bonheur, and the county government had agreed that there should be a single standard of care for children. At the University, the enrollment of African-American medical students increased to 44 in 1984 and to 63 in 2006. Together with traditionally black medical schools, Meharry and Howard, the University of Tennessee, in 2009, was among the top ten medical schools in the United States in graduating African-American students.

ST. JUDE HOSPITAL, MEDICARE, AND SOCIAL CHANGE

A reunion of West Tennessee University graduates. Memphis, unknown date. Seated, left to right: Drs. Norman M. Watson, unknown, possibly A. A. White, Christopher M. Roulhac, and B. F. McCleave. Standing behind Dr. White in a bow tie is Dr. Wheelock Bisson.

"unalterably opposed" to counties' denying membership in medical societies on racial bases, but included no provision for enforcement. The AMA and the National Medical Association formed a Liaison Committee that year. In 1965, the Department of Health, Education, and Welfare proposed requiring compliance statements, but the American Medical Association refused. Then, Title VI of the 1964 Civil Rights Act made discrimination in hospitals receiving federal funds illegal. In 1965, Medicare and Medicaid, which required participating hospitals to sign a statement of intent not to discriminate on the basis of race, color, or national origin, essentially mandated integration.

One physician who lived through these changing times was Dr. Clara Arena Brawner. The only practicing African-American woman physician in Memphis in the mid-1950s, she graduated from Meharry Medical College in 1954, following in her father's footsteps (M.D. degree, Meharry, 1926). Over the next 30 years, she moved into local politics, healthcare administration, fundraising, and theology. A role model and a mentor, she practiced family medicine with an emphasis on pediatrics. She was on the staffs of St. Joseph, Baptist, and Collins Chapel hospitals, where she chaired pediatrics and guided the scientific program. She held several leadership positions in the African-American Bluff City Medical Society (BCMS), was a homecare physician for the VA, and was medical director for Goodwill Homes for Children. In 1966, she was elected to membership in the Memphis and Shelby County Medical Society. Later, she studied at Memphis Theological Seminary. Her sister, Alpha Brawner, was a well-known soprano. Dr. Clara Brawner died in 1991.

The Bluff City Medical Society, begun about 1907, was kept alive in these dark days by the leadership of Dr. Brawner, its first woman president. The group met at her home, and the physicians, whose number had declined to 20 in the 1960s, kept their strong sense of duty to patients

Above: Dr. Clara Brawner, 1967. She donated her time to examine children who lived at Goodwill Home. Dr. Brawner's mother was a nurse. Her father, Dr. J. Brawner, came to Memphis in 1927 to intern at Jane Terrell Hospital. Right: Dr. William Oscar Speight, Sr., 1959. He came to Memphis in 1922 and earned his medical degree at West Tennessee Medical School. Regarded as a "working doctor," he delivered more than 4,000 babies. Dr. Speight died in 1964, age 79. His son, Dr. William Oscar Speight, Jr., an EENT specialist, graduated from Meharry Medical College, and was a founder of the Memphis Health Center. He died in 1975, age 54.

and the community. In 1958, a group of members of the Society were photographed in front of Eli Lily and Company in Indianapolis, where they were honored for their skills and achievements by the company and by Indianapolis physicians. The group included Drs. T. H. Watkins; L. A. Johnson; W. Oscar Speight, Sr.; W. A. Bisson; W. O. Speight, Jr.; B. F. McCleave, Sr.; Clara Brawner; Arthur E. Horne; N. M. Watson; and Leland L. Atkins, president of the Society and vice chairman of staff at Collins Chapel Hospital.

With Dr. Brawner, Dr. C. O. Daugherty and others began the Memphis Health Center Clinic, 360 E. H. Crump Boulevard. Incorporated in 1973, the clinic was a federally qualified health center for the underserved; it provided primary medical care for the African-American community and enabled young physicians to establish their practices. In the early 1980s, the clinic added the Chelsea-Watkins Health Center Clinic (1230 N. Watkins) and the Greenlaw Health Center Clinic (278 Greenlaw). In 1986, the Rossville Health Center Clinic opened (Highway 57, Rossville), and in 1987 the clinic established a program for the homeless.

In the mid-1970s, 15 African-American physicians practiced in Memphis. As opportunities increased in medical education and practice in Memphis, additional African-American physicians came to the city reflecting the growing sociological diversity of Memphis medicine. Dr. Brawner remained president of the Bluff City Medical Society, with Dr. Wheelock A. Bisson as secretary and Dr. Edward Reed as treasurer. Each served for more than ten years.

The first African-American medical graduate at the University of Tennessee was Dr. Alvin Crawford (M.D., 1964). He is now a professor of pediatric orthopedic surgery at the University of Cincinnati and director of the spine center at Cincinnati Children's Hospital. Also a classical clarinetist, he laughingly anticipates a "new career in music" in retirement.

ABOVE: *Dr. John W. Runyan, Jr., 1977, at the time he received a $10,000 Rockefeller Public Service Award for his work in "promotion of health, improved delivery of health services and control of health costs."* RIGHT: *Dr. Lester R. Graves, obstetrician and gynecologist, was the son of Dr. Lloyd M. Graves, director of Memphis's Department of Health.*

The Bluff City Medical Society was recognized by the National Medical Association as "Chapter of the Year" in 2004 because of its superior programming, recruitment of physicians, and service to the community. Currently Memphis has more than 250 African-American physicians. Memphis radiologist Dr. Albert W. Morris, Jr., was national president of the 25,000-member National Medical Association in 2008, after presiding over the Bluff City Medical Society. Dr. Morris was a member of the Committee on Homeland Security and was the lead author of a consensus paper on "Environmental Terrorism," presented in 2002 at a colloquium sponsored by the National Medical Association. As president, Dr. Morris said that he could "look forward to advancing the National Medical Association's agenda of eliminating disparities in our current medical system and creating a healthier America."

DISEASE CONTROL

More new vaccines (measles and chicken pox in the 1960s) demonstrated the benefits of medical research on the quality of life. In 1961, massive administration of the Sabin oral polio vaccine virtually eliminated polio in America, with the last transmission of "wild" (not vaccine related) polio in the United States in 1979. Memphis pediatrician Dr. Hubert Dellinger, Jr., recalled that when he was a resident in the 1960s, the Isolation Hospital (renamed the Jefferson Pavilion) that had formerly housed polio patients had become quarters for the house staff. In 1968, the Centers for Disease Control declared that smallpox was no longer endemic in the United States.

In 1963, Dr. John W. (Bill) Runyan, working with Dr. George Lovejoy, published an article describing the Visiting Nurse Program at the Health Department; the two won a Rockefeller Award for their work. From this beginning grew the University of Tennessee's nurse-practitioner program.

MEMPHIS PRACTICES

Internal Medicine. Dr. J. D. Young (1962) and Dr. I. Frank Tullis (1966) served as presidents of the Tennessee Society for Internal Medicine. Dr. Henry G. Rudner, Jr., joined his father in practice.

Dr. Maury W. Bronstein established a new group, Internal Medicine and Cardiology, and became a prominent internist by the 1960s. A University of Tennessee graduate (1951) who did his residency at Louisiana State University Medical Center in New Orleans, he practiced at Baptist and St. Francis hospitals and taught at the University for many years. He was recognized with an Outstanding Physician Award from the Tennessee Medical Association in 2009. An endowed chair in cardiovascular physiology bears his name, and the medical library at Baptist Memorial Hospital Memphis is named for him.

Obstetrics and Gynecology. Dr. Lester R. Graves, son of longtime Health Department director Dr. L. M. Graves, began practicing obstetrics and gynecology in Memphis in

DR. SHELDON KORONES AND THE NEWBORN CENTER

Dr. Sheldon Korones grew up in Manhattan and received his M.D. degree from the University of Tennessee. After doing his residency at Boston City Hospital, he practiced general pediatrics in Memphis, but his interest in academic medicine led him into part-time teaching at the University for ten years. At Gaston, he saw the contrast between the "more personalized medicine" that children of affluent families received in East Memphis and the level of care given to poor children at the city hospital. In 1968, he joined the University faculty full time.

With Dr. Korones as its first director, the Newborn Center opened July 1, 1968. "I was not a marcher with placards," he said in an interview. "I recognized my professional talents, and I wanted to take care of the most helpless patients." When he went to Washington seeking money, he asked for money for people, not equipment. "If I could train more nurses," he said, "the mortality in the nursery would decrease. It fell by 40 per cent the first year."

From a $100,000-per-year project, the Center's budget grew to over $1.1 million a year in 2008, with funding from state and federal governments. When Dr. Korones needed a van, he obtained a converted bread truck, which enabled nurses to go out 250 times a year to transport critically ill babies from around the Mid-South to the newborn center.

"A more loyal and effective group of people I have never seen," said Dr. Korones of his staff. In more than 40 years, he cared for more than 50,000 sick babies. Recently he met an 18-year-old boy who said, "I was a 460-gram baby when you took care of me." No one can predict which ones will survive. "But," said Dr. Korones, "not knowing who

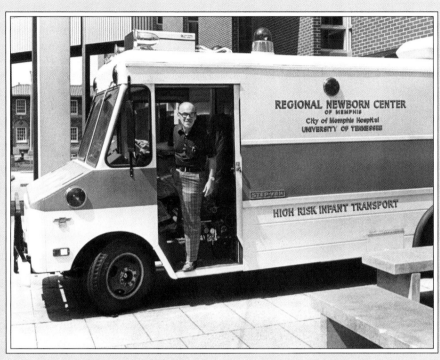

Pediatrician Dr. Sheldon B. Korones at the wheel of his converted bread truck, 1976. The van was specially equipped so that medical personnel could perform delicate procedures while parked or in motion. The Regional Newborn Center of Memphis was organized by Dr. Korones on the statistic that Shelby County had the worst infant mortality rate in Tennessee and one of the worst in the country.

will be impaired, I consider it my obligation to opt for life." He credited "gathering good people" for the success of the Newborn Center, currently directed by Dr. Ramasubbareddy Dhanireddy.

Dr. Korones wrote several books, including *High Risk Newborn Infants*, which was translated into Russian and Spanish. His *Neonatal Decision Making* was translated into Polish. He went to Russia four times on teaching missions after writing the first book. His work has touched millions of lives around the world.

Dr. John K. Duckworth, director of Methodist Hospital laboratories, checks samples at the blood bank, 1975. Methodist had completed its first year of a volunteer blood program.

1961 and still maintains an office in 2010. The founder of his group in the 1950s was Dr. William P. Maury, nephew of Dr. John Maury, longtime obstetrics and gynecology professor at the University of Tennessee. Dr. Maury's first partner was Dr. Henry Leigh Adkins, and Dr. Graves joined them in 1961. Later, Drs. Jack C. Sanford and Sam J. Cox joined the group.

Interviewed in 2009, Dr. Graves remembered obstetrics and gynecology in the 1950s and 1960s, at a time when obstetrics and gynecology specialists treated their patients for everything from minor illnesses to diabetes to deep vein thrombosis to breast masses. Graves recalled working at Gaston as a resident in the 1950s, when there was no air-conditioning, and windows had to be open even in the operating room or the delivery suite. "Nurses with flyswatters were a necessity," he said. A colleague, Dr. William K. Payne, recalled in a memoir that medical residents ordered oxygen tents for heart patients for the cooling effect.

By the mid-1960s, Memphis had four women obstetrics and gynecology specialists—Drs. Sara E. Abbott, Mary E. Bouldin, Martha A. Loving, and Betty Schettler.

Oncology. Although the American Society of Clinical Oncology met first in Philadelphia in 1965 in response to the complexity of cancer care, Memphis had first-generation medical oncologists during the decade. At first, the medical treatment of cancer was primarily provided by hematologists and surgical oncologists. Drs. Charles Neely, Jefferson D. Upshaw, and Gerald Plitman were the earliest hematologists to treat solid tumors with anti-cancer drugs. Outstanding as teachers and physicians, they influenced several University of Tennessee residents to become medical oncologists.

Drs. Reed Baskin, Barry Boston, William West, and Barry Neal were residents under the tutelage of Dr. Gene Stollerman, who had a unique ability to encourage his residents to aspire to be the best they could. Stollerman's work in immunology led them into medical oncology.

Pathology. Dr. Douglas Sprunt remained a leader in the Memphis Society of Pathologists until he retired in 1968, and Dr. Cyrus C. Erickson succeeded him as chair at the University.

A major pathology laboratory, Duckworth Pathology, was a microcosm of the specialty's growth, as was Dr. Thomas Moss's laboratory. When Dr. John Duckworth finished pathology training in 1964, he became chief of clinical pathology at Methodist Hospital. He selected the staff, added pathologists who brought specialties to the lab, and accepted "no second rate work." In 1967, Dr. Duckworth began the independent Duckworth Pathology Group, which built a reputation as a meticulous lab.

Rheumatology, a new subspecialty. Dr. Glenn M. Clark came from the University of Colorado in 1957 to head a developing subspecialty of internal medicine—rheumatology—at the University of Tennessee, soon after cortisone was developed in the 1950s and became a boon to both clinical and academic rheumatology. He helped recruit the philanthropists for whom the Bowld Hospital and its Dobbs Research Institute were named.

Important breakthrough researchers in immunology of inflammation were Drs. Andrew Kang and Arnold Postlethwaite at the University, and Alex Townes at the VA. By 2009, Kang and Postlethwaite had developed a drug for scleroderma that was in advanced stages of testing, having been declared an orphan drug in Europe and America.

Dr. Glenn M. Clark also practiced rheumatology at Bowld Hospital. Bright and charming, he was popular as Memphis's first practicing rheumatologist. In the late 1950s, he was joined in practice by one of his first graduate fellows, Dr. Stanley B. Kaplan, who practiced and taught rheumatology until his death in 2008. Dr. Laura Carbone of the University of Tennessee and the VA Hospital wrote a poignant tribute to the widely known Dr. Kaplan, published in *Annals of Internal Medicine*, which played on the Hebrew meaning of his middle name, *Baruch*, 'blessing,' and concluded, "I need to finish now, for your patients are here and I know what you would want. For you were both their doctor and their blessing, and you have taught us well."

ABOVE: *Dr. and Mrs. Glenn Clark on their first European trip, 1958. Dr. Clark, Memphis's first practising rheumatologist, was an assistant professor of medicine at the University of Tennessee.* RIGHT: *Dr. Stanley Kaplan watches technician Jacqueline Carter as she examines a slide as part of the Memphis and Shelby County Arthritis Research Program supervised by Dr. Alfonse Masi, 1968.*

Dr. Clark was also chief of staff at the City of Memphis Hospitals; as he became increasingly busy, he recruited Dr. Al Masi from Johns Hopkins to chair the Department of Rheumatology at the University. Shortly thereafter, Dr. Gene Stollerman came as chair of medicine, and with his strong immunology and rheumatic disease background, greatly strengthened the rheumatology section. The University's Rheumatology Department collectively trained most of Memphis's clinical rheumatologists, including Drs. H. P. Blumenfeld; Howard Marker; Hugh T. Holt, Sr. and Jr.; Marshall Koonce; R. Franklin Adams; Lowell B. Robison; and Jane Alissandratos. Dr. Charles Arkin trained outside the Memphis program.

CHAPTER TWELVE

TRANSPLANTS, NEW IMAGING, AND CHRONIC DISEASES

1970-1979

From psychedelic colors and Woodstock to Republican president Richard Nixon, from rebellion against the Vietnam War and the "establishment" to "stagflation," from Kent State to oil shortages (1973, 1979) and interest rates of over twenty-one percent (1979), world changes were echoed in Memphis. As the decade began, few in East Memphis or Whitehaven or Raleigh could anticipate busing of public schoolchildren to distant schools to achieve racial balance or the variety of private, often church-supported, schools that sprang up. No one could anticipate that the King of Rock and Roll would turn a spotlight onto Memphis's medical community.

Medicine in Memphis became subspecialized as knowledge expanded beyond the ability of any single individual to master it all. Human kidneys and livers were transplanted.

Medical treatments for cancer were discovered. The most common childhood leukemia was regularly cured in Memphis. And east of I-240 in Memphis, Sister Rita's vision of a hospital easily accessible to suburbia opened.

Milestones

In the 1970s, Memphis medicine became increasingly subspecialized. In internal medicine, subspecialties included pulmonology, endocrinology, rheumatology, nephrology, cardiology, gastroenterology, and infectious diseases.

The 1970s began an era of more precise diagnostic medicine, led by developments in imaging: computerized axial tomography scan (CAT), magnetic resonance imaging (MRI), and ultrasound (US). Drs. J. Cash King at Methodist Hospital, Jack Whiteleather at Baptist Hospital, and Barry Jerald at the University of Tennessee

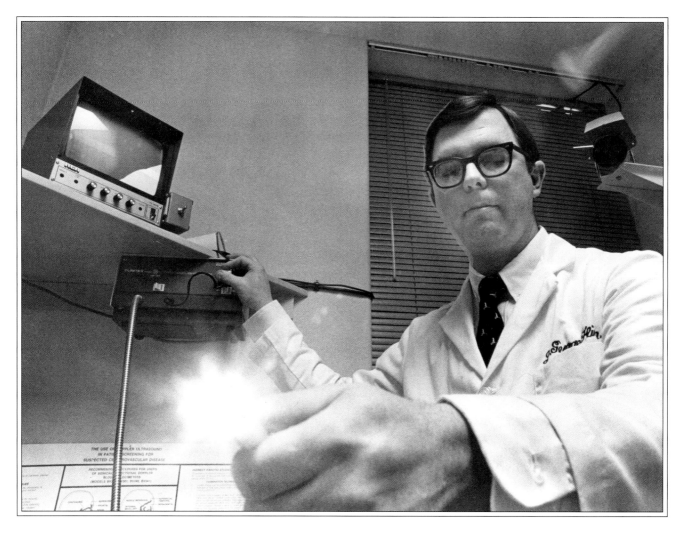

ABOVE: *An important development in medicine during the 1970s was the use of imaging techniques for diagnosis. In this 1983 photo, Dr. George Flinn, Jr., employs a light-scanning device (diaphanography) in the detection of breast cancer. His office was one of only five in the country which used three different tests to diagnose breast cancer: thermography (heat), ultrasonography (sound), and diaphanography.* OPPOSITE PAGE: *Dr. Rex Amonette, 1973, the first physician to use Mohs micrographic surgery in Memphis.*

headed radiology departments that spearheaded imaging technology in Memphis. Drs. George Flinn, Jerry Grise, and Chapman Smith were early advocates for ultrasound, begun in 1975 in Memphis.

Early in 1974, radiologic technologist Bill Robinson from Methodist Hospital and physicist John Kersey from Baptist Hospital attended a course in England on operating and repairing the CAT scan unit. The subsequent purchase of these units, at Semmes-Murphey's urging, brought to Memphis two of the first ten in the United States. Despite competition between the two hospitals, Robinson and Kersey worked together to set up the scan units for Memphis patients. Drs. Bill Mitchum from Methodist and Randy Ramey from Baptist were the first radiologists involved. Drs. Fred Hamilton and Colby Gardner soon wrote a paper on diagnosing a brain abscess via CAT (or CT) scan.

Breakthrough diagnostic procedures changed cardiovascular medicine. Angiography improved the success of vascular surgical procedures of all types. Pioneers included Drs. Thomas Stern, Frank Kroetz, and J. Leo Wright at Baptist, with Drs. Art Sutherland and Dwight Clark at Methodist.

Dermatology. Memphis was the first city in the South to have Mohs micrographic surgery. The city also had a substantial training program in Mohs surgery, introduced by Dr. Rex Amonette, who developed the training program through the University of Tennessee in the early 1970s. The program drew dermatology residents from around the world to learn from Dr. Amonette, the first Mohs surgeon in Memphis, one of perhaps three in the country.

Dr. Frederick Mohs developed his procedure in 1936 while still a medical student at the University of Wisconsin.

First called *chemosurgery* because it used a chemical fixative, the procedure froze tissue for microscopic examination. Dr. Mohs's approach gradually traced out the tumor until all was gone. The procedure required on-site pathological analysis to enable the surgeon to remove only diseased tissue. Mohs micrographic surgeons were specially trained in both pathology and reconstruction to leave the smallest possible scar. By 2010, Dr. Amonette had trained hundreds, and Mohs was the procedure of choice for non-melanoma skin cancer.

St. Joseph East. In the early 1970s, patients from East Memphis drove half an hour to hospitals clustered downtown. In October 1968, Sister Rita, administrator at St. Joseph Hospital, and Sister Rolandina, assistant administrator, discussed purchasing the Lichterman-Loewenberg property at Park and Ridgeway for a satellite hospital. When Sister Rita and associate administrator Jesse L. Luton went to William A. Loewenberg's office to discuss a purchase, Mr. Loewenberg showed them a feasibility study conducted in 1964 for the Sisters of Mercy Hospital of Knoxville. Mr. Loewenberg had offered the property at that time, but the Sisters had not pursued the project.

Groundbreaking ceremony for St. Joseph Hospital East, March 13, 1974. The event symbolized the movement of hospitals to suburbia. Those holding shovels are, left to right, Tommy Price, Edward Barry, Sister M. Rita Schroeder, Dr. Edward C. Segerson, Monsignor Merlin F. Kearney, L. K. Thompson, Jr., and Jesse Luton.

Mr. Loewenberg was executor of the 300-acre estate of Piggly Wiggly founder Clarence Saunders, bought 30 years earlier in a bankruptcy sale by his sister, Lottie Loewenberg Lichterman, and her husband, Ira Lichterman, who built a luxurious log house on the property. Mr. Lichterman had wanted a hospital on the property so that injured children in the area "would not bleed to death before they could get to a hospital."

After negotiations, the Sisters of St. Francis of Mishawaka, Indiana, authorized Sister Rita to purchase 30 acres for $600,000. Although the hospital would face the railroad track despite Sister Rita's objection, patients often asked for rooms on the front so that they could count the passing cars. Later, the Loewenberg-Lichterman Foundation donated seven additional acres to St. Joseph Hospital East

DR. JAMES W. PATE

A MAJOR PRESENCE IN MEMPHIS AND NATIONAL medicine since the 1960s, Dr. James Pate continues to work part time, traveling around the world as a lecturer and consultant. With a shock of white hair and a sense of a life lived well, he projects his focus on the profession to which he has given much. Born in Wedowee, Alabama, Pate graduated from Emory University and from the Medical College of Georgia, leading his class at the age of 20. After an internship and surgical residency at Bethesda National Naval Medical Center, where he helped create the process for freeze-drying arteries used during the Korean conflict, Pate helped discover how bioelectrical impulses caused blood clots. After an additional residency at the University of Alabama School of Medicine, he came to Memphis for two years of thoracic surgery residency at the VA Hospital.

Dr. Pate joined the full-time faculty of the University of Tennessee in January 1959, with a full professorship in 1965. Working with the VA Hospital in 1957, he introduced the aortic valvulotomy procedure in Memphis. He established the first tissue banks at the University and the VA system in 1956. He founded open-heart surgery programs at the City of Memphis Hospital and the VA Hospital (1959) and helped establish the programs at Baptist and Methodist hospitals (1961). Pate was also codeveloper of the first surgical intensive care units with remote electronic monitoring in 1960, for which the American College of Surgery awarded him a Certificate of Honor. One of the founders of the National Society of Thoracic Surgeons, past president of the Memphis Heart Association, and past president of the Southern Thoracic Surgical Association, he also served on the Tennessee Heart Association's board of directors. He was president of the Tennessee chapter and member of the Board of Governors of the American College of Surgeons.

Dr. Pate was the first surgeon in the United States to implant an artificial valve in a pediatric patient at the City of Memphis Hospital (described in *Journal of Pediatrics*, 1967). He also replaced a valve and implanted a pacemaker, as an emergency procedure, in a patient with a gunshot wound to the heart (described in *JAMA*, 1969). The first heart transplant program in the Mid-South began at the University of Tennessee-Bowld Hospital and later moved to Baptist. Dr. Pate performed the first operations in 1985 and directed the program, working with Dr. Glen Crosby. Later, Dr. H. Edward Garrett, Sr., extended the program with the help of Dr. Darryl Weiman.

In 1974, Pate became chair of surgery at the University of Tennessee. He led in merging general surgery resident programs at St. Joseph, the VA, and Baptist into the University's

Surgeon Dr. James W. Pate with a machine which determines heart function, 1972.

surgery program. He spearheaded the statewide trauma system, the Elvis Presley Memorial Trauma Center, and the Firefighters Regional Burn Center (both at The MED). Drs. Pate and Louis G. Britt developed the Mid-South Transplant Foundation, which procures area organs for transplantation, through the University of Tennessee Department of Surgery. Dr. Pate has published over 200 articles and book chapters, focusing on heart surgery and transplantation, and is coauthor of *Science of Surgery*.

In 1982, Dr. Pate was honored, with Dr. Louis Britt, by the staff at the City of Memphis Hospital, Dr. Pate for work with general and thoracic surgery residents, leadership in establishing the Trauma Center, and surgery for cardiovascular disorders. He was named Outstanding Alumnus by both the Medical College of Georgia and the University of Tennessee.

DR. LOUIS BRITT

Grandson of a general practitioner in Parkersburg, West Virginia, who studied at Edinburgh, transplant surgeon Dr. Louis Britt still has his grandfather's 1887 identification card from New York's Bellevue Hospital with the rules, "No riding the elevator, no examining females, and no spitting on the floor."

Fascinated by the flowering of surgery in the 1900s under meticulous surgeon Dr. William Halstead, father of America's surgical residency system at Johns Hopkins, Dr. Britt received his M.D. degree at the University of Tennessee and interned at Cook County Hospital in Chicago. After returning from service during the Korean conflict, he completed his surgical residency under Dr. Harwell Wilson at the University from 1959 to 1963, after which he joined the faculty, practicing at Bowld and at Baptist hospitals. He remembers a deep sense of camaraderie among physicians in Memphis and regards Baptist's Frank Groner and Maurice Elliott (later at Methodist) as great administrators who attracted extraordinary practitioners in the 1950s and 1960s and were friends to doctors.

In 1968, after a talk by Dr. David Hume, part of the Boston team that performed the first kidney transplant, Dr. Britt began developing a transplant program at the University. He used his lab for animal transplantation to perfect organ preservation and the surgical technique of transplantation. Charles Miller, his lab technician, learned tissue-typing techniques in California, and Britt trained with Dr. Thomas Starzl in Denver.

Meanwhile, the University's Dr. Fred Hatch extended the dialysis program begun in the early 1960s, with Drs. Jim Gibb Johnson at the University of Tennessee, Shane Roy at Le Bonheur, and Bobby Kelley at Baptist. Dialysis extended the lives of kidney-failure patients needing transplants. In the early 1970s, Memphis's Dr. McCarthy DeMere (who was also a J.D.) helped write the first brain-death law for Tennessee, later adopted nationally. The definition allowed physicians legally to keep a patient's bodily functions working after death until donor organs could be removed.

At the first International Congress of Transplantation in New York, immunologist Sir Peter Medawar introduced keynote speaker Professor Jean Hamburger (pronounced 'ahm boo zhay'). Dr. Britt's resident whispered, "Either we're rednecks, or we're in the wrong place." When the University's transplant program acquired a horse to make antilymphocyte serum after a human spleen product was injected under the skin, Britt named the horse "Ambu" for the French professor. The horse produced serum until drug companies began manufacturing it.

On April 9, 1970, Dr. Britt performed the first kidney transplant in Memphis at Bowld Hospital. Supporters of the program included Dean Richard Overman and and donated additional acreage off Quince Road to the city of Memphis for a children's park. Named the Lichterman Nature Center, the park opened in 1983, with the house and other buildings as classrooms and laboratories.

On December 8, 1974, St. Joseph Hospital East was dedicated as a 693-room, full-service hospital. On her regular walks on the hospital grounds, Sister Rita carried a grocery bag and picked up trash. Later a Methodist administrator reported to his colleagues that one reason the grounds of the new hospital looked good was that Sister Rita picked up trash in the evening.

After a St. Francis Hospital physician died of a heart attack in an ambulance waiting for a passing train, Sister Rita urged the City Council to expedite completion of an overpass at Park and Ridgeway and to install traffic lights at the hospital. Sister Rita seldom lost a battle. The only overpass on Park Avenue in East Memphis was completed, and traffic lights were installed. An administrator at another hospital once commented, "Sister Rita can run circles around any hospital administrator in Memphis." In 1980, the name of the hospital was changed to St. Francis to specify the connection between the hospital and the Sisters of St. Francis in Mishawaka, Indiana, who supported it.

Community Blood Plan. In the 1970s, years after Dr. Lemuel Diggs had advised such a program, Memphis opened its first community blood plan center using all-volunteer blood. In 1974, Dr. E. Eric Muirhead received a federal grant of $183,000 to establish a separate, independent facility, not based in a hospital, for the Community Blood Center, later the Mid-South Regional Blood Center. In 1979, the new Mid-South Regional Blood Center opened at 1040 Madison Avenue. In 1984, the Center adopted the

Dr. Louis G. Britt receives a check from Mrs. William B. Ashworth of the Nineteenth Century Club to buy equipment for his work in transplants, 1973.

Dr. James Etteldorf. In 1982, Dr. James Williams began a liver transplant program at Bowld. Later figures in the program included Drs. Osama Gaber, Santiago Vera, and Hosein Amiri.

Dr. Britt identifies Dr. James Pate as one of the fathers of heart surgery in Memphis and recalls a power outage when a technician cranked the disk oxygenator by hand while Dr. Pate performed surgery by flashlight until the hospital's generator came on. Britt also recalls the expertise of pediatric cardiac surgeon Dr. Robert Allen, who once, during an aortic-renal bypass, found that the vein to be used was inadequate. "No problem," said Allen, who took a piece of pericardium, wrapped it around a small catheter to form a tube, stitched the tissue, removed the catheter, and continued successfully.

The major development during his career, Dr. Britt says, was Dr. John Gibbon, Jr.'s invention of the heart-lung machine in Philadelphia. He adds that new sciences, notably immunology, grew in tandem with transplantation.

Known nationally for his leadership, Dr. Britt has received many awards, including the L. M. Graves Award and the University of Tennessee Alumni Service Professor Award. In 1982, he was honored, with Dr. James Pate, by the staff at City of Memphis Hospital for his work with surgery residents, renal transplants, and surgery for endocrine disorders.

Lifeblood® logo and by 2009 had blood collection centers around the Mid-South.

Memphis Women in Medicine

Not only was the first graduating class at the University of Tennessee led by its first woman graduate, Dr. Sara Conyers York, but Memphis medicine has included many distinguished women physicians, and the 1970s brought increasing numbers of women to medicine. Some remarkable women from earlier years served as role models.

Eugene S. Forrester, Lifeblood executive director, takes note of the shortage of type O-positive blood, 1981. Shortages were not due to lack of donors, but to increased demand by hospitals.

ABOVE: *Dr. Evelyn Ogle, the first woman to become president of The Memphis Medical Society.* RIGHT: *At the Strep Disease Center in Gailor Clinic, Dr. Iris A. Pearce, center, looks on as nurse Unice Weaver gives a physical check-up, 1970. The clinic, with Dr. Pearce as director, was trying to prevent the occurrence of rheumatic fever and nephritis, two results of strep infection.*

Dr. Evelyn Ogle received her M.D. degree from the University of Tennessee in March 1947 as the only woman in her quarterly class of 36. She married neurosurgeon Dr. William (Billy) Ogle. Named as a prominent woman in medicine by the National Library of Medicine, Dr. Ogle commented that her sojourn at the University was not tough as much as it was lonely, and that "the boys in the anatomy class were always trying to get me to blush." She persevered, however, and became a leader in medicine, serving in 1992 as the first woman president of The Memphis Medical Society. Trained in pediatrics, she developed an interest in electroencephalography (EEG) and became a pioneer in freestanding EEG laboratories. She has remained active with the Medical Society in retirement, especially as a mentor to women medical students.

Dr. Iris A. Pearce, a fourth-generation physician and only child of Dr. Robert Pearce, joined the Navy after graduating from Vanderbilt University during World War II. She received her M.D. degree from the University of Tennessee in 1950. Always committed to those in need, Dr. Pearce served in many administrative capacities at Gaston and at the University. At her death in 2005, her estate endowed the Dr. Robert S. Pearce Chair in Community Medicine at the University.

Dr. Alys Lipscomb came to the University during World War II to study physiology and afterward received her M.D. degree, graduating in 1945. She was one of the first to study medical uses of radioisotopes made available after the war and was the first physician in Memphis to treat hyperthyroidism with 131-I.

Dr. Clara A. Brawner received her M.D. degree from Meharry and was the only African-American woman practicing medicine in the mid-1950s. Dr. Brawner practiced family medicine with an emphasis on pediatrics. She was known as an excellent physician, a mentor, and a champion in civic affairs.

Dr. Alice Deutch was an ophthalmologist and a professor at the University of Tennessee actively engaged in research.

Dr. Margaret Halle was a prominent Memphis ophthalmologist, later a member of the Memphis Literacy Council and a community benefactor.

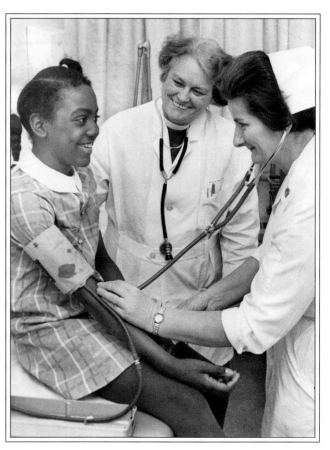

Later, Memphis became home to similarly prominent women physicians. Dr. Nancy Chase received her M.D. degree from the State University of New York College of Medicine and completed her residency in pediatric cardiology at Long Island Jewish Medical Center. Growing up with travel and music because her mother was an opera singer in Europe, Dr. Chase blends her passions for medicine and music. She is an associate professor of pediatrics at the University of Tennessee and maintains a private practice in East Memphis. She also serves on the board of Opera Memphis, plays bagpipes, and works for human rights.

Dr. Valerie Arnold received her M.D. degree from the University of Alabama at Birmingham and completed her residency at the University of Tennessee in child and adolescent psychiatry. She is board certified in both adult and child/adolescent psychiatry and has been acting chief of child and adolescent psychiatry at the University of Tennessee. She is a researcher with Clinical Neuroscience Solutions and was president of The Memphis Medical Society in 2007.

Dr. Val Vogt, a graduate of Rush Medical College in Chicago (1990), interned at the University of California at Los Angeles and completed obstetrics and gynecology training in Chicago and Indiana (combined, 1994-1997). She was named a prominent woman in medicine by the National

LEFT: *Nuclear medicine expert Dr. Alys H. Lipscomb, 1953, was the first director of the Radioisotope Unit at the University of Tennessee in 1948. That year, she was also instructor of medicine, but was promoted to assistant professor in 1950.* ABOVE: *Pediatric cardiologist Dr. Nancy Chase is currently on the board of The Memphis Medical Society.*

Library of Medicine. At the University of Tennessee, she focuses on post-menopausal care. A strong researcher, she was also vice chair of the committee that wrote the obstetrics and gynecology board examinations.

Dr. Claudette Shepard is an associate professor of obstetrics and gynecology at the University of Tennessee. After receiving her M.D. degree from Loma Linda University, she completed her residency at SUNY-Brooklyn, followed by a fellowship in pediatric and adolescent gynecology. She served as a volunteer contact for the Mid-South in the Empire State Medical Association's Health Access/Katrina Relief Plan. Dr. Shepard was president of the Bluff City Medical Society (2005-2006) and has served on the Legislative Committee of The Memphis Medical Society.

Dr. Marion Dugdale received her M.D. degree from Harvard University (Alpha Omega Alpha, 1954) and com-

Hematologist Dr. Marion Dugdale is also director of the University of Tennessee Hemophilia Clinic, Van Vleet Cancer Center.

pleted fellowships in hematology at Duke University (1957-1958) and the University of Tennessee (1958-1962). She has worked with the Tennessee Hemophilia Program and received the Humanitarian Award from the Memphis Chapter of the National Hemophilia Foundation, which also established a scholarship in her honor. As professor of medicine in the Division of Hematology, she is the hematologist whom surgeons call when they encounter bleeding problems. As a teacher, she has won multiple awards, and she received The MED's Distinguished Service Award.

Drs. Susan Murrman and Mary McDonald established their own clinic in obstetrics and gynecology. Sisters Drs. Luella and Mary Ashley Churchwell had their own dermatology clinic in Germantown. In the 2000s, Dr. Karen Chancellor was the medical examiner. Dr. Ellen Kang, wife of Dr. Andrew Kang, is a pediatrician retired from teaching at the University of Tennessee and the author of two books.

By 2008, about half the students in the University of Tennessee College of Medicine were women. So many women specialized in obstetrics and gynecology that the specialty's national societies, concerned that the specialty would be identified as a "woman's" field, organized special initiatives to lure men into the field.

Public Health

Although Surgeon General Dr. William Stewart's reported statement (1967) that "the time has come to close the book on infectious diseases" may be apocryphal, medical historians agree that the idea was widely held in the 1960s and 1970s, in recognition of the need to reallocate funds to heart disease and cancer. At the time, no one could foresee the emergence of new organisms and the resurgence of old ones with resistance to antibiotics. Diseases that would quell some of the optimism were Lyme disease and Legionnaires' disease, which struck in 1976 at a Legionnaires' convention in Philadelphia and was eventually traced to bacteria in the convention hotel's cooling tower. Nevertheless, a major triumph was the 1979 announcement by the World Health Organization (WHO) that smallpox had been eradicated worldwide. The enormous successes of vaccines (rubella, or German measles, 1970; chicken pox, 1974; pneumonia, 1977; and meningitis, 1978) and antibiotics freed researchers to focus on old enemies—cardiovascular disease, cancer, and diseases of the immune system.

Cost Containment

As medical costs increased with technology, the federal government urged states to control healthcare costs by managing the growth of services and facilities. In 1973, the Tennessee General Assembly created the Health Facilities Commission (HFC) to administer the certificate of need (CON) program. An omen of measures to come, the CON is a permit to establish or modify a healthcare facility or service at a specified location. The CON program, according to the Tennessee Health Services and Development Agency (successor to the HFC), ensures that healthcare "projects are accomplished in an orderly, economical manner."

Memphis Subspecialty Practices

Internal Medicine. New discoveries, including the synthesis of insulin (1977), continued to give impetus to internal medicine. At the new St. Joseph East, the names of Drs. Jack Halford; Ed Taylor; Robert Kirkpatrick; Eugene Spiotta; Albert Grobmeyer; and Robert Kerlan stand out. Dr. Stephen Miller arrived in the late 1970s from Johns Hopkins and expanded teaching programs in general internal medicine at the University of Tennessee. Drs. Hall Tacket, Charles Deere, Fred Strain, and Pervis Milnor, plus the Otis Warr Clinic (with eventually four generations of internists in the family) and the John D. Hughes Clinic, were among major names at Baptist Hospital. When Dr. Tacket died in 2009, Dr. James Lewis, in a letter to *The*

Commercial Appeal, called him "the best bedside teacher I have ever witnessed and . . . a true mentor to the generations of residents who trained under him." Dr. Lewis added that, in retirement, Dr. Tacket not only continued to teach residents, but "took basic science courses at the University of Tennessee to help the school improve its curriculum," making the top grade in pathophysiology. Internal medicine stalwarts at Methodist included Drs. John Conway (a cardiology pioneer), J. B. Witherington, Bill Weber, and Hamel B. Eason. Dr. Jean Hawkes moved from internal medicine into the subspecialty of endocrinology, focusing on diabetes.

Neurosurgery. The Harvey Cushing society became the American Association of Neurological Surgeons (AANS) in 1967, now the largest neurosurgical society in America and the active voice of neurosurgery. Dr. Semmes had been president of the society in 1940, and four of his associates in the Semmes-Murphey Clinic were subsequent presidents: Drs. Francis Murphey (1965-1966); Richard DeSaussure (1975-1976); James T. Robertson (1991-1992); and Jon H. Robertson (2007-2008). The Robertson brothers trained at Semmes-Murphey/University of Tennessee and were the first brothers serving as president of AANS.

Orthopedics. In the 1970s and later, new therapeutic interventions such as arthroscopic surgery and direct injections into joints were driven by new imaging techniques. Cortisone had been the primary weapon against arthritis for two decades, but nonsteroidal antiinflammatory drugs (NSAIDs) were introduced in the 1970s and 1980s. Both hip and knee replacements became available in the 1970s. Dr. Harold Boyd at Campbell Clinic performed the first total hip replacement in Memphis in the early 1970s after studying with the procedure's inventor, Sir John Charnley, in England. Before retiring in 1974, he developed the hip replacement program at Campbell Clinic. Several others helped to bring the procedure to the forefront in Memphis, including Drs. James Harkess; Leon Hay; Michael Lynch; Owen Tabor, Sr.; Tom Morris; and Michael Neal.

ABOVE: *Dr. Hall S. Tacket, left, was one of the 83 charter members of the Memphis chapter of the Society of Sigma Xi in 1956. Chapters were selected based on the quality and scope of medical research submitted for membership. Although chartered as the University of Tennessee Medical Units Chapter, it included researchers from several other institutions in the city.* LEFT: *Orthopedic surgeon Dr. Harold B. Boyd, c.1990. He joined Campbell Clinic in 1938, and became president of the American Academy of Orthopaedic Surgery in 1954.*

Immunology, a new specialty. From the breakthroughs of the 1960s and 1970s, immunology broadened knowledge and insight for other subspecialties. The science of immunology and immune-mediated diseases brought knowledge that had been lacking in all internal medicine specialties before the mid-1960s. Dr. Gene Stollerman was on the cutting edge in his investigation of rheumatic fever, which he found to be not only an infectious disease secondary to streptococcal infection, but also one of the first discovered autoimmune diseases in which antibodies to fight the streptococcus also cause damage to the patient's heart. This initial concept of autoimmunity had never been described or perceived before his research.

When Dr. Stollerman arrived at the University of Tennessee, he called the 1960s "the age of the lymphocyte," the one white blood cell that was not understood until recognition of how autoimmune diseases operated. The lymphocyte was the key player in producing antibodies to protect an individual from infections and to provide natural immunity. However, it was the underlying culprit in diseases in which an individual's immune system goes awry and produces antibodies against its own tissues (hence, "autoimmune"), as in systemic lupus, rheumatoid arthritis, many allergic conditions, and many pulmonary, renal, and neurologic diseases. Immunology thus opened a whole world of understanding and treating such diseases.

Oncology, a new specialty. As immunology continued to open new paths for research, it led to another major specialty. For many years, surgery had been the first line of defense against accessible tumors that seemed defined. However, chemotherapy was becoming widely used as an adjuvant treatment, sometimes combined with radiation therapy, although the many variables in cancer barred uniformly applicable protocols.

By the 1970s, bone marrow transplants made possible new approaches to treating cancer. Powerful radiation and chemotherapy could be combined with autologous transplants (self as donor, with removal and freezing of the patient's own marrow), followed by high-dose drug or radiation therapy, after which the marrow was replaced. By 1978, six specialists in hematology/oncology were with the University of Tennessee Medical Group, the practice group for the University's teaching physicians: Drs. Pat Adams-Graves; Steven R. Deitcher; Marion Dugdale; Antonius Miller; Brent Mullins; and Randall Rago. Other Memphis hematology/oncology specialists included Drs. A. Earle Weeks; Margaret Gore; Martin K. Barnett; Kathleen D. Spiers; Jarvis D. Reed; and C. Michael Jones (Methodist and St. Francis); Ronald D. Lawson; Furhan Yunus; William K. Walsh (Methodist); and James M. Holbert (Baptist). Bone marrow transplants were performed at St. Jude and Bowld hospitals. In 1986, Baptist physicians created the first adult bone-marrow transplant center in the Mid-South. This procedure was highly effective in fighting lymphoma and Hodgkins disease.

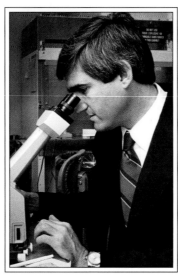

Top: *Oncologist Dr. Reed Baskin currently serves on the Executive Committee of the University of Tennessee Cancer Institute and is codirector of the university's Hematology/Oncology Training Program.* Above: *Dr. William H. West, 1997. The West Clinic, founded in 1979, has combined clinical practice in oncology with research.*

A survey by the World Health Organization (WHO) in 1975 showed that death rates from breast cancer remained the same as in 1900. Intensive research found that some combination of surgery, chemotherapy, and radiation was far more effective than the old, dreaded radical mastectomy. Combination therapy became the treatment of choice, and surgery was simplified to the "modified radical" mastectomy or simply "lumpectomy" followed by chemotherapy and/or radiation.

During the 1970s, Dr. Reed Baskin began practicing at Methodist at a time when medical oncology through chemotherapy was not widely accepted even in the medical community. He was the first fellowship-trained medical oncologist in Memphis, says colleague and noted surgical oncologist Dr. Irvin Fleming, whose son, Dr. Martin D. Fleming, is also an oncologist. Dr. Baskin was also director of internal medicine training at Methodist. Dr. Barry Boston started at Saint Francis, working at the VA Hospital for several years before going into private practice, and Dr. William West went to Baptist. Throughout the 1970s, new treatments and more oncologists came to Memphis.

The West Clinic was founded in 1979 by Dr. William West, son of Dr. Thomas C. West and brother of two physicians. With an M.D. degree from Johns Hopkins and further study at the National Cancer Institute, West left his appointment as an NIH researcher to bring world-class care to Memphis cancer patients. Chemotherapy was new, but he believed in providing it through community oncology clinics. Now located at 100 N. Humphreys Boulevard, the clinic participates in cancer research and community-level treatment with the motto "No stone unturned." By 2009, the organization had six clinics and 30 physicians. It is home to the WINGS Cancer Foundation and received the Hematology and Oncology Practice Excellence (HOPE) award in 2008 from *Hematology and Oncology News and Issues*. Physicians include Drs. Kurt W. Tauer; Guy J. Photopulos; Lee S. Schwartzberg; Linda M. Smiley; Alva B. Weir, III; Benton M. Wheeler;

DR. GENE H. STOLLERMAN

IN THE 1960S, ONE OF THE STRONGEST RESEARCHERS and administrators in the city's history arrived—a man whose mission looked backward to devastating infectious diseases and forward to the body's immune response to infectious agents. Dr. Gene Stollerman, a dynamo of a man with a zeal to eradicate streptococcal infections and their complications, was recruited by Dean M. K. Callison as part of the sea change at the University of Tennessee from graduating mainly primary care physicians to an emphasis on research in all areas. Stollerman chaired and transformed the Department of Medicine from 1965 to 1981.

Dr. James Gibb Johnson, respected physician, professor, and administrator at the University, spoke highly of him: "Stollerman was a dynamic, colorful, and occasionally outspoken leader and teacher," he said, adding that, as a resident, he could call at night to get Stollerman's opinion about a difficult case, and Stollerman would often come in even though he was not on call. He was such a dedicated teacher and builder of the Department of Medicine that it became known throughout the country as a top program for residency in internal medicine. "At one point," Dr. Johnson recalled, "there were eight straight [full] internal medicine first-year residency positions, and the department received 200 applications. One-third of these were from Alpha Omega Alpha honor graduates from medical schools outside Tennessee."

Dr. Stollerman also recruited great prospects for the VA hospital and helped to get it moved to the medical center, where it opened in 1967. He recognized the "huge indigent pediatric population" that was medically underserved, and he worked to improve treatment for this group.

When medical students graduated, Dr. Stollerman often placed them outside Memphis in residencies. Dr. Phil Lieberman (M.D., University of Tennessee; fellowship in allergy and immunology, Northwestern University) helped write a textbook on allergies and served as president of the American Academy of Allergy, Asthma, and Immunology and as president of the American Association of Certified Allergists. Dr. Howard Horne, chief resident under Stollerman, went to Brigham and Women's Hospital in Boston and became a partner of Dr. Bernard Lown, inventor of the cardioverter and winner of the Nobel Peace Prize for his work with Physicians for Peace.

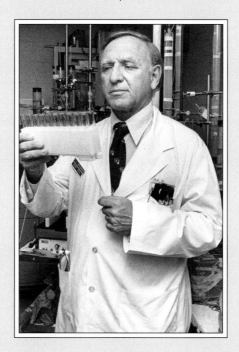

In 1978, Dr. Gene H. Stollerman, head of a team at the University of Tennessee, reported the results of a successful test vaccine against a virulent strain of Streptococcus.

Dr. Stollerman's research, however, defined his contributions to medicine. "It was . . . an era in which the Department of Medicine unified basic and clinical science," he once said. Internationally known for research on Group A Streptococcus and its complications, Dr. Stollerman worked on a vaccine that would prevent streptococcal infections, rheumatic fever, and glomerulonephritis. At Mt. Sinai Hospital, he was one of the first to use penicillin. Before coming to the University of Tennessee, he held a research chair at Northwestern University. Also coming to Memphis in 1965 from Chicago was Dr. Edwin Beachy, who worked with Stollerman on the streptococcus vaccine. Dr. James Dale worked with Beachy and Stollerman and continued the work, finishing preclinical studies by 2009. Since 2006, Dale has held the Gene H. Stollerman Endowed Professorship in Medicine at the University of Tennessee.

Before leaving for Boston University in 1981, Stollerman made a strong, positive imprint on medical education and research in Memphis, and his influence continued even as his research extended to geriatric care. His publications represented the best research in infectious diseases and immunology, his administrative feats enriched the faculty at the University of Tennessee, and his training of students and residents helped build the University's strong reputation in teaching and research.

THE END OF A LEGEND: MEMPHIS'S MOST FAMOUS CITIZEN

On August 16, 1977, a hot, stifling day, Memphis's most famous citizen passed from celebrity into mystery and legend. Elvis Presley—who had originated and popularized a new form of music, toured the world, and starred in movies—was a good and generous citizen of Memphis. He always returned to Graceland, which he had purchased from the family of the late Dr. Thomas Moore, who had built the mansion on property that his wife inherited.

Later, Elvis became a frequent inpatient at Baptist Hospital, receiving treatment for digestive and other conditions while blocking his hospital windows with aluminum foil because of insomnia. His eating habits were notorious—hamburgers, fried peanut-butter-and-banana sandwiches. In later years, Elvis was different from the youngster who had sung on Ed Sullivan's show. Pursued by fans and the press, he avoided crowds by renting theaters and bowling rinks at night for his entourage, then slept through the day. Performances took their toll because he always gave all he had, and more. He took pills to keep him alert so that he could record, pills to put him to sleep, with prescriptions from different physicians in different cities.

During one hospitalization, he formed a friendship with his nurse, Mrs. Marian Cocke. The *Commercial Appeal* featured a photograph of Mrs. Cocke in Baptist's parking lot beside the new Pontiac whose keys Elvis had just given her. Mrs. Cocke published *I Called Him Babe* about her friendship with Elvis. He gave away cars and jewelry, and he visited and supported St. Jude, Le Bonheur, and other charities. Elvis's close friends, including noted ophthalmologist Dr. David Meyer, said that he had a brilliant mind.

On August 16, 1977, Elvis was found slumped on his bathroom floor at Graceland, not breathing. His personal physician, Dr. George Nichopoulos, rushed to the scene and tried to resuscitate him. In the emergency room, doctors labored in vain and time of death was pronounced at 3:30 p.m. Four hours later, three certified pathologists gathered in the autopsy room, led by Shelby County Medical Exam-

William A. Lankford; William H. West; and Paul Lee, Jr.

Medicine in the 1970s was further influenced by "genetic engineering." In 1975, DNA sequencing began. Genentech cloned human insulin in 1978. In the next two decades, the computer revolution combined with these technologies to revolutionize oncology.

Ophthalmology. Advances in ophthalmology also led to subspecialization. Dr. Roger Hiatt, Sr., was first, subspecializing in pediatric ophthalmology. He joined the University of Tennessee as the first full-time ophthalmologist in 1964, became professor and chair in 1969, and retired in 1994, after which he was a Mormon missionary in the Philippines with his wife, Nancy.

Dr. Ralph S. Hamilton returned to Memphis in 1959 after training at Wills Eye Hospital in Philadelphia, where he also received a master's degree in opthalmology and completed a fellowship in oculoplastic surgery. An outstanding ophthalmologist capable of many procedures that in a later day would be referred to multiple subspecialists, he introduced retinal buckling surgery to Memphis in the 1960s. Winner of a variety of awards, he also served as president of the Tennessee Academy of Ophthalmology and served in the 11th Field Hospital, U.S. Army, in Augsburg, Germany. He was elected to the prestigious American Ophthalmological Society and received the University of Tennessee's outstanding teaching award multiple times. Dr. Hamilton's work in ophthalmology began at the age of 15, when he would shine a light into the eyes of his father's cataract patients. The mature cataracts, he said, looked like "a piece of paper." He went on to train almost 100 ophthalmologists at the University. The endowment made by Hamilton and his wife began the Hamilton Eye Institute. He noted that giving must be learned, adding, "But once you do it, you like it."

In 1968, Dr. David Meyer became the first retinal subspecialist in Memphis, joined in 1975 by Dr. Steve Charles, who added vitrectomy to their partnership.

had happened. His heart was enlarged, with some blockage of the coronary arteries. The final results of blood and tissue tests would take days, but the gross examination was inconclusive about the cause of death.

Dr. Jerry Francisco—who had also performed the autopsy on Dr. Martin Luther King, Jr., in 1968—signed the death certificate. For years he was a professor of pathology at the University of Tennessee and medical examiner for the city and county. In the room with Dr. Francisco were Dr. Noel Florendo from Methodist Hospital, Dr. E. Eric Muirhead, pathology chief at Baptist, and Baptist administrator Maurice Elliott. In 2009, as he has since that day in 1968, Dr. Francisco stood by his statement that an arrhythmia of the heart caused Elvis's death, adding that the toxicology screen revealed multiple drugs in Elvis's body, but "even the combined drug levels were sublethal." No clots were present, but two of the arteries were 50 percent blocked. "Today," he adds, "there is a term for this kind of death—*sudden coronary death*." Although questioners attempted to establish a different cause of death, Dr. Francisco remains firm in his conclusion.

Elvis graciously offered the Baptist School of Nursing the use of Graceland for their end-of-the-year party in the 1950s. A student nurse shows him the student page from an issue of a hospital publication.

iner Dr. Jerry Francisco. Outside, crowds gathered, some standing silently, others sobbing. On the autopsy table lay the King of Rock and Roll, a big man with dark hair and sideburns down to the mandible. No one knew exactly what

Continued subspecialization brought Drs. Tom Wood in cornea; Richard Drewery in neuroophthalmology; Jim Wilson and Natalie Kerr in pediatrics; Alan Mandell and later Audrey Tuberville and Peter Netland in glaucoma; Leo Beale and James Chris Fleming in eye plastic surgery; and John Linn, John Elfervig, and Rick Sievers in retina. Dr. Jerre Freeman, also trained in cornea, became the first and only ophthalmologist to limit his practice to cataract surgery.

Malpractice Crisis

Physicians nationwide faced a crisis of malpractice insurance in the mid-1970s. Large numbers of claims and exaggerated judgments in Tennessee led commercial insurors to withdraw coverage. In 1976, Tennessee physicians formed a physician-owned organization, State Volunteer Medical Insurance Company (SVMIC) to provide malpractice insurance, thus resolving the crisis, although the cost of premiums and of defensive medicine remained problematic. A mutual company, it was owned by its members rather than by stockholders. One Memphis physician group even sold its lake house to raise capital. Nashville's Dr. Kelly Avery, president of the Tennessee Medical Association, was a leader in forming the company, with help from Memphis orthopedic surgeon Dr. Allen Edmondson, who was chair for several years. Dr. Jim Gibb Johnson of the University of Tennessee was a highly effective and important vice chair for many years. Memphis surgeon Dr. Hugh Francis, III, has been vice chair for several years.

In 1989, SVMIC began offering coverage to physicians in neighboring states, and practice management services were added in 1996. By the late 2000s, SVMIC had become one of the largest and most successful physician liability companies in the United States. Significant legislative change in 2009 added the requirement that a plaintiff must arrange for expert testimony before filing a malpractice lawsuit.

CHAPTER THIRTEEN

SURGERY CENTERS, MANAGED CARE, AND MEDICAL BUSINESS

1980-1989

From conservatism with Ronald Reagan and Margaret Thatcher, to Mikhail Gorbachev's policies of openness and reform, followed by the break-up of the Soviet Union, to the end of the Cold War that began with the end of World War II, to upheavals in Eastern Europe and China in 1989, to citizens carrying off pieces of history as the Berlin Wall came down, the 1980s were years of readjustment.

Medicine embraced more new technologies and was surprised by a new epidemic. Acquired Immune Deficiency Disease (AIDS) appeared in 1981, and finding drugs to extend the

lives of the infected would take years. In 1981, a vaccine for hepatitis B was developed. By 1984, there was even a vaccine for leprosy. In 1982, the American Medical Association lifted its ban on physician advertising. Managed care, spurred by federal support of health maintenance organizations (HMOs) under Title XIII of the U.S. Public Service Act of 1973, affected all of medicine.

In Memphis, imaging continued to refine specialties, and new delivery of surgical care, through freestanding outpatient surgery centers, affected specialties as diverse as ophthalmology and gynecology, otology and urology. With managed care,

ABOVE: *The Medplex and the Jesse Turner Tower, part of the Regional Medical Center at Memphis, 1990s. The Medplex houses the outpatient services, while the Turner tower contains the Firefighters Regional Burn Center.*
OPPOSITE PAGE: *An AIDS poster produced by the U.S. Department of Health and Human Services, Centers for Disease Control, between 1987 and 2003, for their series "America Responds to AIDS."*

physicians were drawn into medicine as a business. Memphis hospitals morphed into regional "healthcare systems."

MILESTONES

New Infections. As if to shatter the notion that medicine had conquered infectious diseases, a mysterious condition appeared first among homosexual men in New York and San Francisco. In 1978, young, apparently healthy men began presenting with infectious diseases uncommon in their age group. Illnesses progressed quickly, and patients died of opportunistic infections that their weakened immune systems could not overcome. By 1982, the Centers for Disease Control had linked AIDS to blood, and French scientists had identified the human immunodeficiency virus (HIV). In 1985, the Food and Drug Administration approved the first test to identify antibodies in the blood. At the VA Hospital in Johnson City, Tennessee, Dr. Abraham Verghese first recognized a pattern of migration from large cities back to small-town homes as the illness progressed, and published a study on the migration. Later he wrote *My Own Country,* detailing his experiences with AIDS patients. During 1985, 5,636 individuals died of AIDS in the United States.

In Memphis, physicians in every specialty knew of the epidemic. New safety measures were enforced. Memphis was one of those smaller cities to which some individuals returned when they became unable to survive without care, although Memphis had its own cycle of AIDS transmission, particularly from bisexual men to female partners and through prostitutes to clients to others. Memphis infectious disease specialists Drs. Mack A. Land and Bryan P. Simmons urged education and prevention as keys to fight-

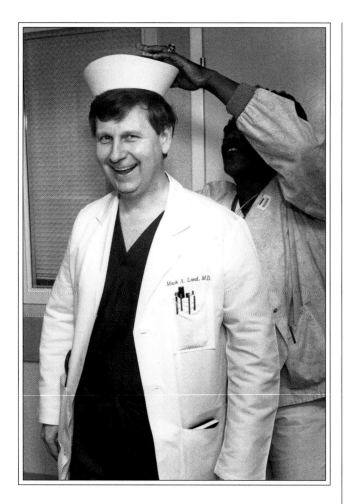

ing the disease. In 1988, forty percent of Tennessee's AIDS cases were in Shelby County.

In 1980, Dr. Richard Kelley, a pathologist in the Microbiology and Immunology Department at Baptist and his colleague Wilton Rightsel, director of Microbiology at Baptist, studied a form of pneumonia that was first identified as Legionnaire's disease in Philadelphia in 1976. They became the first in the United States to isolate and identify the causative organism while an outbreak was occurring.

Blood and Testing. In 1986, Dr. Edward P. Scott joined the Mid-South Regional Blood Center (Lifeblood®) as president and medical director. A University of Mississippi graduate with fellowships in hematology in California and Seattle, Scott worked in San Francisco when the AIDS epidemic began and knew firsthand the evolution of safety measures. In Memphis, he increased donation of blood products, taught at the University of Tennessee, and expanded biotech research.

Imaging and Radiology, and Developing Subspecialties. Imaging became part of treatment as well as diagnosis. Radiologists used radiopharmaceuticals for critical diagnoses such as heart disease, and new nuclear medicine cameras and computers were developed. Digital imaging was added to vascular radiology, and magnetic resonance imaging (MRI) continued to revolutionize specialties, especially neurology and orthopedics. Computer power was added to ultrasonography. In 1982, the University of Tennessee and Baptist combined to operate a nuclear magnet resonance (NMR) unit (the name soon changed to magnetic resonance imaging [MRI] because some feared the term *nuclear*). Drs. Jim Langston, Jim Acker, Frank Eggers, and Bob Lasker were early leaders in this field. Gradually, developments in interventional radiology in the 1980s overlapped specialist lines, from cryoablation of liver tumors (1980) to self-expanding stents (1985).

Cardiac ultrasound became the tool of choice for pediatric cardiologists in diagnosing congenital heart disease. Transesophageal echocardiography gave immediate post-surgical information, and fetal echocardiography was so accurate that it could screen for heart lesions. Leaders in Memphis included Drs. David C. Wolford, Charles M. Jost, Ernest C. Madu, K. B. Ramanathan, and William Walker. The newer CT and MRI were additionally helpful in diagnosing congenital heart disease.

In adults, percutaneous transluminal coronary angioplasty (PTCA) via balloon-tipped catheter essentially defined the overlap between diagnosis and intervention because expanding the balloon could treat as well as identify blockages in the coronary arteries. Leaders in Memphis included Drs. Joseph K. Samaha, Jay M. Sullivan, John E. Strobeck, and Darryl Weiman. By the mid-1980s, an equal number of balloon procedures and bypass surgeries were performed. Shortly thereafter, a number of groups reported using balloon catheterization to administer streptokinase as a treatment of myocardial infarction. These procedures were not without complications, particularly rupture and late restenosis of the artery. In 1986, intracoronary stents appeared, giving lower restenosis rates than balloon angioplasty. Quickly adopted at major hospitals, stents were employed by interventional cardiologists from the Stern Clinic, the Memphis Heart Clinic, and others. Stents would provide reliable intervention in the coming decades.

Electrophysiologist Dr. James Porterfield of Arrhythmia Consultants, a University of Tennessee medical graduate, implanted the first cardiac defibrillator in Memphis in 1986. Described in the *Memphis Business Journal* as "the size of a whiskey flask," it could be felt through the abdominal skin.

Ambulatory Surgery Centers. The 1980s marked Tennessee's first freestanding ambulatory ophthalmological surgery facility, Memphis Eye and Cataract Associates' Ambulatory Surgery and Laser Center, which opened on Poplar Avenue near Germantown in 1985. Other freestanding ambulatory surgery sites and diagnostic clinics followed,

ABOVE: *Recipients of the University of Tennessee Medical Units student awards for 1974, left to right: James G. Porterfield, Outstanding Community Service Award, Ronald E. Christensen, Charles C. Verstandig Award, and Ken Etheridge, Most Promising Physician Award.* OPPOSITE PAGE: *Dr. Mack A. Land celebrates Nurses Day in the ICU at Baptist Memorial Hospital in the medical center, 1994.*

with a profound effect on how Memphis medical care was delivered—treatment or surgery at an ambulatory center, discharge the same day, and no overnight hospitalization. By 2004, approximately twenty-five percent of all procedures in Memphis were done in ambulatory centers.

MEMPHIS VOLUNTEERS: CHURCH AND SECULAR
MEDICAL MISSIONS AT HOME AND ABROAD

Memphis physicians continued a long tradition of charity work, both caring for the poor in Memphis and taking up-to-date medicine and surgical procedures to developing countries. So numerous are these volunteers that no list can be complete.

Christ Community Health Services began as a vision of four physicians—Drs. Richard Donlon, David Pepperman, Karen Miller, and Steven Besh. All studied medicine at Louisiana State University in the late 1980s and envisioned a plan to integrate their medical training with their desire to serve God by serving the medically underprivileged. After residencies at different institutions, they selected Memphis, where there was no primary care physician for more than 50,000 residents in the southwest part of the city.

The four physicians moved to Memphis in 1993 and, with a grant of $200,000 from the Baptist Memorial Health Care Foundation, opened on Third Street in 1995. Since then, additional clinics have opened: in Frayser (2005); in Orange Mound (2006); on Broad Street (2008); and at Central and Hollywood (2009). Christ Community is federally qualified and sees 90,000 patients annually, with a third uninsured. Many patients have TennCare or Cover TN. Fees are based on a sliding scale.

With the assumption that treating a minor problem can prevent a serious illness later, St. Andrew African Methodist Episcopal Church opened a free walk-in clinic at the church on South Parkway, staffed by volunteer doctors, mostly members of the congregation. Led by Dr. Kenneth Robin-

ABOVE: *The founders of Christ Community Health Services, left to right, Drs. Richard Donlon, Steven Besh, Karen Miller, and David Pepperman, 1995. Baptist had given them the start-up funds for their clinic.* RIGHT: *Dr. A. Roy Tyrer reading* A Ship Called Hope *by Dr. William B. Walsh, 1964.*

son (M.D., Harvard), St. Andrew offered preventive health services and healthful living ministries. Dr. Robinson became the first African-American Commissioner of Health in Tennessee in 2006.

St. Joseph's MedWise opened to provide primary care in the Evergreen neighborhood of Midtown for those 65 years of age and older. North Mississippi Health Services dedicated a mobile clinic, Work Link, which traveled from one employer to another providing basic services and screenings.

Other Christian missionaries have represented Memphis abroad for many years. Plastic surgeon Dr. Louis Carter (M.D. degree, University of Tennessee, 1964) and his wife, Anne, from Chattanooga, were medical missionaries in the African bush for 40 years. Son of a Memphis physician who practiced at Methodist Hospital, Dr. Carter worked at three mission hospitals in Kenya, repairing birth deformities and burn scars in children who fell into fire pits.

Dr. William (Chubby) Andrews, a well-known, influential surgeon at Baptist, made frequent missionary trips to Africa.

Otolaryngologist Dr. Coyle Shea made annual trips to South America with a team of nurses, physicians, and dentists, working in remote parts of the jungle.

Project HOPE, a global charity for the last 50 years, which took its name from the S.S. *Hope*, the first peacetime civilian hospital ship, was begun in 1958 by Dr. William B. Walsh. It is now land based. Drs. Roy Tyrer, Douglas Hawkes, and Nicholas Gotten, Sr., participated in its program. Dr. Julia Hurwitz of St. Jude worked through Project HOPE to set up clinical trials of an HIV vaccine in Zimbabwe, and her daughter, Liz Coleclough, worked there with HIV-infected children.

As a member of the Christian Medical and Dental Association, Dr. Al Weir (internal medicine and hematology/oncology) served at Eku Baptist Hospital in Eku, Nigeria. Interested in palliative care and the role of faith in medicine, he also made multiple trips to Albania with the Albanian Health Fund and published two books.

Working with the Memphis-Afghan Friendship Summit (MAFS), Memphis gastroenterologist Dr. Zach Taylor and his wife, Cindy, made repeated trips to Afghanistan, where they tested and treated underserved people. Anesthesiologist Dr. Ghany Zafer joined them, with ministers from the Germantown Baptist Church, which helped support MAFS.

DR. SCOTT MORRIS AND THE CHURCH HEALTH CENTER

AS A YOUNG MAN READING HIS BIBLE, SCOTT Morris noted references to caring for the sick. When hospitals were built with church funding, however, he saw a distance between those who gave and those who needed care. So concerned was he that he studied both medicine and divinity. Then, in 1986, armed with an M.D. degree from Emory and a master's degree in divinity from Yale, Dr. Morris chose Memphis as his place of work because of its high poverty rate and the number of working poor who could afford neither medical insurance nor medical care. As associate pastor for St. John's Methodist Church, he began raising money to build his dream—a clinic where the working poor, a large, underserved group, could receive medical care at minimal cost based on income. Christ's commandments, Dr. Morris believed, were to minister through preaching, teaching, and healing. Reverend Frank McRae, long a highly respected service minister at St. John's, became Dr. Morris's mentor, with connections that were invaluable in gaining support and raising funds for the project.

Situated in a renovated house near St. John's, the Church Health Center (CHC) had a few paid positions, although most services were provided by volunteers from doctors, dentists, and nurses. The working poor, said Dr. Morris, "have fallen through the health-insurance safety net." They are the people who "cook our food, shine our shoes, and will one day dig our graves."

Dr. Scott Morris, founder of the Church Health Center, a volunteer organization which serves those who cannot afford health insurance. The Hope and Healing Center, affiliated with this institution, offers programs for dealing with various medical conditions, a gym, and a swimming pool.

Their options had shrunk with cutbacks in state-funded insurance for the poor and cutbacks at public hospitals such as The MED.

These patients, Dr. Morris emphasized, do not ask for free care. They pay a minimum of $10 and, if they are behind, they are asked to pay at least $2 per week. The referral work is donated by hospitals and by physicians who see the center patients without charge in their private offices. All patients at the Church Health Center are among the 118,000 uninsured in Shelby County. Physicians there try to emphasize prevention—in Dr. Morris's words, "to keep workers healthy" because they are "the economic backbone of a city."

The Church Health Center accepts patients of any race, creed, or color. Receiving no direct government support, it relies on donor congregations (numbering at least 200), plus individuals, groups, foundations, and businesses for funding. "Hospitality," a concept deeply rooted in our cultural heritage, used to underlie the concept of church-related hospitals, but, as medicine changed, too many have become separated from this concept. Dr. Morris's life's mission is to end this separation, to bring churches back to ministering to both body and soul. Medical directors at the center have included Dr. Morris (1989-1998), Dr. Mary Nell Ford (1998-2002), and Dr. David K. Jennings (2002-present). The work of the center is has become a world model for medical service to the working poor.

DR. WILLIAM NOVICK

IN 1996, TEN YEARS AFTER THE CHERNOBYL EXPLOSION, Dr. William Novick of Le Bonheur Hospital in Memphis began taking teams of medical volunteers to Belarus to repair children's congenital heart defects caused by exposure *in utero* to radiation from the Chernobyl incident. At the time, approximately 6,000 children were on the waiting list in Belarus and perhaps 10,000 in Ukraine. Dr. Novick's work was featured in *Chernobyl Heart*, a documentary by Mariann DeLeo about the effects of radiation in Belarus. The film won an Academy Award for Best Documentary Short Subject in 2004. For his work with children in Belarus, Dr. Novick received the Frankskaya Scorina Humanitarian Presidential Medal.

In 1994, Dr. Novick founded and became medical director of the International Children's Heart Foundation, a non-profit organization that focuses on improving the care of children in developing countries. In the late 1990s, Variety Children's Charities purchased the Loewenstein mansion for $550,000 to provide a permanent home for Dr. Novick's International Children's Heart Foundation. Through this organization, thousands of children with congenital heart defects in 17 different countries have received life-saving surgery.

A Memphis resident and holder of the Paul Nemir, Jr. Endowed Professorship in International Child Health at the University of Tennessee, Dr. Novick also received the Church Health Center's "Good Samaritan Award" (1995), the American Philanthropy Association's "Jefferson Award" (2001), the National Federation of Croatian Americans' "Humanitarian Leadership Award (1996), and a special White House reception recognition by Hillary R. Clinton, "Work with Bosnian Children" (1996). In 2004, he was awarded the Red Star of Croatia and, in 2007, he was a cowinner of the Inaugural Frederique-Constant "Passion Award" in Geneva, Switzerland.

Dr. William M. Novick cares for a child in Bejing, China, 2009. Founder and medical director of the International Children's Heart Foundation, Dr. Novick travels around the world as lead surgeon of a team which operates on children with congenital heart defects.

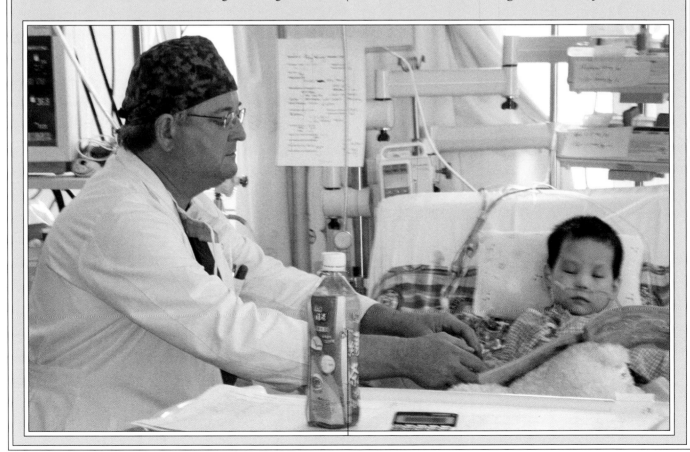

In 1997, Dr. Gordon J. Kraus began Haiti Medical Missions of Memphis (HMMOM), which opened a permanent free clinic in the Port-au-Prince area for the needy, established a long-term healthcare program managed by Haitians, and trained Haitian healthcare providers. Begun as an outreach program of East Memphis's Church of the Holy Spirit, HMMOM welcomed participation regardless of religious affiliation, race, or gender. After the major earthquake in 2010, the organization sent medical and surgical volunteers to aid the injured.

Secular efforts also continued the tradition of volunteerism. In 1994, the Junior League of Memphis created Hope House Day Care Center in Midtown for children with HIV/AIDS.

Ophthalmologists were particularly active in international efforts. Established in 1978 by Dr. Jerre Minor Freeman, the World Cataract Foundation sponsored cataract removal and lens implantation in many countries. Dr. Freeman's wife, Anne, worked with the foundation to build communities across boundaries. Dr. Freeman has traveled annually to Ometepec, Mexico, for 40 years, taking a University of Tennessee resident each year and introducing a generation of ophthalmologists to international volunteerism. Dr. Cathleen Schanzer went to Sierra Leone in Africa; Dr. Roger Hiatt went to the Philippines; and Dr. Tom Gettelfinger traveled annually—to countries from A to Z, Afghanistan to Zaire, over more than 25 years. Drs. Bill Hurd, Ralph F. Hamilton, and John Freeman were also frequent volunteers. Many ophthalmologists have volunteered with the Orbis Flying Eye Hospital, prominently Dr. Steve Charles, but including Drs. Fred Hidaji, Tom Gettelfinger, Chris Fleming, and Barrett Haik.

Ophthalmolgist Dr. Jerre M. Freeman receives the DAR Medal of Honor from Ethlyn Burt, regent of the Commodore Perry Chapter, 1981.

Healthcare Systems: Regionalism Comes to Memphis

Methodist Hospital opened Memphis's first satellite facility—Methodist South—in 1973, followed by Methodist North in 1978. St. Joseph East opened in 1974 and Baptist East in 1979. Responding to patients' enthusiasm for conveniently located, uncongested facilities in the 1980s, Memphis hospitals embraced the concept of "healthcare systems," with both suburban and regional facilities. In 1981, Baptist became Baptist Memorial Health Care System, Inc., a multi-hospital system with Baptist Memorial Hospital Central in the Medical Center as its flagship facility. In 1982, Methodist incorporated as Methodist Health Systems, and in 1986 Methodist Germantown came under the umbrella corporation. St. Francis was an independent hospital after its name change in 1980 until it was purchased by Tenet in 1994 and became part of Tenet Healthcare Systems.

Managed Care in Memphis

In the 1980s, managed care seemed a panacea for the increasing costs of medical care because it offered lower rates to large groups. Under the "preferred-provider" (PPO) concept, patients paid less if they used network providers, which offered discounts and in return gained patient volume. For providers outside the network, patients paid full price. BlueCross BlueShield, Cigna, Aetna, and smaller companies offered these plans, which attracted some of

Saint Francis Hospital, originally named St. Joseph Hospital East, 1980s.

Memphis's largest employers and became the most popular plans.

Other providers offered "managed care organizations" (MCOs), under which the patient selected a primary-care physician (PCP) as a gatekeeper whose written referral was required to visit a specialist; without referral, no payment was made. Point-of-service (POS) organizations also required a gatekeeper, and patients without a gatekeeper's recommendation received lower reimbursement.

An alphabet soup of providers' organizations evolved that remained confusing into the next decades. Shifts in reimbursement, failure to keep payments in line with inflation, and increasing costs of malpractice insurance led to the first decrease in physicians' annual income since the Great Depression.

One physician response was an Independent Physicians Association (IPA), which helped doctors interact with hospitals and insurance companies with a common voice. Begun in 1983 by Dr. Richard Raines and several others, including Drs. Robert Richardson and Riley Jones, MetroCare was a physician-based organization. Although collective bargaining was disallowed by antitrust laws, MetroCare educated its physician members, discussed contracts with members, and sent contracts to members for individual consideration, with a goal of providing high quality and cost effectiveness. Drs. Reed Baskin, Carter Towne, and James West followed in Dr. Raines's footsteps, and the organization successfully represented employers, insurance companies, and HMOs.

In 1985, MetroCare joined with the Methodist Healthcare System, forming Health Choice, a physician hospital organization (PHO) that allowed the hospital to negotiate for its services and gave physician members the right to reject or accept an insurance plan's proposals after a review by Health Choice. Technology solutions and support services were also available through Health Choice, which served more than 1,250 physician members, plus ancillary healthcare professionals in the Mid-South.

Baptist Healthcare System similarly formed Baptist Health Services Group (BHSG), a provider-owned healthcare network that, by 2007, contracted with 3,200 physicians, more than 50 hospitals, and 160 ancillary facilities in the Mid-South. BHSG offered diverse services such as PPO, HMO, and POS plans, plus workers' compensation plans and occupational services.

St. Francis Hospital was the last of the three major institutions in Memphis to join the contracting bandwagon. Comfortable in its East Memphis niche, St. Francis Hospital did not embrace discount networking until the early 1990s, when its occupancy declined from almost one hundred percent to about sixty-three percent (reflecting shorter hospital stays and competition), according to the president, Sister M. Rita Schroeder. Discounted contracts accounted for less than seven percent of its private insurance business. St. Francis then expanded outpatient care, actively sought contracts, and emphasized its family-practice residency program. St. Francis also joined a BlueCross BlueShield discount healthcare network with St. Joseph and the private Brannon-McCulloch Primary Health Care Center (with locations in Orange Mound, Binghampton, and Horton Gardens).

As a nonprofit facility, however, St. Francis struggled until the hospital and its nursing home were purchased in 1994 by American Medical International (AMI), which merged with National Medical Enterprises six months later to form Tenet Healthcare Corporation. The purchase price was not revealed, but a cash payment of $45 million was recorded. St. Francis became a for-profit institution, but retained the ethical position of its Catholic history. Through the Tenet company, St. Francis established HealthNow and contracted with several provider networks.

Purchase funds from the nonprofit St. Francis Hospital were used to establish the Assisi Foundation, whose mission is to respond "to the diverse needs of our community to support health, lifelong learning, social problems, and responsible use of resources with respect and compassion for all." The core values of the foundation are respect for life, compassionate concern, joyful service, and stewardship. Its interests include health and human services, education and literacy, social justice and ethics, cultural enrichment, and the arts. Grants are available to tax-exempt organizations and to public charities for improving the well-being of people and institutions in Memphis and the Mid-South. By 2009, the foundation had awarded over $100 million to non-profit organizations.

Memphis's Regional Medical Center, The MED (the state's oldest hospital), remains today the "safety-net" hospital for the Mid-South region, of whose physicians and staff it is often said, "They can bring you back once, anyway."

BELOW: *The Regional Medical Center at Memphis (The MED), 2010. It contains the Elvis Presley Trauma Center, the Emergency Department, and inpatient and intensive care units.* RIGHT: *The entrance to Birthplace, part of the Regional Medical Center at Memphis.*

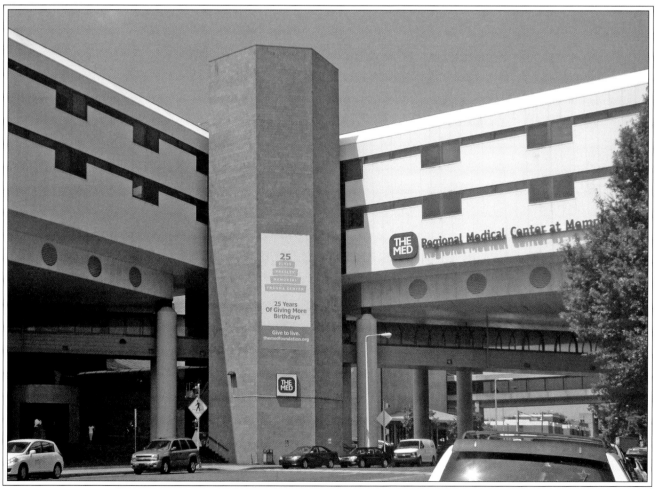

SURGERY CENTERS, MANAGED CARE, AND MEDICAL BUSINESS

Firefighters Regional Burn Center, 2010. It is part of the Regional Medical Center at Memphis and is housed in the Jesse Turner Tower.

With its Elvis Presley Trauma Center (1983); Burn Center (1985, rededicated in 1993 as the Firefighters Regional Burn Center); Wound Center (1992); MedPlex Ambulatory Care Center (1994); and primary-care network called the Health Loop (until 2009, this consisted of six health department clinics, plus four MED-owned community-health clinics), The MED is vital to the city's healthcare.

Retaining its connection with the University of Tennessee for medical education but operated by the Shelby County Healthcare Corporation, The MED began offering TLC Managed Care in the 1980s. Widely misperceived as Memphis's hospital for the poor, The MED was much more. Although thirty percent of its inpatients had no source of payment, The MED received only about twelve percent of its annual revenue from Shelby County, yet it offered one-of-a-kind services to the community. Further, it served many patients from neighboring states with no compensation. The emergency room was characteristically swamped with patients who could not pay, postponed seeking help, and arrived with exacerbated, often multiple, illnesses.

In 2009, financial problems led to The MED's closing five clinics by December 31. Officials planned to transfer the remaining clinics to the Memphis Health Center (with similar clinics and eligibility for significant federal funding), but when Dr. Reginald Coopwood became chief executive officer of The MED in March 2010, he questioned the decision and delayed divestiture of the Health Loop clinics.

DRGs

Reimbursement schemes earlier had used measurements such as teaching status and number of beds to account for the case mix in a hospital, but they were replaced in the 1980s by diagnosis related groups (DRGs), which provided a schema for monitoring quality of care and utilization of services in hospitals. The DRGs described all types of patients in an acute hospital setting. By limiting the hospital's share of Medicare reimbursement, DRGs effectively shortened a patient's time in the hospital.

MEMPHIS PRACTICES

Anesthesiology. Self-assessment and self-improvement characterized anesthesiology. In the early 1980s, the mortality rate of patients under anesthesia was about one in 5,000, and malpractice insurance rates were double those in other specialties. Although physicians often resist outside regulation, the American Society of Anesthesiologists established the Anesthesia Patient Safety Foundation in 1985, and anesthesiologists at Harvard University compiled a set of standards that required careful monitoring of anesthesia patients following minute-by-minute guidelines. By 1995, malpractice premiums for anesthesiologists had dropped almost fifty percent. Technology also contributed to patient safety by monitoring blood gases in expelled breath, but anesthesiologists led the way in safety.

By 2006, the mortality rate for anesthesiology was approximately one in 250,000. Dr. W. Roger Funderburg, Jr., commented for the *Memphis Medical Journal,* "Anesthesiologists have taken it upon themselves to critically review practice procedures and. . .make anesthesia safer." Dr. Gary Kimzey, another prominent anesthesiologist in Memphis, endorsed the changes.

Gastroenterology. In Memphis, Drs. Mike Gompertz, Alva Cummings, and Richard Bicks had been among the first physicians to practice and train younger physicians in gastroenterology in the 1960s. Their tools included liver-spleen scans, blood chemistries, and rigid scopes to allow a peek at the lower and upper gastrointestinal (GI) tract (students and residents called the rigid sigmoidoscope the "silver stallion"). Dr. Bergin F. "Gene" Overholt, a University of Tennessee graduate, designed an early version of the fiberoptic colonoscope in 1961, and in 1967 he reported 40 successful sigmoidoscopies. During the next 10 years, new scopes allowed examination of the entire length of the large bowel and extraction of polps or tumors for biopsy. Other advances included knowledge of different hepatitis viruses and long-term consequences of viral infections that could produce liver failure and cirrhosis.

Treatment of peptic ulcer was simplified after Australian physicians Barry Marshall and Robin Warren discovered in 1982 that *Helicobacter pylori* in the stomach was linked to peptic ulcers, long assumed to have been caused by stress, diet, and/or excess gastric acid. In Memphis, as elsewhere, the medical community gradually accepted this possibility and began to test routinely for the bacterium in cases of suspected ulcer. In 2005, Drs. Marshall and Warren received the Nobel Prize for Medicine for their discovery.

In the 1980s, gastroenterologists were among subspecialists who established freestanding clinics with labs for endoscopies. Drs. Lawrence Wruble; Robert Kerlan; Jeremiah Upshaw; Lee Wardlaw; Myron Lewis; and Dan Griffin were among the first to bring the new technology to Memphis. By the new century, most large groups maintained their own freestanding clinics.

Oncology. During the 1980s, the major oncology groups became more defined. Dr. Ron Lawson joined Dr. Reed Baskin. Drs. Lee Schwartzberg, Kurt Taur, and Al Weir joined Dr. William West to form the West Clinic. Dr. Earle Weeks joined Dr. Barry Boston. These medical oncologists pioneered outpatient chemotherapy, which helped patients avoid hospitalization and enabled them to receive their treatment in a physician-supervised infusion center. Subspecialties evolved during the time, including gyneco-

Dr. Raza Dilawari, oncologic surgeon at Methodist and a professor of surgery at the University of Tennessee.

REVOLUTIONS IN OPHTHALMOLOGY

ADVANCES IN EYE SURGERY IN the 1980s were nothing short of revolutionary. Vitreoretinal surgery by Drs. David Meyer and Steve Charles at the Vitreoretinal Foundation and intraocular lens implantation with cataract surgery, pioneered in Memphis by Drs. Jerre Freeman and Thomas Gettelfinger at Memphis Eye and Cataract Associates, were noteworthy.

Vitreoretinal Foundation

In 1968, Dr. David Meyer was the first retinal subspecialist in Memphis, joining Drs. Melvin DeWeese and Tom Grizzard. He founded the Vitreoretinal Foundation, the first angiography, sonography, and laser center for retinal care in the region. Dr. Steve Charles joined him in 1975, adding vitrectomy to the treatment modalities for retinal repair, and together they operated at the Mid-South Hospital. The practice was extremely busy, with operating rooms in constant use. In 1969, Dr. Paul McNeer of Richmond, Virginia, completed the first retinal fellowship, named for Dr. Roland Myers, a pioneer ophthalmologist and friend to Dr. Meyer. In 1972, Dr. Meyer established the Crippled Children's Vitreoretinal Research Foundation (CCVRRF), which trained more than 50 retinal fellows and contributed clinical and basic science articles to the eye research literature. The foundation was later led by a former CCVRRF fellow, Dr. Larry Donoso, of Wills Eye Hospital in Philadelphia, Pennsylvania. The practice grew, staffing retinal clinics around the region, with travel by private plane. The foundation built and opened the first vitreoretinal outpatient surgery center in the United States in 1988.

Both men have rare and strong personalities: Dr. Meyer is an imposing man, friend of Elvis, breeder of cattle, hunter, entrepreneur, and developer of Spring Creek Ranch Golf Course in eastern Shelby County. Trained in major clinics in the United States and Europe, Dr. Meyer, at St. Jude Hospital, coauthored the protocol for staging and treatment of retinoblastoma, the most common child eye cancer of, and

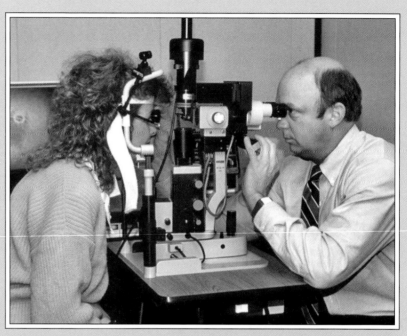

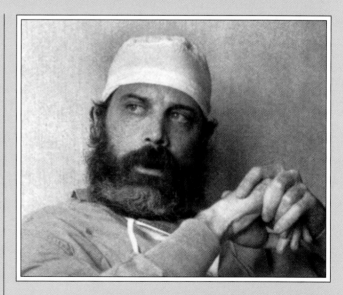

TOP: *Dr. Steve Charles peers through a laser slit lamp used to treat the retina. Dr. Charles, an ophthalmologist, a vitreoretinal surgeon, and an electrical/mechanical engineer, has nearly 50 patents for medical devices.*
ABOVE: *Dr. David Meyer, cofounder with Dr. Steve Charles of the Vitreoretinal Foundation, was the first retinal specialist in Memphis.*

pioneered the use of radioactive plaques to treat ocular melanoma, a procedure now used across the world.

Dr. Charles is known internationally. Requiring little sleep, and with a frenetic work schedule, he is a developer of multiple surgical instruments, once housing a virtual machine shop at his home. Charles, who is also an engineer, began using computer guidance for robotics that made tiny cuts—smaller even than the tremor of a human hand. Early in the 1990s, he realized that more vision could be saved by minimizing cutting movements and created MicroDexterity Systems to combine new imaging with robotics to devise microsurgical techniques used not only in the eye, but in the brain, the inner ear, and other sites in the body.

In 1985, Dr. Charles opened the Charles Retina Institute, and Dr. Meyer continued to direct the Vitreoretinal Foundation. In 2005, he established the Meyer Eye Group in Memphis, with four regional offices in Arkansas, Mississippi, and Tennessee.

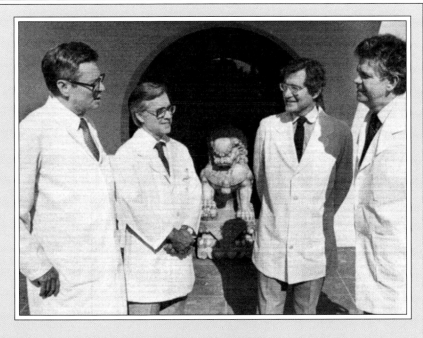

At the opening of Memphis Eye and Cataract Associates (MECA) Clinic and Ambulatory Surgery Center, July 22, 1985, are, left to right, Drs. Joseph Scott, Joseph Weiss, Tom Gettelfinger, and Jerre Freeman.

Memphis Eye and Cataract Associates and the Ambulatory Surgery Center Revolution

Dr. Jerre Freeman saw his practice shift from cataract removal in the early 1970s, requiring five days in the hospital and wearing Coke®-bottle glasses afterward, to a streamlined procedure that patients often call "miraculous."

The intraocular lens marked a revolution in eye surgery that began with Sir Harold Ridley in England in 1949 and required many years, many surgeons, and manufacturing innovations to achieve perfection and wide acceptance. First, Dr. Melvin DeWeese, then soon after Dr. Freeman, brought to Memphis the technique of small-incision cataract surgery with phacoemulsification and implantation of intraocular lenses. Dr. Freeman did the first lens implant in 1974. Dr. Thomas Gettelfinger, in 1976, and later Dr Joseph Weiss joined Dr. Freeman, forming Memphis Eye and Cataract Associates (MECA).

MECA not only revolutionized techniques of advanced cataract surgery for the region but also revolutionized how surgery was delivered, through Tennessee's first freestanding ophthalmological ambulatory surgery center in 1985. The cost savings were significant for patients and insurers. With cataract surgery as the most common procedure for Medicare-age patients, Dr. Freeman estimated that savings in healthcare costs during the first decade of MECA's operation were in the tens of millions of dollars, plus reduction in patient stress and in time consumed for the physicians.

Over time, MECA has expanded its services. Dr. Freeman performed the first laser vision correction surgery in the Mid-South there in the early 1990s. Advanced corneal transplants, corneal prostheses, and retinal procedures have been added. In addition to Drs. Freeman and Gettelfinger, the MECA group, which now has a 60-person staff, includes Drs. Hal Wright, James and John Freeman (who joined their father in practice), and retinal surgeon David Irvine. Dr. Joseph Scott, now retired, practiced with the group for many years.

Dr. Freeman has been on a number of national ophthalmology boards, president of some, and president of The Memphis Medical Society. His innovations in eye surgery and foresight in developing ambulatory surgery in Memphis are matched by his holding a number of medical patents and his election to the Memphis Society of Entrepreneurs. Dr. Gettelfinger, a former president of the Tennessee Academy of Ophthalmology, pioneered training and certification of ophthalmic technicians in Tennessee. Almost all of MECA's technical staff are credentialed as Certified Ophthalmic Assistant (COA) or Certified Ophthalmic Technician (COT).

Other independent ambulatory centers appeared after MECA, especially ambulatory gastrointestinal centers. Today there are more than 25 ambulatory centers in Memphis.

logic oncology. Drs. Bobby Hughes, Guy Photopulos, and Linda Smiley were among the earliest to focus on tumors of the cervix, uterus, and ovary. Dr. Raza Dilawari joined the University of Tennessee Department of Surgery during this time as a full-time surgical oncologist. Later, he became director of surgical residents at Methodist Hospital.

In the late 1980s and into the 1990s, most major medical disciplines developed advanced training for physicians interested in cancer. Dr. Robert Heck with the Campbell Clinic and Dr. Michael Neal with OrthoMemphis became the first orthopedists to concentrate on cancer. Advances in chemotherapy also included modulating the 30-year-old fluorouracil treatment for colorectal cancer with leucovorin in 1985. Autologous stem cell transplantation was used in the 1980s as a treatment for multiple myeloma. The latter advances were incorporated by Memphis's oncology groups.

Psychiatry. By 1980, Memphis had more than 100 psychiatrists, including some practicing part time. Among these were Dr. Frank Lathram, who was active locally and nationally in creating stronger psychiatric disciplines; Dr. Garo Aivazian, who headed the Gailor Unit and was chairman of psychiatry at the University of Tennessee, helping train hundreds of students, interns, and residents; Dr.

Drs. James C. Hancock, left, and Garo Aivazian, Gailor Clinic, 1965. Dr. Aivazian was also chairman of psychiatry at the University of Tennessee.

Theron S. Hill, chair of psychiatry at the University and head of the Gailor Unit (1946-1970); and Dr. Fontaine S. Hill, a pediatrician and a minister who completed a degree in child psychiatry.

Psychiatrist Dr. Frank Lathram noted in his history the following fellow psychiatrists active at Methodist in 1980: Drs. Elizabeth Gehorsam (formerly chief of the department); Kenneth Tullis (acting chief of the department); John Harris (formerly acting chief); Nancy Duckworth; Bob Davison; and David Moore. He also emphasized important work done in allied fields, especially by physiologists Dr. David Knott and James Beard, Ph.D., in alcohol and drug treatment. St. Joseph Hospital operated a care unit for alcohol and drug patients for several years during this time.

Urology. Long a center for advances in urology, Memphis, in 1985, saw the introduction of laparoscopic assistance in removing kidney stones percutaneously from pelvic kidneys, often eliminating lengthy surgery and painful recovery. Laparoscopic assistance in staging prostate cancer in 1989 opened the way for a new era in operative urology.

Drs. Lynn Conrad, Richard M. Pearson, John R. Adams, Jr., Rair Chauban, Paul Eber, Robert Hollabaugh, Jr., and Michael McSwain of the Conrad-Pearson group were among the leaders in this surgery, as was the Mid-South Urology Group. Dr. Samuel Raines continued as a prominent urologist long after his term as president of the Southeastern Section of the American Urological Association in 1955. Dr. Fontaine B. Moore, Jr., was president of that organization in 1985.

The Hospice Movement

Although the hospice movement is relatively new in U.S. health care, the concept is ancient, linked to the tradition of family caring for the dying and to the concept of shelter for tired or ailing travelers. Through her work with dying patients in the 1940s, English physician Dame Cicely Saunders taught, "We do not have to cure to heal," and came to the United States in 1964 to teach at Yale University. Dr. Elisabeth Kubler-Ross published *On Death and Dying* in 1969, which brought death into public consciousness and changed the way Americans thought about death.

In 1982, Congress provided a hospice benefit under Medicare, and the hospice movement flourished, with grants and legislation supporting research and improvements. Hospice care was designed to provide relief from whole-body suffering, anxiety, and loneliness to patients whose life expectancy was six months or less. Rather than provide disagreeable treatments in a last effort to provide a cure when none was deemed medically possible, hospice care offered relief from pain, a relaxed atmosphere, and compassion. Volunteers were trained to work with patients, playing music the patient liked and listening to the patient's concerns. Specially trained nurses oversaw the patient's medical needs.

Melissa Russell, left, Hospice of Memphis nurse, and Frank Boatwright, executive director of Hospice, interview a prospective volunteer during a 1981 recruiting drive.

MEMPHIS'S ORTHOPEDIC DEVICE INDUSTRY

By DR. JAMES CALANDRUCCIO

Orthopedic commerce has thrived in Memphis. Musculoskeletal disease is the major cause of disability worldwide, affecting 500 million individuals. The orthopedic industry sought to provide devices that lessened periods of morbidity and promoted improved lifestyles, as well as rapid return to work. Memphis was fortunate to have orthopedic device companies that began here and have grown, including Smith and Nephew, Richards; Wright Medical Technology; and Medtronic. Startup companies included Active Implants, TiGenix, Regenisys, and Expanding Orthopaedics.

The first American orthopedic manufacturing company was begun by Revra DePuy in 1895 in Warsaw, Indiana. In 1934, Don Richards, a salesman for this company from North Mississippi, opened his own company, Richards Manufacturing, in Memphis. Its association with the Campbell Clinic greatly benefited the company, which expanded its market share and global presence by purchasing Rhohr Pharmaceuticals in 1968, Coopervision in 1986, and most recently Smith and Nephew in 1987. The present company emerged as a leader, often by collaborating with local orthopedic surgeons.

With Memphis's Elvis Presley Memorial Trauma Center ranking as one of the three busiest trauma centers in the country, innovations in trauma repair and fracture management were natural consequences of surgeons' experience.

Another Richards employee, Frank O. Wright, left Richards to begin Wright Manufacturing in 1950, first making rubber heels for plaster casts. Since its beginning, this company has been the leader in the extremity market, adding small-joint implants as well as large-joint arthroplasty devices. Controlling interest by Warburg Pincus Equity Partners in 1999 took the company public. The company, headquartered in Arlington, Tennessee, had 1,100 employees, 850 of whom were local, in 2009. In 2003, Wright Medical Technology was the fastest-growing local orthopedic company. Wright also collaborated with local orthopedic surgeons; one design was an expandable femoral replacement prosthesis allowing noninvasive femoral lengthening in pediatric limb salvage procedures.

Begun in 1980 by Alan Olsen as Danek Medical, sold to Biotechnology, Inc., a company with a spine focus, in 1985, and sold in 1993 to Sofamor, the Sofamor Danek Group was bought by Medtronic in 1999, becoming Medtronic Sofamor Danek. After purchasing Kyphon in 2005, Medtronic

---

Social workers and chaplains visited if the patient wished.

In Memphis, hospice care was available through all three major hospitals and a variety of other providers, and the University of Tennessee had a fellowship program. Both home and institutional care were available. A leader in this area was Dr. Clay Jackson, medical director of Methodist Hospice and Palliative Services and director of Methodist University's fellowship program. Dr. Jackson also holds a diploma in theology and is double board certified in hospice and palliative medicine and in family medicine.

Memphis's First Astronaut/Doctor

Originally from Murfreesboro, Margaret Rhea Seddon graduated from the University of Tennessee with an M.D. degree in 1973 and completed her surgical internship and three years of general surgery residency in Memphis. Her research interest was nutrition in surgical patients. She also worked in the emergency room in hospitals in Tennessee and Mississippi between internship and residency.

Dr. Seddon became an astronaut in 1979, after intensive training with the shuttle's medical kit, launch and landing rescue, and crew communications. A veteran of three flights—one on *Discovery* and two on *Columbia*—she logged more than 722 hours in space. She was payload commander on *Columbia*'s Spacelab Life Sciences-2 in 1993, when the crew performed nine Extended Duration Orbiter Medical Projects; the flight was recognized by NASA as the most efficient and successful Spacelab flight to date. Sent to Vanderbilt University Medical School in Nashville in 1996, she

Frank Ozanne Wright displays his company's orthopedic heels and other devices, 1966.

tion and surgical demonstration. Of the 804 courses the institute offered in 2008, 270 were on orthopedic surgery.

Orthopedics has experienced technological advances once unimaginable and was uniquely positioned because of its overlap with specialties including pediatrics, neurosurgery, sports medicine, vascular surgery, and oncology.

UT–Baptist Research Park was planned, in part, as a collaborative campus of orthopedic surgeons and affiliate medical partners, research scientists, and partners in industry. The musculoskeletal piece of this effort was InMotion, founded by the Assisi Foundation, Baptist Memorial Health Corporation, the Campbell Foundation, Hyde Family Foundations, Medtronic, and the Plough Foundation. The hospital-clinic-university-industry partnership followed a translational research model whereby fundamental research, applied research, and clinical research led to improved patient care. In 2009, InMotion studied biologics and biomaterials with a focus on growth, repair, and regeneration of musculoskeletal tissues.

further established its position as an international leader in spine care. In 2009, Medtronic had 40,000 employees worldwide, 2,000 of whom were in Memphis, with 600 more planned, and had its own airport parking facility. The company's spine focus was headquartered in Memphis.

These companies benefited from their associations with the Memphis Education and Research Institute, a unique cadaver laboratory offering courses in anatomic dissec-

Globally, Memphis has been synonymous with orthopedics, thanks to the pioneering efforts of Dr. Campbell, the care delivered by the city's orthopedic groups, and the burgeoning orthopedic industry.

helped prepare cardiovascular experiments for *Columbia*'s April 1998 flight. She retired from NASA in 1997, joining the Vanderbilt Medical Group in Nashville.

Medical Business

In the last half of the twentieth century, Memphis became the center of a highly innovative medical-device industry, enhanced by FedEx's outstanding distribution capabilities.

In Janaury 1978, Dr. Margaret Rhea Seddon was notified that she was one of 35 people accepted into the United States astronaut training program. She eventually made three voyages into space aboard the space shuttles Discovery *and* Columbia.

CHAPTER FOURTEEN

BIOTECHNOLOGY, MINIMAL INTERVENTION, AND INSURANCE

1990-1999

FROM THE GULF WAR, TO THE TAKEOVER OF Hong Kong by the People's Republic of China, to the splitting of Yugoslavia, to peace in Ireland, to genocide in Bosnia and Rwanda, to flourishing market economies, to the World Wide Web, the 1990s brought contrasts. Unprecedented prosperity was shadowed by the Al Qaeda car-bombing of the World Trade Center and the 1999 arrest of an Al Qaeda militant intent on bombing the Los Angeles airport. In medicine, technology flourished and minimally invasive surgery came to several specialties. Despite new treatments, the AIDS epidemic continued, bringing renewed infections. Health care was an increasing burden as the population aged, and end-of-life medical issues were rivaled only by beginning-of-life issues. In 1997, the Federal Drug Administration loosened restrictions on direct-to-consumer advertising of prescription drugs, unleashing a bonanza for pharmaceutical companies. New vaccines became available in the 1990s for hepatitis A (1992) and for Lyme disease (1998).

Major changes came to Memphis's healthcare delivery through insurance reimbursement and migration of hospitals, a veritable hospital hopscotch. Major insurance carriers asserted their influence by contracting with large corporations and adding thousands to HMOs, bringing insurance companies into examination and operating rooms. Poor patients were squeezed out of the system by high insurance costs, and many addicted and psychiatric patients found doors closed. Changes in Medicare, the beginning of TennCare, evolving managed care programs, and the increase in malpractice insurance changed physician reimbursement. More patients developed chronic healthcare problems: obesity, cardiovascular diseases, and diabetes, plus emphysema and chronic obstructive pulmonary disease (partly the legacy of cigarette ads and World War II-era free cigarettes to soldiers). Even meticulous and caring physicians expected to be sued.

In 1993, Congress's Omnibus Budget Reconciliation Act modified reimbursement to Medicare providers via the resource-based relative value scale, which established new standards for payments to physicians, replacing "usual and customary" fees and attempting to factor into reimbursement the physician's work based on relative value units

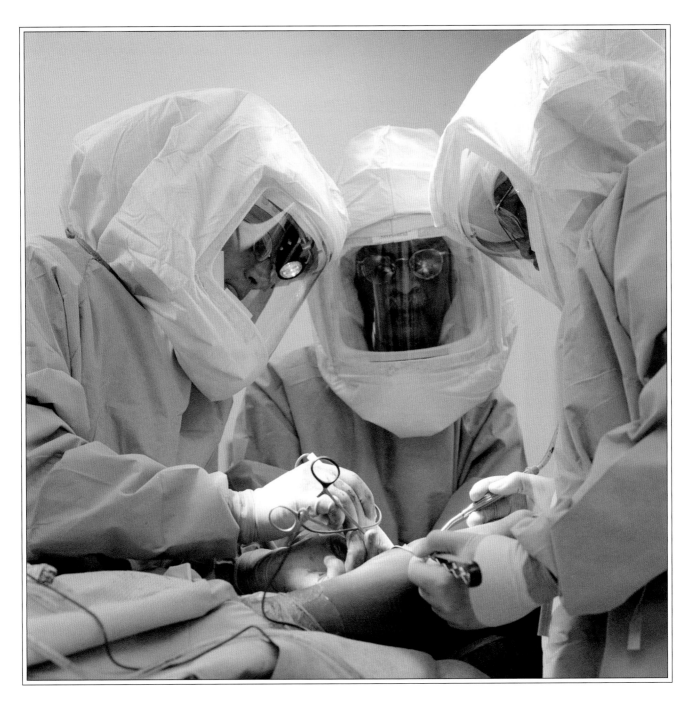

Drs. Bhaskar Rao, center, Mike Neel, left, and Veerland Thompson, right, perform a limb-saving surgery at St. Jude Children's Research Hospital, 2001.

(RVUs) reflecting the time, skill, and training required. Additional modifiers related to overhead and geographical location were added, plus a code that was translated into dollars and modified yearly to control costs. The geographical modifiers mean that payments to physicians in Manhattan are twenty-three percent higher than payments in Tennessee for comparable work; further, forty-four percent more is spent on Medicare patients in Manhattan than in Tennessee, even though federal tax rates are the same in all locations. For years, the Tennessee Medical Association has presented resolutions to equalize payments, but Congress has not responded.

Despite attempting to differentiate monetarily between advanced cognitive functions required in different diagnoses and treatments, the payment system produced a disparity in reimbursement, with physicians who performed procedures rewarded more than primary care physicians (a large percentage gain from 1993 to 2003 went to invasive cardiologists). Overall, reimbursements dropped as Medicare failed to keep pace with inflation. Private insurance

consultation when needed. Hospitalists may work from an office or at a single institution. The Memphis chapter of the Society of Hospital Medicine, established in 1997, included Drs. Tapan Thakus, Chris Sands, Michael Puchaev, Kamal Mahan, and Ayesha Mugammil.

TennCare

Introduced in 1994, TennCare marked another change in healthcare delivery. Seeking to provide managed health care as a substitute for Medicaid, Governor Ned McWherter obtained a waiver from Medicaid to move federal and state

LEFT: *Dr. Wiley Robinson, c. 2005. Dr. Robinson, along with Dr. Mark Hammond, founded the hospitalist group Inpatient Physicians of the Mid-South, which now includes 14 physicians. These physicians practice strictly in a hospital setting and have been quite effective in lowering healthcare costs and providing excellent care for their patients.* BELOW: *Clinical psychologist Dr. Allen O. Battle holds the Lifetime Achievement Award given to him by the* Memphis Business Journal *in 2009.* OPPOSITE PAGE: *The Memphis Mental Health Institute, 1961. Built just north of Le Bonheur, it was torn down to make way for the new Le Bonheur Children's Hospital.*

companies based their reimbursements on Medicare rates, and payments for almost all medical care were affected.

Physicians in family medicine, internal medicine, and general surgery were affected adversely by Medicare changes and by managed care. Graduates of American medical schools sometimes avoided the primary-care specialties such as family medicine, internal medicine, general surgery, and obstetrics/gynecology because of their extended work hours in relation to reimbursement. Medical students often express their motivation for choosing subspecialization as a desire to master a particular field, but poor reimbursement for primary care has also been suggested as another factor.

As a subspecialty of internal medicine, hospital medicine (practitioners are called "hospitalists") developed as many primary care physicians found it impossible to spend time in several hospitals and maintain their offices. In 1993, Memphian Dr. Mark Hammond began working solely for hospital inpatients while still a member of Family Physicians Group. Dr. Wiley Robinson joined the group in 1996. In 1999, Drs. Hammond and Robinson formed the first group of hospitalists, Inpatient Physicians of the Mid-South, entering agreements with physicians rather than hospitals, and serving all the major hospitals in Memphis. Dr. Robinson explains that hospitalists may also serve as intensivists or provide hospital

dollars into a fund administered by independent managed care organizations (MCOs), which would contract with physicians and hospitals to provide care. With Vice President Al Gore's help, the state received the waiver, and TennCare began trying to provide for the uninsured and the working poor while maintaining budget neutrality. The state shifted 800,000 individuals from Medicaid and added an additional 500,000 individuals previously ineligible for Medicaid. Twelve private MCOs were established to manage the program and deal with providers and hospitals. TennCare was clouded by confusion, with many MCOs lacking the necessary experience and the medical community unprepared to deal with them. Payments were often late or denied, and reimbursement rates were unacceptable to most physicians. Some MCO board meetings were not businesslike, and some funds allocated to one MCO were never accounted for. As costs rose, the state tightened enrollment requirements.

In 1996, TennCare separated behavioral health benefits from the program. Even with reduced numbers in the program, the budget for TennCare increased from $1.8 billion in 1995 to $7.1 billion in 2007 and remained a drain on Tennessee taxpayers. Although many physicians at first declined to participate, with one in four Tennesseans enrolled, more doctors accepted the lower rates.

The combination of Medicare regulations, limited TennCare reimbursements, and psychotropic drugs prescribed to outpatients led to an unprecedented shrinking of psychiatric facilities. Mental patients were often left to fend for themselves, with many of them homeless. Public Defender A C Wharton suggested that TennCare—by refusing reimbursement for inpatient psychiatric care and relying instead on mental patients themselves to take prescribed medicines—was indirectly responsible for a huge increase in the number of petty crimes, and for an increase in the number of inmates diagnosed with mental illness. As mental health facilities closed across the state, deinstitutionalization of mentally ill people accelerated. In January 1997, 242 new inmates were diagnosed as mentally ill. The second floor of the county jail seemed more like a mental institution than a jail, filled with people moaning, screaming, and shaking. Those unable to post bond remained in jail until arraignment.

Common crimes included trespassing, urinating in public, minor thefts, and verbal assault (yelling at strangers). Allen O. Battle, Ph.D., longtime psychiatry professor at the University of Tennessee, formerly in charge of psychiatric emergencies at The MED, commented: "The transfer from inpatient to outpatient care was a blessing for some and for others a nightmare. The situation was like musical chairs

St. Joseph Hospital, 1960s. After the property and buildings were acquired by St. Jude, the hospital was torn down in 1999. In 2000, St. Jude began an expansion of its facilities, starting with the Integrated Research Center.

without the chairs; it was a curse. Many patients do not take their meds; they live in the gutter, and then policemen have to take them in, although Memphis has a number of psychiatric care facilities."

In 1998, community volunteers, mental health professionals, and county officials formed a task force, the Memphis and Shelby County Mental Health Summit. Convinced that care for the mentally ill had declined severely, the group studied how to enhance care systems for the mentally ill and for substance abusers. At the time, *Memphis Medical News* reported that about 230,000 uninsured or uninsurable individuals were without care. Although the poverty rate in Memphis suggested a high percentage of eligibility for TennCare, a significantly smaller number of people were enrolled. The task force helped the needy reach out for help and facilitated their enrolling in established programs.

In March 2010, a new era began for all American medicine with the signing of the Patient Protection and Affordable Care Act, a large, controversial, and poorly understood measure designed to reform the health system in the United States. Far-reaching effects of this healthcare reform act will unfold gradually in the next few years.

Hospital Migrations

Hospitals followed the population eastward. In 1995, Methodist and Le Bonheur hospitals merged to form Methodist Le Bonheur Healthcare (the umbrella organization for a growing number of facilities). Methodist assumed Le Bonheur's debt, but gained a connection with University pediatrics and a stronger negotiating position in competing for managed-care contracts. Methodist Healthcare purchased a small, for-profit hospital in Germantown and added obstetric services to draw preferred provider organization patients.

Downtown, the aging St. Joseph Hospital at 220 Overton Avenue, once the site of a flourishing nursing school and an active hub of medical practice, had fallen on difficult financial times. It merged with Baptist in a three-year process beginning in 1997, with most of its 1,100 employees absorbed by Baptist. For $20 million, St. Jude acquired the land and buildings from the Sisters of St. Francis of Perpetual Adoration of Mishawaka, Indiana, who had founded

St. Joseph in 1889. The hospital closed in December 1998 and was subsequently demolished to make way for new St. Jude facilities.

In addition to their suburban facilities, the two largest private hospitals also continued to upgrade facilities in North Memphis, Southaven, Mississippi, and East Memphis and to add locations in West Tennessee, North Mississippi, and East Arkansas.

Combining Practices

By the end of the 1990s, physicians began adapting to the realities of diminishing professional fees and the shift of patients to the suburbs. They formed new, larger practices by merging and adding new physicians and services. Some, such as Dr. Randolph Turner, who had earned a reputation as a highly respected general surgeon at Methodist Hospital, realized earlier that physician practices should consolidate to have clout when dealing with pressures from insurance companies. Dr. Turner had expanded in 1966, combining with Drs. Scott King, Leonard Hines, Hugh Francis, Jr., Ned Laughlin, and Carter McDaniel to form a single large practice that worked out of multiple hospitals. They recruited young surgeons—Drs. Justin Monroe, Ben Gibson, and Will Gibson—with similar interests in general as well as oncologic surgery. By the 1990s, many of the larger groups offered services at multiple locations as diagnostics, procedures, and complex therapies were shifted from the hospital to freestanding complexes. Such changes would continue in the 2000s.

As healthcare delivery changed, rapid technological growth and advances in understanding the biology of disease brought improvements in patient care and further changes. The relationship between hospitals and physicians changed as physicians joined multiple hospital staffs and were no longer identified with Baptist, Saint Francis, or Methodist.

Technological Influences on Memphis Medicine

Minimally Invasive Surgery. Early minimally invasive surgery (MIS) had used imaging devices to assist the surgeon in visualizing the field. In the 1990s, robot-assisted surgery miniaturized the surgeon's touch and brought new innovations.

Dr. Guy Voeller was one of a few early surgeons interested in the procedure when it was first publicized in France in 1988. Gynecologists had done tubal ligations for years through the laparoscope, but the scope occupied one hand. In 1990, a new type of laparoscope opened the door for

Dr. Guy R. Voeller, an early proponent of minimally invasive surgery, has been an innovator in the use of laparoscopic procedures. He is a founding member and past president of the American Hernia Society.

MIS, allowing surgeons to operate through multiple small incisions instead of the traditional larger incision. Such procedures decreased postoperative recovery pain and reduced operating time for many procedures.

In Memphis, a group of University of Tennessee-affiliated surgeons were leaders. Dr. Richard Patterson and Dr. Guy Voeller, after much work in the lab, performed the first laser laparoscopic gallbladder removal in Memphis in 1989. The two surgeons held a series of courses beginning in 1990, teaching laparoscopic cholecystectomy to physicians from Memphis and around the United States. Dr. Jim Green began using the procedure at Methodist, along with Drs. Mark Miller and Charlie Frankum at Baptist.

Dr. Gene Mangiante assisted Dr. Voeller in the first laparoscopic inguinal hernia repair in Memphis in 1990, and the two surgeons directed one of the world's first courses on that procedure in Memphis in 1992.

With Dr. Mangiante assisting, Dr. Voeller performed ulcer surgery through the laparoscope, followed by the fourth

Dr. Eugene Mangiante worked with Dr. Guy Voeller using laparoscopic surgery.

laparoscopic highly selective vagotomy in the country at The MED in 1991. Urologist Dr. Bob Wake and Dr. Voeller performed the region's first laparoscopic adrenal surgery at Bowld Hospital in 1994. It was followed by laparoscopic colon, hiatal hernia, achalasia, spleen, and incisional hernia repair, with work in the lab before actual surgery. Dr. Voeller did the first course in the United States on endoscopic saphenous vein harvest for coronary bypass grafting at Medical Education Research Institute (MERI) in Memphis. He taught a course in laparoscopic surgery for about 60 Japanese surgeons who came to Memphis. His group was one of the world's first to use a 3D laparoscope in 1995 and a robot in 1997. In 2000, Dr. Voeller received *Memphis Business Journal*'s Health Care Innovation award.

Other MIS surgeons, including Drs. Mark Miller, Leonard Hines, and Alan Ellis, also performed advanced laparoscopic procedures. Dr Hines was president of the University of Tennessee National Alumni Association and a leader in surgery in the Saint Francis system. Pediatric surgeon Dr. Thom Lobe refined MIS hernia repair and videotaped the procedure for those who questioned his six-minute procedure.

After the early 1990s, laparascopic surgery became standard for gallbladder removal and for hernia, bowel, gastrourinary, and gynecological procedures in Memphis. In many cases, it turned a several-day hospital stay and weeks of recovery into an outpatient procedure with a shortened recovery period.

In the mid-1990s, pediatric surgeon Dr. Bhaskar Rao at St. Jude reached the milestone of 100 successful limb-salvage surgeries, using a technically demanding procedure to remove bone tumors and to implant a prosthesis that could be adjusted as the child grew. Children as young as five were candidates, with the most common procedure involving bone tumors in the knee.

In another milestone for Memphis, leading transplant surgeon and University of Tennessee professor Dr. Osama Gaber performed Memphis's first segmental liver transplant in 1992. Because liver tissue regenerates itself, a segmental liver transplant can come from a live donor.

In 1997, the University's Bowld Hospital initiated a bloodless surgery option, beginning with abdominal surgery. Driven partly by concern about AIDS and hepatitis, the procedure also appealed to some religious groups. Surgeons used plasma expanders and drugs to stimulate red cell growth. In appropriate patients, blood was salvaged during surgery and returned to the patient. The procedure was further aided by the argon beam laser that seals as it cuts. Dr. Jeff Woodside, urologist, and Dr. Louis Britt, chair of surgery, led the initiative, which included 40 physicians..

New programs demonstrated the growing emphasis on research in Memphis medicine. In 1994, the University of Tennessee established its Pancreas Islet Transplant Laboratory. Dr. Gaber performed the first pancreatic islet cell transplant in Memphis in 1995. The Diggs-Kraus Sickle Cell Center was rededicated in 1994 to honor the three people who had spent their lives studying and treating this disease—Drs. Lemuel Diggs, Alfred P. Kraus, and Lorraine Kraus. In 1997, the University established its Transplant Clinical Pharmacotherapy Research Institute.

In 1997, St. Jude announced plans for a $115-million Integrated Research Tower, which was part of a ten-year, $1 billion plan to expand its research capacity. The new Department of Chemical Biology compared tumor cells with normal cells in an effort to design drugs for each patient in an individualized approach matched to a person's genetic makeup. In 1998, the American College of Microbiology (ACM) cited genetic research at the University of Tennessee and St. Jude as examples of the future of immunization. Both facilities had projects in which modified genes were delivered by a virus vector to treat disease.

PETER C. DOHERTY, Ph.D., NOBEL LAUREATE AT ST. JUDE

TOLD IN HIS OWN WORDS IN *Les Prix Nobel* (the yearbook of the Nobel Foundation edited by Tore Frangsmyr, Stockholm, 1996), Dr. Peter Doherty's autobiography reveals a man who learned early to recognize opportunity, an unconventional scientist steeped in learning but always ready to think in a new way, a thinker who by his own admission followed his own drumbeat.

Born in Brisbane, Australia, and drawn to literature, history, and music, Peter Doherty was good at science and conscious of the need to earn a living. At the University of Queensland, he studied veterinary science, and later he joined the Queensland Department of Agriculture and Stock, where his laboratory research led him to the Animal Research Laboratory in Yerongpilly. There, his work with bovine leptospirosis led to his master's degree, and he met his wife, Penny.

In Melbourne, while studying virology techniques at the John Curtin School of Medical Research (JCSMR), he focused on research. In 1967, he went to Edinburgh to the Moredun Research Institute, which trained graduate students at the University there. While in Edinburgh, he became a part-time graduate student at the medical school, but he returned to Melbourne, Ph.D. degree in hand, accepting a position at the Commonwealth Scientific and Research Organization to learn basic immunology. In 1971, the family moved to Canberra, where he met collaborator Dr. Rolph Zinkernagel, who shared the 1996 Nobel Prize in Medicine.

At the Second International Immunology Conference in Brighton, England, in 1973, the "single T-cell receptor altered self" hypothesis was almost in place. Immunologists realized this was an important discovery, though counter to accepted wisdom. In 1975, Dr. Doherty joined the Graduate Immunology Group at the Wistar Institute in Philadelphia. He headed the Department of Experimental Pathology at JCSMR from 1982 to 1988, but research was his *métier*, and then-director of St. Jude Children's Research Hospital, Dr. J. V. Simone, lured him to Memphis in 1988 to "a superb, open research environment" with significant funding. Through the efforts of Dr. Alan Granoff, head of the Virology Department, and Dr. Robert Webster, he also joined the University of Tennessee, though Memphis proved too sunny for his Irish complexion and too far from the Pacific

Dr. Peter C. Doherty holds the Albert Lasker Basic Medical Research Award which he received in 1995. Others, from left to right, are Drs. Arthur Nienhuis, Robert Webster, and James Ihle.

for full-time residence. He divides his time between St. Jude and the University of Melbourne.

Dr. Doherty characterizes the origins of his scientific quest as "a non-conformist upbringing, a sense of being something of an outsider, and looking for different perceptions in everything...." Drawn to complexity and influenced by Karl Popper and Thomas Kuhn, he received the prestigious Albert Lasker Award for Basic Medical Research in 1995 and was Australian of the Year in 1997.

The prize-winning discovery of Doherty and Zinkernagel concerned the way that killer T-cells recognize infected cells and destroy them to prevent virus reproduction.

MEMPHIS PRACTICES

Bariatric (Weight Loss) Surgery. In 1991, as president of the American Society for Bariatric Surgery, Dr. George Cowan, with Dr. Mervyn Deitel, a surgery professor in Toronto, co-founded the first bariatric surgery journal, *Obesity Surgery*. Now in its twentieth year, it is among the top ten most frequently cited surgery journals. Dr. Cowan also founded the International Federation for the Surgery of Obesity (IFSO) in 1995 and wrote the by-laws. In 1997, as incoming president, Cowan presented a document to the IFSO Council defining bariatric surgery and providing a model to groups awarding "Center of Excellence" designations to bariatric surgical practices meeting the high standards required. In Memphis, Dr. Stephen Behrman and Dr. Virginia Weaver became bariatric surgeons.

Neurosurgery. In 1994, a respected group, the Canale Clinic, established by neurosurgeon and medical historian Dr. Dee J. Canale, consolidated its offices in a freestanding facility on Germantown Road, across the street from the Campbell Clinic.

In 1995, the merger of two distinguished clinics—Baptist's Semmes-Murphey Clinic and Methodist's Neurosurgical Group of Memphis—brought together some of the top neurosurgeons in Memphis. Four physicians, plus office and practice connections, were added to the Methodist

ABOVE: *Dr. Russell Chesney is associated with the Department of Pediatrics at the University of Tennessee and Le Bonheur Children's Hospital. His research has been valuable in the management of kidney disease in children. In recognition of his work, he was the recipient of the Ira Greifer Award, in 2010, given by the International Pediatric Nephrology Association. His wife, Dr. P. Joan Chesney, has focused on pediatric infectious disease in her research.* LEFT: *Dr. George Cowan, Jr., a leader in the field of bariatric surgery. Recognized internationally for his work, he has published prolifically, including* Intravenous Hyperalimentation, *the first book on intravenous nutrition.*

Healthcare system. With connections to both Baptist and Methodist, the Semmes-Murphey Clinic was a key sponsor in establishing the Medical Education Research Institute (MERI) in 1995.

Neurosurgical progress was striking in children. In 1997, neurosurgeons at Le Bonheur introduced a new microscope that showed where tumor ended and brain tissue began, enabling surgeons to remove tumors with minimal damage to healthy tissue. In the late 1990s, St. Jude reported a ninety percent survival rate of low-grade astrocystoma, the most common brain tumor in children, in a ten-year study of 142 patients.

Orthopedics. In 1990, the Campbell Foundation joined the University of Tennessee's Department of Orthopaedic Surgery in a formal relationship for managing orthopedic residencies through a fully accredited joint program requiring five years, with 40 residency slots, and known as the University of Tennessee-Campbell Clinic program. The Campbell Clinic moved its main facility from 969 Madison in the Medical Center to a new complex at 1400 Germantown Road in 1993, while retaining an office at 1211 Union Avenue. This move proved prophetic for Memphis medical clinics.

Pathology. Laboratory director Dr. John Duckworth, after first retiring in 1980, organized a clinical research department at Methodist and later directed the pathology residency program at the University of Tennessee. He credited former student Dr. Charles Handorf with building the pathology residency at the University. A pioneer in laboratory quality control, Dr. Duckworth spearheaded efforts to computerize lab results and transmit them electronically.

Pediatrics. In the 1990s, Methodist and the University of Tennessee collaborated in studying Memphis's high rate of prematurity, which continued despite technological advances and aggressive outreach for prenatal care to prevent lifetime health problems and high medical costs associated with prematurity. Drs. Joan and Russell Chesney were prominent figures in University pediatrics. Dr. Russell Chesney chaired the Department of Pediatrics, and both were active researchers.

Pulmonology. In 1998, Dr. G. Umberto Meduri at the University of Tennessee and his team published in the *Journal of the American Medical Association* their finding that administering methylprednisolone improved hospital survival by fifty percent compared with conventional management of acute respiratory distress syndrome. The research was supported by the Baptist Foundation, the University of Tennessee and its Clinical Research Center, and the Assisi Foundation.

Radiology. By the mid-1990s, ultrasound could produce a remarkable image of a fetus's cleft palate. Memphis's first trained pediatric radiologist (who did his residency at Boston Children's Hospital and studied in London and Stockholm), Dr. Webster Riggs, Jr., continued to train radiologists through the University of Tennessee and Methodist Le Bonheur. The cost of a CT scan fell so that some physicians' groups added CT imaging to their offices. With the cost below the threshold requiring a certificate of need, physician-owned diagnostic centers appeared in the city.

Urology. In the 1990s, drug treatment became an alternative to transurethral resection for benign prostatic hypertrophy. Nondrug options such as needle ablation and

Dr. Webster Riggs, Jr., examines an x-ray in his office in Le Bonheur East.

microwave thermotherapy were also available. One drug was available for urinary incontinence.

Once testing for prostate-specific antigen (PSA) was recommended for routine screening for prostate cancer in 1993, the number of new cases of prostate cancer increased as prostate screening and biopsy became common.

At the University of Tennessee, Dr. Clair Cox chaired the Department of Urology until 1999, training 88 residents who all became board certified. In 1999, Dr. Mitchell S. Steiner, a 1986 University graduate, became chair, after residency at Johns Hopkins, teaching at Vanderbilt, and serving the University of Tennessee as a professor and director of the Urologic Research Laboratories since 1994. In 1996, Dr. Steiner treated the first two prostate cancer patients with gene therapy, using a retrovirus mixture that kills cancer cells and stimulates the immune system to produce more tumor-suppressing cells. Building on awareness that cancer cells fail to get the signal to die, Dr. Steiner used a modified gene to deliver the message. His research led to the development of two promising drugs.

THE NEW NEUROSURGERY

Neurosurgery has been especially prominent in Memphis medicine, from the Semmes-Murphey Clinic, to the Neurosurgery Group and Dr. Dee J. Canale's group, to significant recent developments. In the mid-1990s, two important facilities continued that tradition of excellence—the Gamma Knife facility at Methodist Central and the Medical Education and Research Institute (MERI) on Cleveland Avenue.

Gamma Knife. Dr. David L. Cunningham, a member of the Neurosurgical Group at Methodist with a doctorate in pharmacy and an M.D. degree from the University of Tennessee, spearheaded the Memphis Regional Gamma Knife facility, which opened in 1995 at Methodist Central. Cunningham trained in the Gamma Knife procedure at the University of Pittsburgh and eventually became medical director of Methodist's Gamma Knife Center.

Developed by Swedish neurosurgeon Dr. Lars Leksell, the procedure gives a one-time application of high-dose radiation rather than a fractionated delivery of lower-dose radiation over several weeks. Unique equipment enables radiation to pinpoint a single spot in the brain without cutting through healthy tissue. The procedure is highly effective in controlling certain tumors, arterio-venous malformations, and trigeminal neuralgia. Because the brain does not feel pain, patients usually receive only a sedative. With normal tissue spared, intensive care and lengthy hospital stays are unnecessary. The facility is used by neurosurgeons from the major hospitals with credentials in its use. By 2009, more than 2,400 cases had been treated.

In addition to his innovation as a neurosurgeon, Dr. Cunningham served on the Tennessee Board of Medical Examiners from 1993 to 2007 and was president in 2005.

Medical Education and Research Institute (MERI). Memphis's unique MERI began with a meeting of Dr. Cunningham, neurosurgeon Dr. Kevin Foley, pathologist Dr. John Duckworth, and Methodist administrator Maurice Elliott. Dr. Foley originated the idea of MERI and helped to bring the reality to Memphis. Foley of Semmes-Murphey and Cunningham of the Neurosurgical Group represented the two physician groups that conceptualized the facility and purchased the

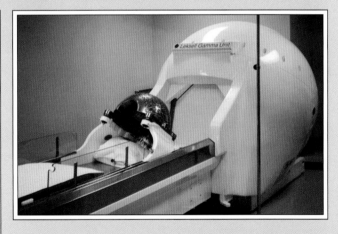

TOP: *The first Memphis class to train at the University of Pittsburgh in the use of the Gamma Knife, February 1995. Standing, left to right: Drs. Douglas Konziolka from the University of Pittsburgh, unknown visitor, Michael Muhlbauer, David Cunningham, Milo Solomito, Seated, left to right: Thomas Monaghan, D. Canale, and unknown visitor.* ABOVE: *An early version of the Gamma Knife. Note the collimator helmet attachment.*
OPPOSITE PAGE: *Dr. Kevin Foley teaching at MERI, 1996, where he is also medical director. Note the imaging machine in the background. In addition, he holds appointments at the University of Tennessee as professor of neurosurgery, in the School of Biomedical Engineering, and at Semmes-Murphey Neurologic & Spine Institute.*

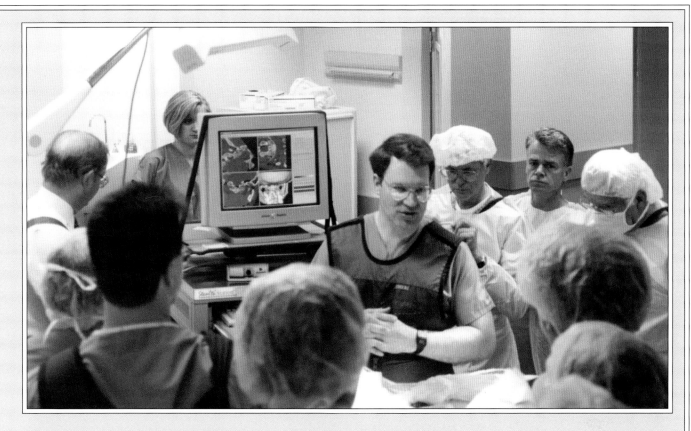

former Crosstown Post Office building at 44 S. Cleveland. The facility opened in 1995. Dr. Foley is medical director and chairman, and Dr. Elizabeth Ostric is executive director.

The only facility of its kind in the United States, MERI uses fresh-frozen, not embalmed, cadavers. Treating each specimen with the same respect given to living patients in an operating room, physicians from various specialties are trained in using new instruments or performing new procedures, particularly endoscopic ones. Originally the facility purchased specimens, but the Genesis Anatomical Donor Program at MERI reduced the need for purchases. Each year, under the direction of Shirley Woodard at Methodist Medical Education, more than 5,000 physicians from all 50 states and 26 foreign countries come to MERI for training. In 2008, the facility added a Simulation Center, which is used for training firefighters and other emergency personnel.

Founded jointly by Semmes-Murphey, Baptist, and Methodist, the facility also offers animal research. In 1998, MERI conducted 12 off-site seminars, including one in Palm Springs, California, for the American Association of Neurosurgeons. In 1999, the institute was named as the training site for a $26-million NIH study on carotid stents.

Minimally Invasive Spine Surgery and Biomechanical Innovations. The major force behind MERI, Dr. Kevin Foley, is known worldwide for neurosurgical innovation. He is a professor of neurosurgery at the University of Tennessee and holds additional appointments in the School of Biomedical Engineering and at the Uniformed Services University for the Health Sciences in Bethesda, Maryland. In addition, he is director of complex spine surgery at Semmes-Murphey and medical director for both MERI and the Image-Guided Surgery Research Center.

After receiving his M.D. degree from the University of California, Los Angeles (UCLA), Foley trained at Letterman Army Medical Center and at UCLA, after which he served in key positions in the U.S. Army Medical Corps at Brooke Army Medical Center, Tripler Army Medical Center, and Walter Reed Army Medical Center in Washington, D.C.

In 1992, Dr. Foley left the military for a career in academic and research medicine. In addition to conceptualizing and helping found MERI, he maintains an active practice in spine surgery while conducting research on minimally invasive spine surgery, image-guided spinal navigation, and spinal biomechanics. An author of many professional publications, he lectures frequently at scientific meetings and workshops at universities in the United States and abroad.

As a pioneer in minimally invasive spine surgery, Foley holds a number of patents on devices such as artificial cervical and lumbar discs. Noting that Dr. Foley's inventions have revolutionized spine surgery, a colleague commented that he has helped innumerable patients through new implants and instruments. His new techniques have lessened surgical trauma and have brought quicker recovery times and improved outcomes for patients.

CHAPTER FIFTEEN

ICONIC CHANGE, TRANSLATIONAL MEDICINE, AND ECONOMIC IMPACTS

2000-2010

THE NEW CENTURY BROUGHT ANXIETY—from fear of a worldwide computer crash, to the terrorist attacks of September 11, 2001, to wars in Afghanistan and Iraq, to the disintegration of the space shuttle *Columbia,* to the technology bubble and collapse of 2002, to the housing and market collapses of 2008. The turbulent first decade also brought scientific advances at the molecular level and new treatments for many diseases. The first vaccine for human papilloma virus (linked to cervical cancer) was developed in 2006, and surgery to correct refractive errors in vision became popular.

In Memphis, much research focused on basic science and "translating" discoveries into marketable drugs and devices. The 2000s also brought heightened awareness of how the medical community is intertwined with the economic well-being of the Mid-South, as healthcare costs increased, reimbursements dropped, and many were priced out of the health insurance market.

TennCare and Medicare

In 2005, TennCare2 began with another waiver from the Department of Health and Human Services. Patients ineligible for Medicaid were dropped from TennCare rolls. In Memphis, The MED, with twenty-five percent of its patients already uninsured, was hit hard, with the same number of patients needing care but fewer eligible for TennCare. Geographically, The MED was near a high number of uninsured patients who chose to go there. Trauma cases continued to come to The MED from adjoining states that contributed nothing for their care (a complaint registered regularly since 1898).

Further complicating the problem was patients' delay in seeking medical attention until their illnesses were serious,

An architectural rendering of the UT-Baptist Research Park, 2005. Currently being constructed on the site of the old Baptist Memorial Hospital in midtown, it will be part of the future of medicine in Memphis.

and often multiple illnesses were involved. The MED regularly saw the "pathologies of poverty"—i.e., obesity, diabetes, high blood pressure, and congestive heart failure.

In 2005, TennCare cuts shifted The MED's patient profile. Uninsured patients increased twenty-two percent, and often they came from middle-income areas—East Memphis, Fox Meadows, and Hickory Hill. Many reported lack of access to primary care or prescribed medications. More of the patients were sicker than before the cuts and needed expensive care.

As the roll of eligible patients was trimmed, the number of managed care companies decreased. Then, in 2008, the state capped the amount allocated for the program and put management companies at risk, making the companies responsible for costs exceeding the cap. By 2010, The MED's financial woes had led to the closing of some clinics and other cost-cutting measures.

In 1998, Medicare changed further. The Omnibus and Reconciliation Act of 1998 capped annual Medicare spending and required revenue-neutral provider reimbursement. Caps based on a new calculation, the sustainable growth rate, were used to hold down payments for physicians' services and services "incident to" physicians' charges, including laboratory tests, imaging, and physician-administered drugs. Throughout the decade, caps were regularly exceeded, and Congress threatened to reduce reimbursement rates that were already below those of commercial insurers. Each year, organized medicine lobbied for a reprieve. Physicians were in an increasingly difficult position and, by 2009, the

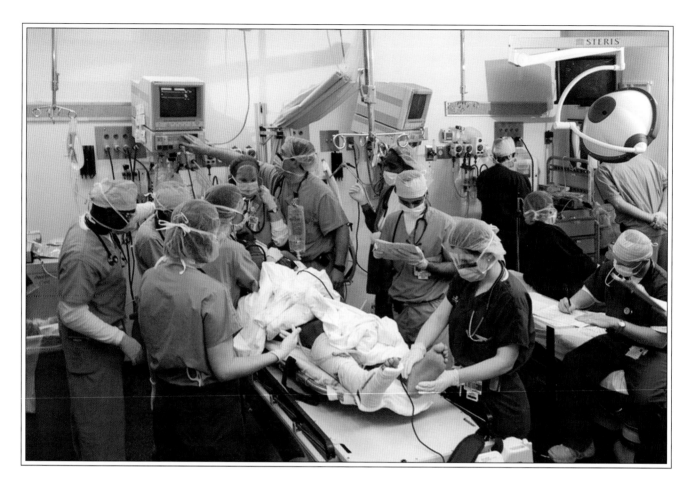

cost of obtaining adequate reimbursement had taken its toll. Offices spent more time verifying insurance, battling denials, and negotiating delayed payments. Physicians had necessarily joined the business and healthcare industries, with much patient contact time replaced by reimbursement-based administrative tasks.

When Medicare added drug benefits in 2006 (Medicare Part D), the bill produced changes for practices delivering cancer chemotherapy or anti-inflammatory medications. Outpatient rates for infusions were reduced, and more patients returned to hospitals for treatment.

Healthcare Delivery Changes

Delivery of health care continued to shift as old Memphis facilities closed, new facilities were built, and regional facilities expanded. The first major event in Memphis medicine came in 2000, with the closing of Baptist Hospital's flagship facility in the Medical Center. A behemoth of a building, the hospital staffed about 500 of its 2,000-bed capacity, and utility costs were staggering. Much delivery of medical care had shifted to outpatient facilities. After proposing a union with the University of Tennessee that failed, decision makers at Baptist committed to its east location. In the Medical Center, Baptist Memorial Hospital, whose first building dated to 1912, was closed in 2000. At 6:45 a.m. on November 6, 2005, in an iconic moment for Memphis, the huge building, once the site of the nation's largest nongovernment hospital by number of admissions, was imploded. Watchers included Memphis dignitaries, individuals who had been born there, physicians who had practiced there, and citizens who remembered the building as a landmark. Prominent general surgeon Dr. Eugene R. Nobles, who practiced at Baptist from 1961 to 1996 and was formerly chair of surgery and of the medical staff, remembered the eagerness of patients to stay in the Madison East tower when it opened. "It was a great place to practice," he said. "The hospital provided the doctors everything they needed."

Turning a potential negative into a positive, Baptist's administrators donated the property to the University of Tennessee for a research park to be named the "UT-Baptist Research Park." After the implosion, most of the land became the site of a planned bioscience center under the auspices of a nonprofit organization, the Memphis Bioworks Foundation®, devoted to research, education, entrepreneurial development, and expansion of bioscience. The foundation developed a ten-year master business plan calling for the UT-Baptist Research Park to comprise a six-

OPPOSITE PAGE: *In that the Elvis Presley Memorial Trauma Center is the only level 1 facility of its kind within a 150-mile radius of Memphis, it is understandable that it is frequently used by non-Tennessee residents. In this image, c. 2007, 14 members of the Trauma team are working on a patient.*
ABOVE: *Implosion of Baptist Memorial Hospital, November 6, 2005. The event sounded the death knell for the largest private hospital in the country, but saw the birth of a major biotechnology research facility, the UT-Baptist Research Park.* RIGHT: *Baptist Memorial Hospital East. Originally built to serve an area with the largest growth potential in Memphis, it became Baptist's flagship facility in 2000.*

story, 165,000-square-foot biotechnology research facility, five modern buildings for laboratories and offices, a regional biocontainment laboratory (RBL), and an incubation center to develop new biotechnology businesses and spur job creation. BioWorks owns the donated Baptist property, with a piece given to the University of Tennessee for the RBL and the new Pharmacy School building, plus the 910 and 930 Madison buildings. Both the pharmacy building and the RBL were nearing completion in 2010.

Memphis's hospitals had grown with sometimes fierce competition, but when Baptist East became the flagship Baptist Memorial Hospital Memphis in 2000, leaders focused on Baptist's four greatest strengths: cardiac care, women's care, cancer care, and orthopedics. In 2000, Baptist

DRS. H. EDWARD GARRETT, SR., AND H. EDWARD GARRETT, JR.

Dr. H. Edward Garret, Sr., performed the first coronary artery bypass graft in 1964, shortly before moving to Memphis to join Dr. Hector Howard. Before his death in 1996, Dr. Garrett, Sr., and his son worked together as thoracic surgeons.

Having witnessed surgery at Baptist Hospital when young, Dr. Garrett, Jr., felt compelled to be a surgeon. With his M.D. degree from Vanderbilt University and surgical internship, general surgery residency, and cardiothoracic surgery residency at Washington University-Barnes Hospital, St. Louis, he took a peripheral vascular fellowship at the University of Tennessee. He followed in the innovative tradition of his father by performing the Mid-South's first heart-lung and double-lung transplants at Baptist.

The mid-1980s development of cyclosporin (which suppressed the immune system and made transplantation feasible) is, for Garrett, the most important development in transplant surgery. For him, heart transplants, involving only four anastomoses, are not technically difficult; the key to successful surgery is selecting both donor and recipient and getting the recipient to commit to lifelong follow-up. A heart transplant takes about two hours once the donor organ arrives, but with the time necessary to secure a donor heart, the whole procedure may require six hours. The time elapsed from the moment the donor organ is removed until blood flow is established in the recipient must be no longer than four hours. Usually, two transplant teams are involved—one to retrieve the organ and one to implant it. When the donor is in the same hospital, less time is required.

In 2009, Dr. Garrett and Dr. Russell Carter were the two surgeons doing the actual transplants, but others were involved in harvesting. Organs for transplant often become available in the middle of the night, and the transplant teams, including residents, had to be available. Transplant teams include not only surgeons, but infectious disease consultants, a physician knowledgeable about immunosuppression, and a psychologist to help in selecting the recipient.

Also crucial in enabling transplantation, says Dr. Garrett, was the heart-lung bypass machine, which made the cardiac bypass graft (CABG) a common procedure. Another major

also restructured and relicensed its Germantown facility as a musculoskeletal hospital and surgery center, offering sports medicine, occupational medicine, workers' compensation, rehabilitation, and diagnostic imaging. In 2001, Baptist opened its Women's Hospital at the east complex, at that time the only women's hospital in Memphis, and announced plans to spend $175 million on its DeSoto County facility. Dr. Lynn W. Gayden had been medical director of the Women's Health Center at Baptist East since 1987 (the center itself was a concept rather than one structure). In 2001, the newly established Baptist Comprehensive Breast Center opened, also directed by Dr. Gayden. Dr. Stephen G. Portera became medical director for the Center for Urinary and Pelvic Disorders in the Women's Hospital. Perinatal specialist Dr. Thomas N. Tabb became program director for the Maternal Fetal Diagnosis Center at the Women's Hospital.

In 2001, the Baptist Heart Institute opened. Led by Dr. H. Edward Garrett, Jr., it offered a full continuum of heart care on the Baptist campus. Dr. David Wolford became medical director of the Cardiac Catheterization Laboratory and performed the nation's first radial brachytherapy, using radiation to prevent scar tissue from clogging the coronary artery after balloon angioplasty or stent placement (arteries other than the radial could be used). Building on Baptist's history of affiliation with the University of Tennessee and a flourishing heart surgery program, surgeons at Baptist, led by Dr. Garrett, performed Memphis's two hundredth heart transplant in 2003. Heart surgeons in Dr. Garrett's group originally included Drs. O. Brewster Harrington and Glenn Crosby, later joined by Drs. Rodney and Brad Wolfe and Russell Carter. Other cardiovascular surgeons in Memphis included Drs. Jerry B. Gooch; Jeffrey B. Gibson; Timothy J.

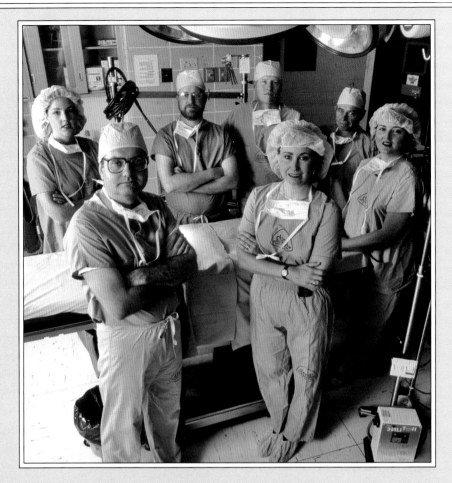

Dr. Edward Garrett Jr., front left, with his transplant team at Baptist Memorial Hospital.

development was the medical therapy of the 1990s and the 2000s, which has treated coronary disease so effectively that fewer CABG surgeries are required.

For the future, Dr. Garrett notes two bright areas: improved heart replacement devices for congestive heart failure (including whole heart replacement) and treatments for atrial fibrillation (AF). A common, usually non-life-threatening condition, atrial fibrillation affects eight million individuals in the United States, limiting quality of life and sometimes leading to clotting that causes strokes. Electrophysiological conversion to a normal rhythm is not always successful. Two decades ago, surgical correction used an open technique. By 2009, the technique was minimally invasive and highly successful.

The effect of robot-assisted surgery, Dr. Garrett says, is "to miniaturize your hand." Three years after he began using the system, he reported performing 60 mitral valve repairs, 75 lung-cancer surgeries, and seven aortic bypasses with the system.

The Garrett family donated funds to Baptist to construct a 282-seat auditorium and five classrooms on the Baptist Memorial Hospital Memphis campus. Named for his father, Dr. H. Edward Garrett, Sr., the facility opened in 2009. At the dedication, Baptist president and chief executive officer Stephen C. Reynolds said, "Dr. Garrett, Jr., has a servant's heart, and many in the Mid-South are living, breathing testaments of his devotion to the community."

Powell; Samuel G. Robbins, Jr.; Alim Khandekar; and Glenn P. Schoettle, Jr.

In 2003, Baptist introduced the first Da Vinci® surgical system, in which the surgeon, sitting at a console, looks through two eyepieces at a 3-D image of the procedure and, in real time, maneuvers three robot arms with two foot pedals and two hand controllers, which perform surgical maneuvers. The Da Vinci Robotic System was developed by the National Aeronautics and Space Administration (NASA) for use on aircraft carriers or space flights, with remote interventions from an on-ground base (hence *tele*-robotics). It began when ophthalmologist Dr. Steve T. Charles, of MicroDexterity Systems in Memphis, collaborated with a team at NASA's Jet Propulsion Laboratory in Pasadena, California, to meet the challenge of extremely accurate and dextrous robotics for laser retinal surgery.

At Baptist, surgeons first learned to do prostatectomies on the Da Vinci. Among these surgeons were Drs. Richard M. Pearson, who performed the first such surgery, Lynn W. Conrad, and H. Michael McSwain. Methodist soon added two Da Vinci systems, one at Methodist University Hospital (MUH) and the other at Methodist North. Urologists Dr. Ravi D. Charles and Dr. Robert A. Donato performed a robot-assisted prostatectomy at Methodist North for an educational videotape.

In 2004, Baptist became one of the first hospitals in the country to use a new treatment for abnormal heart rhythm and other problems. Stereotaxis® used magnets to guide catheters through arteries, allowing physicians in the Heart Institute to access hard-to-reach areas of the heart accurately and quickly. Dr. Eric Johnson of Memphis participated in the initial trial of the system.

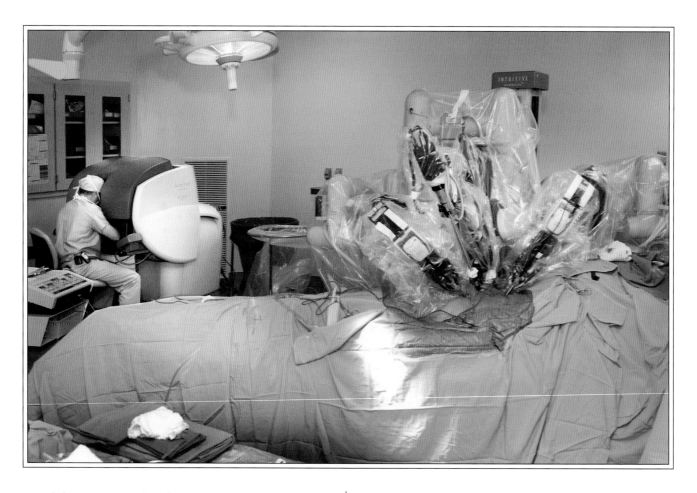

The Da Vinci Surgical System is robotic surgery with a minimally invasive approach. The surgeon sits at a console, as seen in the photograph, and is given a 3-D view of the operative environment.

While Baptist was building on its historical strengths, institutions in the Medical Center established the University Medical Center Coordinating Council (UMCC), whose goal was to support and improve patient safety, quality of care, and efficiency in all institutions remaining downtown. Dr. Robert Waller, University of Tennessee graduate and former chief executive officer of the Mayo Clinic Foundation, chaired the effort, which brought together Methodist, the University of Tennessee Health Science Center, Bowld, The MED, the VA Medical Center, St. Jude, and the Memphis Business Group on Health. The UMCC joined a quality network designated IMPACT, headquartered in Boston and comprising 56 organizations.

Led by CEO Gary Shorb, the Methodist system, which had incorporated in 1982 as Methodist Healthcare, began its own transformation. Emphasizing a central tenet of Methodism, healing the body as well as the soul (the hospital retains its affiliation with the United Methodist Church), Methodist Hospital was firm in its decision to be the only private tertiary care hospital in Midtown. In 2002, Methodist Central assumed management of the William F. Bowld Hospital, part of the University of Tennessee system, and took the name Methodist University Hospital. In 2004, the University's Transplant Program, directed by Dr. Osama Gaber of the Department of Surgery, moved to Methodist University Hospital (MUH). Later, to streamline administration between the University of Tennessee and MUH, Dr. Pat Wall, then dean of the College of Medicine, recommended an institute structure in which the director of the transplant program reported to the dean of the College of Medicine. Thus began the Methodist University Hospital Abdominal Transplant Institute, with an interdisciplinary team of surgeons, hepatologists, nephrologists, and anesthesiologists. Kidney dialysis, noninvasive cardiology, and nuclear medicine programs moved from Bowld to The MED.

Methodist University focused on four areas of established strength: organ transplants, neuroscience, cardiology, and pediatrics. Leaders at Methodist reiterated their promise "to support all of the citizens of this area including those in the urban areas of Memphis."

The Abdominal Transplant Program, directed by Dr. Osama Gaber, together with his prominent colleague Dr. Santiago Vera, emphasized liver, pancreas, and kidney transplants. With four residency positions each year in transplant surgery, Methodist's Institute performed an impressive 238 liver transplants from 2006 through 2008.

Because patients are very sick when they become eligible for a transplant, they often have clotting and platelet problems that make liver transplants difficult. With experience, however, what once took as long as 17 hours has been shortened to five or six hours. Two teams rotate on-call duties, but both teams are needed if a kidney is also available. After preparing the patient, the team removes the diseased liver once the donor organ arrives in ice in a plastic picnic cooler. Meanwhile, the new liver is prepared. Because the liver processes a large volume of blood, a cannula redirects blood flow while the donor organ is sewn in place, and the team watches the organ turn pink as blood flow begins.

Under Dr. Gaber's leadership, the transplant program received $1.7 million in 2001 to establish one of ten islet cell resource centers in the nation. The institute also supported studies of organ rejection, immunosuppressant drugs, and quality of life in transplant recipients. Researchers devised an ever-improving model of post-transplant care to share with other centers.

When Dr. Gaber moved to Houston in 2006, a three-way search by Methodist, UT Medical Group, and the University of Tennessee College of Medicine recruited Dr. James D. Eason, a third-generation University medical graduate who spent ten years in the Air Force and later was chief of the Transplant Institute at Ochsner Clinic in New Orleans. By 2007, the MUH Transplant Institute was one of only 13 centers in the nation to perform more than 100 kidney and 100 liver transplants in one year. By 2009, the institute had performed its one thousandth liver transplant, with a one-year survival rate for liver transplants of 87.57 percent and a one-year survival for kidney transplants of 89.71 percent. Dr. Eason and his team have helped place Methodist University Hospital among the top ten transplant centers in the United States. The choice of Steve Jobs of Apple Computers to have his liver transplant at Methodist University in 2009 reflects the high standards and growing stature of the Methodist University Institute.

In 2006, Dr. Satheesh Nair, Dr. Eason's colleague at Ochsner, became director of hepatology, completing the transplant team. Dr. Timothy Fabian remained chair of the

Top: *Dr. Osama Gaber, second from the right, c. 2004. He was director of the University of Tennessee Transplant Program before moving to Houston in 2006. Dr. Gaber performed the first segmental liver and first pancreatic islet cell transplants in Memphis.* Above: *Dr. James D. Eason, director of the Transplant Institute, is flanked on the left by Dr. N. Nezakatgoo, and on the right by Dr. Santiago Vera.*

University's Department of Surgery, as well as chair of the Division of General Surgery.

As Methodist's transplant program developed, Dr. Christian Gilbert came to Le Bonheur in 1995 as a pediatric heart transplant specialist and associate director of the International Children's Heart Foundation. With Dr. Gilbert's

"I Can Fly," the beautiful mosaic obelisk in the lobby of the new Le Bonheur Children's Hospital by Jeanne Seagle and Lea Holland, 2010. A child sitting on a bluebird is supported by a delightful multicolored design with the words "love," "children, "faith," and "hope" at its base.

Neuroscience was the second area of emphasis for Methodist University Hospital. The Neuroscience Institute, with its Gamma Knife facility, expanded in 2009 with the formation of its Virtual Brain Tumor Board, which makes comprehensive studies of brain tumor cases available via the Internet. Serving on the faculty panel of the board are noted specialists Drs. Michael R. Farmer; Holger Gieschen; Raja Khan; Robert Laster; Ronald Lawson; L. Madison Michael, II; Thomas O'Brien; Jon H. Robertson; Jeffrey M. Sorenson; Clarence Watridge; Jason Weaver; and Alva B. Weir, III.

For its third area of emphasis, oncology, MUH recruited Dr. Dava Gerard in 2007 as administrative director of the Cancer Center. She brought extensive administrative experience from the West Coast to Memphis, plus a determination to inspire cooperation among physician groups involved in cancer care.

Furthering Methodist's emphasis on pediatrics, a new Le Bonheur Children's Hospital was begun. The Memphis Mental Health Institute (MMHI), located at 865 Poplar since 1962, was demolished in November 2007 and was the site of the new Le Bonheur hospital. A 75-bed MMHI was constructed and officially opened on September 27, 2007, on the site of the University of Tennessee's William F. Bowld Hospital (demolished in 2004 because its configuration for half teaching hospital and half research was no longer economically feasible).

arrival, the neonatal ICU at Methodist closed and a 15-bed Neonatal ICU opened at Le Bonheur.

Methodist University Hospital opened a bone marrow transplant unit under the direction of Drs. Furhan Yunus and Donna Prezpriorka. The first adult allogenic transplant (from a donor) in Memphis was completed at Methodist in 2005. Dr. Furhan Yunus, medical director of the Blood and Marrow Transplant Center, completed Memphis's first umbilical cord blood transplant to an adult at Methodist University in 2006.

The third major hospital in Memphis, St. Francis Hospital, retained its location at 5959 Park Avenue in East Memphis. In 2004, St. Francis-Bartlett opened as the first full-service hospital in that community, and St. Francis Healthcare's umbrella system was born. It includes the Memphis and Bartlett hospitals, five St. Francis Sports Medicine and Rehabilitation centers in the Mid-South, the Surgery Center at St. Francis, the St. Francis Nursing Home, and the St. Francis Senior Healthcare Center.

In 2009, St. Francis-Memphis offered advanced endoscopic retrograde cholangiopancreatography (ERCP), which optically visualizes problems in the liver, gallbladder, and common bile duct, to enable efficient diagnosis of obstructions, suspected cancers, and cystic lesions, according to Dr. Mike Lachina.

St. Francis Healthcare continued to offer comprehensive services, including the area's first Chest Pain Emergency Center (opened in 1988 under the leadership of Dr. B. T. Harris). Both the Emergency Department and the Chest Pain Emergency Center were expanded in 2008 to include a Stroke Emergency Center designed to assess patients quickly and restore function when possible. St. Francis offered other centers, including a Bariatric Services Center (surgical weight loss) directed by Dr. Virginia Weaver.

INTERNATIONAL MEDICAL GRADUATES AND STUDENTS

Predicting a shortage of physicians by 2010, especially in primary care, Dr. Stephen T. Miller, senior vice president for research and education at Methodist University Hospital, noted that aging baby boomers need more care than younger people, often from primary care physicians. Although the need in Memphis varies with location, it is growing.

At the beginning of the twenty-first century, primary care specialty training programs increasingly depended on international graduates, with approximately twenty-five percent of overall residencies in the United States and in Memphis held by physicians who received medical degrees abroad. They have brought to this country new insights, inventiveness, a strong work ethic, and a determination to be excellent physicians.

Successful immigration of international physicians actually began in Memphis in the 1960s, when an exodus from Cuba occurred as the educated and the affluent left Fidel Castro's dictatorship. Many, including some of the best physicians who left Cuba, came to Memphis. Dr. Ricardo J. Fuste became a pioneer in therapeutic radiation after arriving in Memphis in 1967.

Dr Raza Dilawari, from Pakistan, became an oncologic surgeon and a professor of surgery at the University of Tennessee. He helped pave the way for other physicians from Pakistan to come here.

Dr Tulio Bertorini, from Peru, became professor of neurology at the University of Tennessee, and started one of the leading specialty groups in the city.

Dr. Sergio Acchiardo, a nephrologist from Santiago, Chile, worked with kidney dialysis at the University and became chief of its Nephrology Division.

Dr. M. Bashar A. Shala received his M.D. degree from the University of Aleppo in Syria and was board certified in internal medicine, with subspecialties in cardiovascular diseases, interventional cardiology, nuclear cardiology, and echocardiography. He completed a cardiology residency with Dr. K. B. Ramanathan at the University of Tennessee.

Dr. David Z. Lan, an electrophysiology specialist at Memphis Heart Clinic, came here from Beijing, specializing in atrial fibrillation ablation.

Aided by friends, Memphis neurosurgeon Dr. Fereidoon Parsioon overcame great difficulties to leave Iran with his family for fellowship training in Memphis.

In addition to international graduates, Memphis also attracted immigrants who became distinguished physicians. Coming to the United States from Cuba in 1960, Dr. Roberto C. Heros became a U.S. paratrooper and was held captive for two years after the Bay of Pigs invasion. Afterward, he studied at the University of Tennessee, leading his class in 1968. Extensively published, he has been president of the American Academy of Neurological Surgeons and the American Association of Neurological Surgery.

Dr. M. Bashar A. Shala is a cardiologist who received his medical training in Syria.

ABOVE: *Born in Gutatemala, Dr. Antonio Cabrera received his medical degree from the Universidad Francisco Marroquin Facultad de Medicina Guatemala and is currently the medical director of Le Bonheur's Cardiovascular ICU.* RIGHT: *Dr. Christopher J. Knott-Craig, chief of the University of Tennessee Division of Pediatric Cardiothoracic Surgery, was given an audience with Pope John Paul II in recognition of his humanitarian service in cardiac surgery.*

Dr. Antonio Cabrera, medical director of the Cardiovascular ICU at Le Bonheur, came from Guatemala for fellowships in critical care cardiology at Children's Hospital in Cleveland, Ohio, and at Arkansas Children's Hospital in Little Rock. He was drawn to Memphis by the need to improve high infant cardiac mortality in the Delta area.

Distinguished international physicians include Russian Dr. Michael Gelfand, an infectious disease specialist who received his M.D. degree from the University of Texas Southwestern Medical School, completed his residency at Vanderbilt, and is currently director of internal medicine at the University of Tennessee. In 2008, Dr. Christopher J. Knott-Craig, whose medical degree and residencies were from Groote Schuur Hospital, University of Cape Town, South Africa, left a similar position at the University of Alabama Medical Center to become chief of the Division of Pediatric Cardiothoracic Surgery at the University of Tennessee and codirector, with Dr. Thomas Chin, of the Pediatric Cardiovascular Institute at Le Bonheur. Dr. Harold Sacks, an internationally known South African endocrinologist, came to the University in the 1970s.

In addition, Memphis has produced two "celebrity" physicians trained in the United States, but with strong international connections. Neurosurgeon Dr. Sanjay Gupta, who completed neurosurgical training at Semmes-Murphey with Dr. Kevin Foley, became affiliated with Grady Memorial Hospital and Emory University in Atlanta. Dr. Gupta has published scholarly and popular articles, and he was an embedded correspondent with American forces in Iraq in 2003. By 2010, he regularly appeared on CNN.

Another Memphis physician, University of Tennessee infectious disease specialist Dr. Manoj Jain, who came to the United States at the age of ten, has degrees in biomedical engineering, public health, and medicine from Boston University. Jain is a leader in quality care improvement, chair of Healthy Memphis Common Table nutritional group, and a regular writer for *The Commercial Appeal*, *The Washington Post*, and *The New York Times.* He travels frequently to India to collaborate with colleagues.

Dr. Manoj Jain is an expert on infectious diseases. He has been a consultant on HIV to the World Bank and has conducted research on HIV epidemiology.

In 2008, Congress passed a bill to help attract physicians to underserved areas, including the rural Mid-South and pockets of poverty in Memphis. Before this law, physicians with American visas for residency training had to return home for two years before reentering the United States and applying for permanent residence. Under the 2008 law, foreign doctors who completed five years in underserved areas could obtain green cards for permanent United States residence.

Stalwarts of Primary Care

In Memphis, as the need for primary care physicians has grown, a core of physicians have carried on the tradition of personal care, including family practitioners Drs. John Avergis; Laurie M. Baker; Oran L. Berkenstock; Tina K. Burns; Cary Finn; Preston G. Givens; Robert C. McEwan; Michele E. Neal; Trung T. Nguyen; Ronald L. Terhune; Graham Warr; Jeffrey S. Warren; Charles J. Woodall; and Carl T. Younger. Internal medicine specialists also served as primary care physicians, including Drs. Eric D. Blakney; Felix L. Caldwell, II; T. Kyle Creson, Jr.; John D. Fleenor; Julian G. Fleming; Nicholas Gotten, Jr.; Jeffery N. Hoover; Oakley Jordan; Robert and Gordon Kraus; Mike Lemond; Kerry Milligan; Carol J. Mitchell; Chuck Munn; Frederick Pelz; Robert L. Richardson, III; Barry R. Siegel; Vincent D. Smith; Sam Veroza; Leonidas N. Vieron; Martin J. and William T. Weiss; Lawrence Whitlock; and Mark S. Young.

Drs. John Conway, Hall Tacket, Barney Witherington, Otis Warr, Jr., Walter Hoffman, and Eugene Spiotta, Sr., were there at the beginning of the new age in the 1960s and 1970s, but probably none of them could have predicted what was to come.

Memphis Specialty Practices

Allergy. At a conference on interdisciplinary medicine in 2003, the National Institute of Allergy and Infectious Diseases focused on the growing prevalence of allergic diseases and their effect on quality of life. The conference evaluated treatment options for allergic diseases, and addressed the clinical implications of allergic reactions, including quality of life and patient compliance. Dr. Michael S. Blaiss of the University of Tennessee was a member of the faculty at the conference.

Dr. B. Manrin Rains, III, was a pioneer in treating fungal sinusitis, which occurs with particular frequency in the Southeast. Director of the Mid-South Sinus Center, Rains performed the first endoscopic sinus surgery in Memphis in 1985 and has made chronic sinusitis the focus of his career. Once considered an allergic condition, fungal sinusitis has been reclassified as an overreaction to the fungus, with frequent secondary bacterial infections. Because this overreaction causes polyps and chronic inflammation, patients sometimes require multiple surgeries. Dr. Rains devised a new protocol using itraconazole to treat the fungal infection. After surgery to remove the fungus, Dr. Rains administers oral itraconazole to kill the fungus, followed by nasal steroid sprays. By 2010, he had refined his protocol to bring ongoing relief to patients.

Cardiology. By 2000, sudden death from a "heart attack" in an otherwise healthy person was less common than it had been a decade earlier. New technologies screened for atherosclerotic heart disease, and invasive cardiologists could dilate and place stents in occluded coronary arteries, changing the outlook for patients. As patients lived longer with a better quality of life, they required ongoing care that increased the need for cardiologists.

Although some of Dr. Neuton Stern's colleagues in the 1930s doubted EKG technology, contemporary heart care

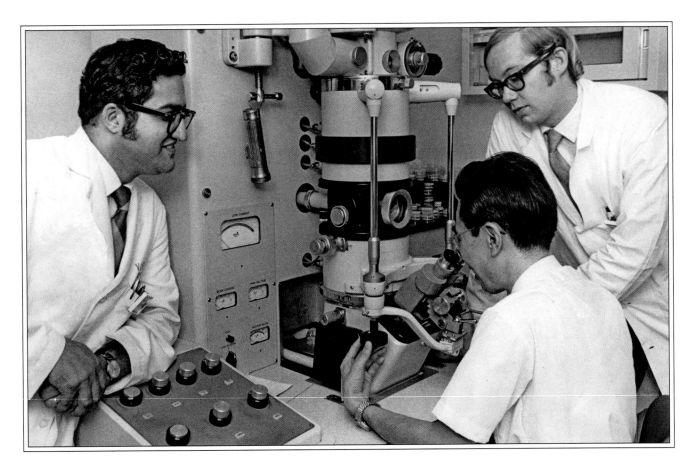

ABOVE: *Drs. Abbas Kitabchi, left, and William C. Duckworth, right, watch Dr. Ken Hashimoto at the electron microscope, 1970.*

LEFT: *Dr. B. Manrin Rains, III, Director of the Mid-South Sinus Center, is also on the staff of Methodist. The Methodist Healthcare Foundation funded his first two studies on the application of itraconazole in the treatment of chronic sinusitis.*

requires technology. The larger heart clinics in Memphis—Stern, Sutherland, and Memphis Heart clinics—expanded throughout the 2000s and opened freestanding facilities offering a multitude of services.

Dr. Dwight W. Clark, Jr., of Cardiovascular Specialists, was the first cardiologist to move cardiac catheterization from the hospital to the outpatient clinic, which he accomplished by contracting with Methodist to provide the equipment and the certificate of need. At first, some hospitals argued that such centers were unsafe for patients, but with careful selection of patients, Dr. Clark proved that most patients could safely have outpatient testing.

The Sutherland Clinic, begun by Drs. Arthur Sutherland, Jack Hopkins, and Frank St. Clair, added Drs. Joe Samaha, David Kraus, and Jeff Kerlan during the 2000s.

Endocrinology. Dr. Abbas Kitabchi, holder of the M. K. Callison Professorship in Medicine at the University of Tennessee, is a leading endocrinologist. His landmark studies have led to the nationally used protocol for treating diabetes with low-dose insulin. With Dr. Samuel Dagogo-Jack, he explored the effects of controlling diabetes on small- and large-vessel cardiovascular disease. Currently he studies how weight loss and lifestyle changes affect cardiovascular health. He was instrumental in establishing a program for Iranian students to study in the United States.

Gastroenterology. Dr. Patrick J. Dean heads GI Pathology Partners, which has had a training program with three

ABOVE: *Thoracic surgeon Dr. F. Hammond Cole, Jr., is a past president of The Memphis Medical Society and is currently professor of surgery at the University of Tennessee.* RIGHT: *When Dr. Christine T. Mroz, surgeon, finished her residency at the Mayo Clinic in 1978, she was only the third woman to complete the program in surgery.*

fellowship positions in Memphis since 2006. Dr. Lawrence Wruble received the University of Tennessee's Outstanding Alumnus award in 2005. Other gastrointestinal specialists in the 2000s include Drs. Isaac Jalfon; Paul Bierman; Myron Lewis; Kenneth Fields; Gerald Lieberman; Terrence L. Jackson; Wesley E. Jones; Mary C. Portis; David D. Sloas; Eugene J. Spiotta, Jr.; W. Zachary Taylor; Daniel E. Griffin; William G. Hardin; Raif W. Elsakr; Michael S. Dragutsky; James J. Upshaw; and Robert S. Wooten.

Oncology. In 2006, the University of Tennessee and St. Jude both became members of the National Association of Cancer Centers. In 2010, Memphis has specialists in every facet of oncology. Dr. F. Hammond Cole, Jr., initially started practice in the 1970s with his father, distinguished chest surgeon Francis H. Cole, Sr., who built a reputation for both integrity and excellence. The two of them provided expertise in lung cancer surgery for more than 60 years. Dr. Steve Behrman at the University of Tennessee focused on cancer of the hepatobiliary tract.

The treatment of breast cancer changed dramatically with new knowledge of tumor biology. All three major hospitals opened breast centers to coordinate diagnosis and treatment, and some surgeons specialized exclusively in the diagnosis and surgery of the breast. Dr. Christine Mroz began specializing in breast care in 1995. Dr. Elizabeth Pritchard at the University of Tennessee emphasized breast cancer. In the 2000s, Dr. Michael Berry and Dr. Susan Hoover came to Memphis as fellowship-trained (University of Texas) surgical oncologists.

The West Clinic, which was opened in 1979. was directed in the 2000s by Drs. Lee Schwartzberg and Kurt Tauer. It was one of the first clinics to add advanced diagnostic equipment in the outpatient setting. In 2001, the University of Tennessee Cancer Institute/Boston Baskin Group was formed as the medical oncology division for the Department of Internal Medicine at the University. Drs. Kirby L. Smith and Peter W. Carter led the Mid-South Cancer Center. Dr. C. Michael Jones established the Jones Clinic, with Drs. Brent A. Mullins and Aleksandar Jankov. Drs. A. Earle Weeks and Don Graevnor formed the Family Cancer Center. All provide the latest in cancer therapy.

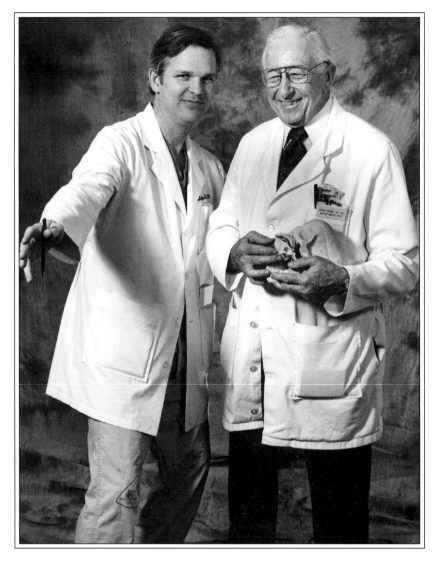

Drs. Jon Robertson, left, and Edwin W. Cocke, Jr., at the time when they developed a new technique to remove brain tumors. Dr. Robertson is professor of surgery at the University of Tennessee, on the staff of Semmes-Murphey Neurologic & Spine Institute, and president-elect of the American Association of Neurological Surgeons. Dr. Cocke is a head and neck surgeon and otolaryngologist. His father, a respected psychiatrist, was head of Western Mental Health Institute, Bolivar, from 1918 to 1933, and author of a 1919 Tennessee law about legal issues in the treatment of psychiatric patients.

Orthopedic specialists, including Dr. Michael Neal with OrthoMemphis and Dr. Robert Heck with Campbell Clinic, regularly were able to remove cancer but salvage patients' limbs.

The first multidisciplinary clinic in head and neck cancer was created at Methodist University Hospital. Each patient was seen by Dr. Sandeep Samant, a head and neck surgeon at the University of Tennessee, along with a medical oncologist and a radiation therapist in the same clinic. After each clinic, the cases were discussed and a treatment plan was developed.

Drs. Alan Sills and Jon Robertson were leaders in the effort against brain cancer and helped establish a brain tumor tissue bank at Methodist University Hospital.

Conferences in Memphis bring together neurosurgeons, particularly those from neurooncology at Johns Hopkins, to make the newest treatments available through Methodist.

By 2010, Memphis physicians stressed early detection, often leading to less invasive treatment and increased chances of cure. Cancer-risk genetic testing and counseling centers were established at the major hospitals to advise patients concerned about risk because of work, lifestyle, and cancer cases of blood relatives. The Trumble and Duckworth Labs performed genetic testing and helped medical oncologists tailor treatment on the basis of a tumor's biology.

Ophthalmology. In the mid-1990s, the National Eye Bank Center, established by Tissue Banks International (TBI), opened its state-of-the-art facility in Memphis, where its proximity to FedEx and to the medical and biotechnology communities enhances access to the country's largest collection of ocular tissues for corneal transplant and other ocular surgery. Tissue collected by eye banks from multiple locations undergoes processing, with stringent requirements for safety and quality, plus quick distribution to surgeons around the world. Fifty percent of the funds needed for the center were raised by companies including the Plough Foundation in Memphis. In 2007, Carl Zeiss Meditec AG donated the first OPMI® VISU 140 (the same microscope used in operating rooms) to the center. The first such machine in an eye bank setting, it helps with tissue evaluation and preparation. Dr. Thomas O. Wood commented, "More precise graft preparation will enhance the quality of TBI corneas, and allow for improved outcomes for transplant recipients."

In 2010, the Mid South Eye Bank was the second-oldest continuously operated eye bank in the country. Other eye

UNIVERSITY OF TENNESSEE HAMILTON EYE INSTITUTE

Situated on property donated by Baptist in 2001, the University's Hamilton Eye Institute (HEI) was inspired by the vision of its first benefactor and namesake, Dr. Ralph S. Hamilton, who trained at Wills Eye Hospital in Philadelphia. After the first endowment from Dr. Hamilton and his wife, Barbara, gifts followed from the Hyde Family Foundation, George Cates of Mid-America Apartments, and others.

Coming to the University of Tennessee as department head in 1995, Dr. Barrett Haik brought a dynamic personality, wide experience in building programs, and a stellar research career. Known internationally for his expertise in diagnosing and managing ophthalmic tumors, he specialized in evaluating diagnostic imaging techniques and interactive computer technology. He was widely known as an educator and received an Honor Award from the American Academy of Ophthalmology. Dr. Haik also participated widely in medical and scientific organizations. He was president of the Association of University Professors of Ophthalmology in 2009 and the American Eye Study Club in 1994.

The HEI treats and manages ocular diseases and complex problems with vision, ranking among the top ten programs in the nation for delivery of care. Opened in three phases from 2004 to 2007, the facility includes laboratory space, examination rooms, space for training and education, and an ambulatory ophthalmic surgery center operated by Baptist and Methodist, plus floors for pediatric and adult vision. The Hyde Foundation funded a glaucoma center directed by ranking glaucoma specialist Dr. Peter Netland (who left in 2009 to assume the chairmanship of ophthalmology at the University of Virginia). Its high-tech, fiberoptically wired Freeman Auditorium, named for noted ophthalmologist Dr. Jerre M. Freeman, can transmit information worldwide. Eight surgical work stations, similar to those at the Medical Education and Research Institute, allow physicians to practice new skills.

At the dedication of the University of Tennessee Hamilton Eye Institute in 2008 were, standing left to right, Barbara Hamilton, Dr. Ralph S. Hamilton, Dr. Barrett G. Haik, Dr. Herschel "Pat" Wall, and Joseph E. Johnson.

Locally, HEI and St. Jude collaborate, bringing together specialists to design effective treatment plans. Their partnership has other centers in Panama (primarily for retinoblastoma), Guatemala, Honduras, and Jordan.

Early in 2009, a joint biomedical project of Oak Ridge National Laboratory and HEI, directed by Dr. Edward Chaum, Plough Foundation Professor of Retinal Diseases at HEI, sponsored diabetic retinopathy and age-related macular degeneration screening for underserved patients, including some from the Church Health Center. The project, which uses special cameras to photograph the back of the eye, is presently being expanded to other underserved communities. Other important work at the HEI has been done by Drs. Chris Fleming and Natalie Kerr.

Dr. Leonard and Nancy Hines, at home, 1987. Dr. Hines's association with Saint Francis Hospital dates to the 1970s when it was named St. Joseph Hospital East. He retired in 2007.

banks organized long ago have either been discontinued or absorbed into different organizations.

Orthopedics. Dr. Robert E. Tooms, a widely recognized expert in rehabilitation and prosthetics at Campbell Clinic, received the University of Tennessee's distinguished alumnus award in 2005.

Plastic Surgery. In the 2000s, the American Board of Plastic and Reconstructive Surgery dropped the term *reconstructive.* Surgeons often offered nonreimbursable cosmetic procedures such as Botox® (with a reported 1.1 million injections each year through 2006). The American Society of Plastic Surgeons required surgery under anesthesia to be performed only in surgical facilities accredited, licensed, or certified by Medicare.

In 2006, the FDA approved the use of silicon breast implants "with conditions" after a moratorium of 14 years. Saline implants were studied again and found to be safe. Society members worked with Congress to support mandatory insurance coverage of breast reconstruction.

Continuing a tradition of excellence in plastic surgery, Memphis has a number of outstanding practitioners, including Drs. Edward Luce; Allen Hughes; Garnett Murphy; Lou Adams (son of Dr. Lorenzo); and William M. Adams, Jr. (son of Dr. Milton). Dr. Bill Hickerson heads the Firefighters Burn Center at The MED. Dr. Robert Wallace heads the University of Tennessee's division of plastic surgery and directs the Cleft Palate Clinic.

The Moving Finger Writes

In 2007, Dr. Leonard Hines, a former president of the University of Tennessee National Alumni Association and a leader in surgery since St. Joseph Hospital East opened, paid tribute at his retirement dinner to the surgeons who taught him. Attendees constituted a *Who's Who* of surgeons: Drs. James Pate, Louis Britt, Roger Sherman, Ken Sellers, and Richard Cheek. Hines commented, "None of us is born knowing how to do this, and without our teachers and role models, there would be no surgeons." Several surgeons who had served Memphis for three to four decades retired during the 2000s, including Drs. Sidney Birdsong, Jim Green, Hugh Francis, Jr., and Van Wells.

From the Lab to the Clinic: Translational Medicine

In the first decade, much discussion in Memphis focused on "translational medicine," shortening the time from research to patient care or "from bench to bedside." Memphis's success in orthopedic devices is as old as the Campbell Clinic and as varied as glue for bones and adjustable rods for lengthening bones. In 2004, the University of Tennessee opened a lab dedicated to studying peripheral neuropathy and bringing discoveries to market, under the auspices of the Neuropathy Research Foundation.

In 2007, the University of Tennessee opened its Translational Science Institute under Executive Director Dr. James Dale. The Clinical Research Unit (CRU), was managed by Dr. Samuel Dagago-Jack, with Dr. Dennis Black in charge of pediatric research. From 2008 to 2009, the University had 86 research projects ongoing through the CRU.

In 2005, three prominent scientists—Drs. Douglas R. Green, Rodney K. Guy, and Leslie L. Robison—came to St.

Dr. Richard Smeyne, left, of St. Jude Developmental Neurobiology, and Dr. Robert Webster, of St. Jude Virology, discuss the effect of avian influenza on the human brainstem, 2010. The collaborators discovered that the H5N1 avian influenza virus produces long-term neurological effects similar to Parkinson's and Alzheimer's diseases.

Jude to translate basic science into clinical treatments. By 2008, researchers led by James Ihle, Ph.D., had published in *Nature* their findings on apoptosis (programmed cell death). The facility expanded the hospital's translational capabilities; AIDS, SARS, and flu virus vaccines are studied there.

Research

In 2008, because of its interactive, interdisciplinary research in all three main types of cancer research—basic science, clinical science, and population science—St. Jude was designated by the National Cancer Institute as a Comprehensive Cancer Center, the only pediatric institution in the United States so designated. The center includes a new Cancer Prevention and Control Program, which developed from St. Jude's long-term patient follow-up.

Robert Webster, Ph.D., was the first to link human flu and bird flu (1957). Working at St. Jude since 1969, he became known worldwide for his work on flu viruses, and he developed the preferred method of flu vaccination. He directs the World Health Organization's Collaborating Center on the Ecology of Influenza Viruses in Lower Animals and Birds, the world's only laboratory focusing on the animal-human link. By 2009, he was the acknowledged world expert on the H5N1 virus.

In the 2000s, St. Jude researchers studied the effects of antibiotics on the inflammatory process, effects that might be the key to survival for patients who develop bacterial pneumonia after a bout of flu. Some antibiotics kill the bacteria but also aggravate preexisting inflammation, while other antibiotics, notably clindamycin, kill the bacteria but do not exacerbate inflammation. Dr. Jon McCullers and his group believe that planning for pandemics should consider these effects.

A program involving collaboration of the University of Tennessee, Methodist University Hospital Neuroscience Institute, and Semmes-Murphey Neurologic and Spine Institute, through the efforts of Valery Kukekov, Ph.D., and Tatyana Ignatova, Ph.D., uses adult stem cells from blood or tumor to study how tissue develops and how injuries lead to

Drs. Valery Kukekov, left, and Tatyana Ignatova in their laboratory. They were recruited from the University of Florida in 2005 to head a new multidisciplinary program on tumor and stem cell biology, working with the Semmes-Murphey Clinic. Dr. Kukekov's expertise is in adult brain stem cells, while Dr. Ignatova is a cancer cell biologist.

tissue regeneration. The two researchers came to Memphis to work with the Semmes-Murphey Clinic.

InMotion Musculoskeletal Institute began in 2005 as an independent, nonprofit laboratory for musculoskeletal research, changing its name in 2009 to InMotion Orthopedic Research Center. It supports collaborative research, bridges the gap between basic musculoskeletal research and clinical treatment, supports development of products for commercial use, and furthers clinical and research education. It is privately supported and is closely linked to the University of Tennessee.

Practice groups, such as the Stern and Sutherland clinics, regularly participate in drug studies and trials. Dr. Frank McGrew supervises the Stern Clinic's research activities. Research is central to the West Clinic's mission.

Medical Business

New imaging devices joined the competition between Memphis's two largest hospitals. Methodist and Baptist installed positron emission tomography (PET) scanners in the early 2000s, each costing $10 million each, including a cyclotron and the services of a physicist. At Methodist, Dr. Michael Fleming first established PET scanning.

Economic Impact of Medicine in Memphis

An economic constant, health care is vital to Memphis's economy. In the 2000s, the University of Tennessee alone provided about $2 billion to Memphis's annual economy, accounting for five to six percent of the $30 billion total personal annual income earned in the Mid-South. In fiscal 2005, the state appropriated $94.6 million to the University. For every dollar of the state appropriation, the University and related operations contributed approximately $20 to the local economy. More than 4,500 graduates of the University of Tennessee Health Science Center lived and worked in Shelby County.

In 2004, the American Hospital Association reported that health services and supplies constituted approximately fifteen percent of the U.S. Gross Domestic Product (GDP). As a major provider of stable jobs and regular incomes in the Mid-South, hospitals did more than provide health care, including holding health screenings and convening support groups. Free or reduced-fee charity care helped relieve the burden on taxpayers.

Cyril Chang, Ph.D., of the University of Memphis, observed that the volume of Shelby County's "hospital market" was essentially flat in the last years of the decade, with annual inpatient admissions between 142,000 and 143,000 and market size shrinking at a 0.2 percent annual rate. This shift, however, may reflect the rising use of outpatient facilities. Following the national trend, government programs insured ever-growing numbers of patients. In Memphis, The MED took about twelve percent of inpatient admissions, with the remaining 88 percent split among the other Memphis hospitals.

The hospital scene in Memphis shifted with population trends. When TennCare first arrived, The MED lost patients, but regained its share by 2004. Methodist and St. Francis gained market share, while Le Bonheur and Delta Medical remained stationary.

Baptist voluntarily reduced its market share by closing Baptist Central. In the period from 1995 to 2004, inpatient admissions in Memphis declined slightly (0.2 percent per year), while gross charges increased 9.8 percent.

LEFT: *Dr. Cyril F. Chang, professor of economics at the University of Memphis, 2009. Dr. Chang is also director of the Methodist Le Bonheur Center for Healthcare Economics. His publications include papers on TennCare and Tennessee Hospitalizations, and a book* Redefining Health Care. ABOVE: *The entrance to the emergency room of The MED, 2010. Chang observed that TennCare and self-pay patients were more likely to overuse emergency rooms.*

In difficult economic times, Dr. Chang observed, hospitals help maintain the local economy and later revive it. During the recession of 2001, for example, education and health care expanded and added jobs in the Mid-South while other sectors declined.

Problematic in the Mid-South is overuse of emergency rooms by patients with nonemergency problems. TennCare and self-pay patients, Chang found, were more likely to overuse emergency room services than patients with other insurance. Because Memphis has approximately twice the number of individuals living in poverty as other cities in the United States of comparable size, intensive efforts are needed to improve the economic picture.

An assessment by the Memphis Area Chamber of Commerce in the late 2000s recommended commercialization of research to develop Memphis's biomedical potential, with a collaborative medical council to bring together hospitals, the medical school, universities, and the business community to refine the biomedical mission and strengthen biomedical goals. Memphis's universities, medical community, and business community have moved rapidly to implement the steps recommended.

MEMPHIS MEDICINE

A HISTORY *of* SCIENCE AND SERVICE

EPILOGUE

COME, NOW, ON A BRIEF JOURNEY THROUGH THE speck of time that is Memphis. We see trappers and traders plying the great river on flatboats. They come alone, without families, French and Spanish mostly, to trap furs in the wilderness, to bring trade to the few settlers and the Native Americans. They stop at the Fourth Chickasaw Bluff because it offers protection from floods, a vantage point against attack, and crude medical care—bleeding, applying leeches, lancing boils, sewing wounds, setting bones, and distributing potions such as "blood purifiers" reputed to cure disease.

Come next to the bawdy frontier town that has sprung up, its muddy streets paved by wooden blocks that harbor waste and filth. Its night soil and garbage are dumped into a wide, open sewer, the Gayoso Bayou. Its residents are subject to typhoid, malaria, and dysentery, and it is called, in derision, "the cemetery on the bluffs." It is named for Memphis on the Nile, ancient capital of Egypt's Old Kingdom and home to Imhotep, perhaps the first ancient physician.

On the outskirts of the city, see the father of the family stretched out on the kitchen table with a bullet festering in his shoulder and a weary physician attempting to remove it, while the man, only dazed by alcohol, cries out and his wife searches Gunn's *Domestic Remedies*.

Come now to growing Memphis, where Union and Confederate troops move by river, leaving their sick and wounded here, where mercantile stores and hotels become hospitals. Gangrene is common among the wounded, and amputations of arms and legs are rampant. The numbers are great, the treatments rudimentary, the suffering horrific.

A few years later, witness the devastation wrought by yellow fever. Erroneously attributed to "bad air," the disease respects none. Crowds gather downtown to read the daily list of the dead. The affluent flee. The poor, the religious, and the doctors remain, many of them losing their lives to the pestilence. The dead are carried via wagon to mass graves, sometimes to private graves, in Elmwood Cemetery. Finally, with the first frost, the disease abates. A massive clean-up will follow, and public health concerns will come to Memphis, though the vector-carried cause of yellow fever will not be found (in Cuba through Dr. Carlos Finley's and Dr. Walter Reed's efforts) until two decades later.

See Memphis physicians leaving, in hospital groups, for World War I, working together, distinguishing themselves in service. Watch the young mother in a nearby town, finally diagnosed with a ruptured appendix, giving her six-week-old daughter to her sister. "Take care of my baby" will be her last sentence as the ambulance wails its way too late to Memphis. See, too, in Memphis, a young African-American man wracked with pain from sickle cell disease. Two decades later, young Dr. Lemuel Diggs will begin a lifetime study of the disease.

Watch, during World War II, the wounded brought to Millington's Naval Hospital, to Kennedy General in Memphis, and the community responding with gifts and companionship for its heroes. See, after the war, the 128 frame buildings of the VA hospital in East Memphis, where crack surgeons discover ways to salvage the wounded and train a new generation of surgeons with their knowledge.

In the 1950s, enter the dining room of a young couple whose two year old lies, wearing a little suit and slightly scuffed shoes, in a tiny casket, months after being diagnosed with liver cancer. Watch as Dr. Gilbert Levy dashes from his car, engine running and door open, into a Midtown home where a month-old boy, seen for a well-baby exam a week ago, does not move his arm. Dr. Levy will rush him to the Isolation Hospital for treatment, and the child will survive.

Fast-forward to contemporary Memphis. See trunks unloaded at FedEx's airport facility, marked "Medical Equipment," returning from a sight-restoring voyage to

The board of The Memphis Medical Society meeting in 2009. At the head table to the left are Don Alexander, chief executive of the Tennessee Medical Association (dark suit), Michael Cates, executive director of The Memphis Medical Society (white shirt and tie) and then Memphis Medical Society president, Dr. Clarence Watridge (white coat). The board has met on the first Tuesday of each month since 1879.

Central America. See the golden dome of St. Jude Children's Research Hospital glowing downtown, against the image of the M-shaped interstate bridge. Check St. Jude's website and see a photo of smiling children saying "Thank you!" to hospital staff. Some are bald, some are in wheelchairs, but all are smiling, and all glow with hope, gratitude, and growing health. Move to the Peabody Hotel, where a young woman and her husband sip champagne. Her dark hair is very short but flattering, and her low-cut dress is not lost on her husband. They are celebrating after the lumpectomy and treatments that eliminated the 3-mm cancer in her breast.

See the new Regional Biocontainment Laboratory near where the imploded Baptist Hospital once stood. There, multiple-level safety procedures enable researchers to study the most dangerous organisms on earth, devising ways to counter terrorist attacks and to prevent pandemics.

See the tower of the new Le Bonheur Children's Hospital, where Dr. Chris Knott-Craig has just repaired the heart of a six-week-old baby. See the complex at Methodist University Hospital where Dr. David Cunningham discharges an executive whose facial palsy was eliminated this morning with the Gamma Knife. See the eastern complex of Baptist Memorial Hospital, where Dr. Ed Garrett, Jr., just repaired a fellow physician's dissecting aneurysm. See the bustling St. Francis Hospital, where Dr. A. Koleyni repaired a cleft lip and palate today.

Distribution center, medical-device manufacturing center, research center, world-renowned treatment center —Memphis has become what it is today through the efforts of many. Not least of those are the practicing physicians, teaching physicians, researchers, technicians, nurses, therapists, and others who have brought medicine in Memphis this far and will continue to inscribe its story on their patients' lives.

MEMPHIS MEDICINE

A HISTORY *of* SCIENCE AND SERVICE

MEMPHIS MEDICAL HISTORIES

THE HISTORY OF MEDICINE IN MEMPHIS and Shelby County stretches back well over 150 years and provides a fascinating picture of the immense changes that have taken place since the first doctors settled in Memphis in the 1830s and 1840s. It shows how hard physicians in Tennessee have worked to establish the sound practice of medicine as they supported health and sanitary laws and high standards in medical education. The following pages offer histories of some of those men and women who carry on the legacy of providing quality medical care in an area now known as Metro Memphis, which includes not only the City of Memphis, but six other incorporated towns in Shelby County (Germantown, Collierville, Bartlett, Arlington, Lakeland, and Millington), a large unincorporated area, parts of North Mississippi (Southaven and all of De Soto County) and Fayette and Tipton counties in West Tennessee.

These profiles offer a glimpse of how the communities within the Metro area continue to be shaped by men and women of medicine. Whether in solo practice or part of a group, whether near the end of their practices or just beginning, they show commitment to their patients and to medicine. These professionals will continue the medical heritage in Memphis and its Metro area well into the future. Both in practice and in research, these physicians have carried Memphis's name far beyond its physical boundaries. The geographical location of Memphis has played an important role in shaping the medical history of this area, while medicine has significantly influenced the culture, organizations, businesses, and communities in which physicians work and live. The Memphis area is better and stronger for its dedicated and talented physicians. In sharing their stories, the physicians in these pages contribute another dimension to a rich tradition.

OPPOSITE PAGE, CLOCKWISE FROM ABOVE: *Outpatient cataract surgery at Memphis Eye and Cataract Associates; the new Le Bonheur Children's Hospital, which opened in 2010; the Semmes-Murphey Neurologic & Spine Institute.*

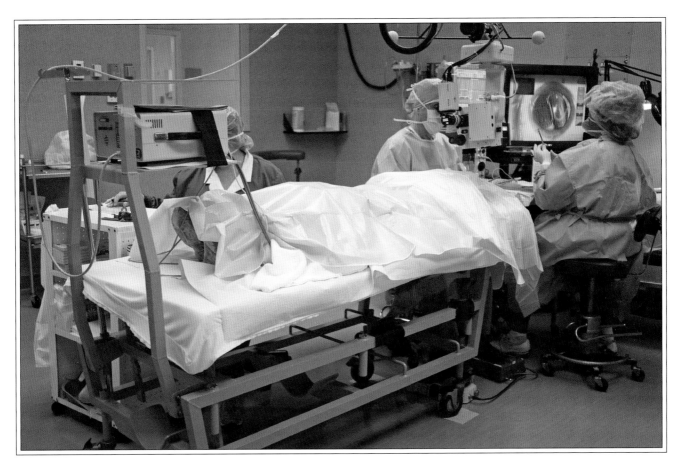

THE MEMPHIS MEDICAL SOCIETY

The Memphis Medical Society (MMS) traces its continuous history to 1876, with the same mission, "to unite the physicians of Memphis and Shelby County into an organization to promote the highest quality of medical practice and the health of our citizens." From the beginning, the Society has been a spokesman for important health issues of the day, speaking on behalf of both patients and physicians.

Chartered by the Tennessee Medical Association (TMA), a federation of 40 medical societies in the state, the TMA in turn unites with similar organizations in the other states and with specialty societies to form the American Medical Association (AMA). State, county, and specialty societies in the AMA comprise the Federation of Medicine.

Organizing local societies throughout Tennessee was encouraged by the Medical Society of Tennessee, organized in Nashville in 1830. The first local society in the state was Nashville in 1843. A predecessor Memphis Medical Society was the fifth, incorporated by the state legislature in 1852 and lasting until the Civil War. The AMA was formed in 1847 to raise medical educational standards and to adopt a code of professional ethics and eliminate quacks and inadequately trained physicians. The oldest continually operating medical society in the United States is the Massachusetts Medical Society (1781), publisher of the prestigious *New England Journal of Medicine*.

Today's medical society, organized in 1876 as the Shelby County Medical Society, has changed its name three times: Memphis Medical Society (1887); Memphis and Shelby County Medical Society (1902) after the Medical Society of Tennessee reorganized as the TMA; and The Memphis Medical Society (2001), under the leadership of President Hugh Francis, III, with a simpler, more modern logo and graphic design.

Since 1879, the MMS has met on the first Tuesday of each month. For much of its history, presentation of interesting medical cases was part of each meeting. The governing House of Delegates would convene, followed by a scientific program for the full membership. Clinical education was a high priority but, as those needs were supplied by other organizations, the scientific programs were discontinued. The Society published its own scientific journal from 1924 until 1969, and still publishes the *Memphis Medical Quarterly*.

The Society's first dedicated office opened in 1952 at 1363 Union Avenue, Apartment 11, when telephone numbers were still five digits, 7-3022. Sometime after 1976, the Society acquired a permanent home at 774 Adams. This building was sold in 1980 to Le Bonheur Hospital for $60,000, and after a debate over whether to purchase the Hill Mansion near Methodist Hospital, the Society moved east to 6264 Poplar in March 1981, reflecting the membership migration from the city center. When the medical society outgrew that site in 1988, it moved to its current home at 1067 Cresthaven.

Four executives have led the Society's business. The first was Robert C. Bird, executive secretary from 1952 until June 1956, whose hiring as a full-time executive was considered the greatest advance for the medical society in many years.

The Memphis Medical Society did not have a dedicated office until 1952. The Society moved to the current location, its fourth, on Cresthaven Road in 1988.

The second executive, Leslie Adams, served from July 1956 until 1979, with the title changed to executive director. John Westenberger was executive director and executive vice president from 1979 until May 1985. Michael Cates, the current executive director and the longest serving, started in July 1985. He is now in his twenty-fifth year. Janice Cooper is his executive assistant, and Victor Carrozza is communications and membership director.

There are four affiliated but separate organizations. The first, the Woman's Auxiliary to the Memphis and Shelby County Medical Society, was formed in 1928 to further friendly relations between doctors' families and between doctors and the community. The name was changed to the Memphis and Shelby County Medical Society Auxiliary sometime before 1984, and in 1994 to the Memphis and Shelby County Alliance. In 1984, the Auxiliary, through its foundation, was responsible for the Health Sciences exhibit at the Pink Palace Museum. This project was coordinated by Ruth Crenshaw (Mrs. Andrew H. Crenshaw, M.D.). The exhibit was accompanied by *From Saddlebags to Science* by Patricia M. Lapointe; both are still great resources for Memphis medical history. For years the group raised more money for the AMA education fund than any similar group in the country. Although the Alliance gave up its charter in 2005, former members still raise education funds through an annual Christmas Sharing Card.

Second is the Memphis and Shelby County Medical Foundation, begun in 1962 as the charitable arm of the Society, with its first project defraying the cost of polio immunization programs. It is still active supporting local charities.

The third affiliated organization, MemPac, a political action committee, was chartered in 1986. The fourth, a for-profit business bureau was formed, also in 1986, encompassing MedTemps, a subsidiary that continues to provide temporary office personnel to members, and MedType, a transcription service no longer offered.

In 2010 there are 819 actively practicing members, 209 retired, and including students, interns and residents, a total membership of 1,706. Current members will be amused to know that dues in 1924 were $10, with $2 for the society's scientific journal and the remaining $8 split between the TMA and the MMS. For $1 more you could "obtain the privilege of legal defense in case of suit for mal-practice".

In its early years, the Society was involved in promoting licensure laws, urging the development of health departments and educational standards, pushing for adequate food and milk regulations, advocating for tuberculosis and polio control, and providing scientific programs and journals. Today the Society promotes healthy lifestyles, advises members on insurance contracts, advocates reform of malpractice laws and, through MemPac, lobbies to secure adequate payment and prevent undue government intrusion into the physician-patient relationship.

The presidents listed in the appendix represent a virtual *Who's Who* of Memphis medicine. Over the years there have been many hundreds of others who have given of their time and energy to work through the Memphis Medical Society and the Federation of Medicine on behalf of the patients and physicians of Memphis. We owe them a great debt of gratitude.

MEMPHIS AND SHELBY COUNTY MEDICAL ALLIANCE

The Memphis and Shelby County Medical Auxiliary was chartered in 1928, one of the several auxiliaries being formed in the South at that time. Mrs. Wilford H. Gragg, Sr., the first president, and Mrs. Willis C. Campbell were leaders in the new organization. At the outset, the Auxiliary had as its primary purposes promoting fellowship among medical families and between doctors and the community and actively furthering the work of the Medical Society.

Through the years these objectives were expanded to include raising funds for medical education scholarships, supporting legislative advocacy for the medical profession, and participating in organizations promoting the health and well-being of the community.

In the early 1980s Auxiliary members researched the medical history of the Memphis area and gathered artifacts and documents that were used to create a permanent health history exhibit at the Memphis Pink Palace Museum. Opened in 1984, this award-winning exhibit was coordinated by Mrs. Andrew Hoyt (Ruth) Crenshaw. The Health Sciences Museum Foundation, the non-profit arm of the Auxiliary, published the book *From Saddlebags to Science: A Century of Health Care in Memphis 1830-1930*, also the name of the exhibit.

In 1994 the Auxiliary's name was changed to The Memphis and Shelby County Medical Society Alliance, a change made at the national level. Through the Holiday Sharing Card, for some 25 years, the Alliance raised a substantial amount of money for the American Medical Association (AMA) Foundation's medical scholarship program. In 1998, the AMA Alliance presented an award to the Memphis Alliance for an outstanding record of fundraising to benefit this program.

Two Memphis Alliance members have been the recipients of the AMA's prestigious Belle Chenault Award for national political participation bestowed in alternate years by the AMA's Political Action Committee at its annual convention in Chicago. Mrs. Jesse C. (Annabel) Woodall, Jr. received the award in 1995, the first Tennessean to be honored nationally. Mrs. Robert J. (Barbara) Trautman, Jr., the 2007 recipient, was the only other Tennessean to be recognized for outstanding accomplishments in this field.

With a grant from the Assisi Foundation, the Alliance implemented "Growing Healthy," a comprehensive health education program for city school students in grades K-6.

Alliance members (left to right) Mrs. John R. Adams, Jr., M.D. (Stephanie), Mrs. Robert J. Trautman, Jr., M.D. (Barbara), Mrs. J. Cameron Hall, M.D. (Terrie), and Mrs. Gary W. Kimzey, M.D. (Jane) pose with Rep. Curry Todd, (center) on Doctor's Day at the State Capitol in 2003.

With state and local cooperation, a video designed to teach children how to resolve conflicts without violence was given to every school in Tennessee along with supporting literature. In 2009, the Alliance, working with a coalition of 13 other groups, organized a one-day health education exposition, "All About Women," at the Cook Convention Center. Local physicians and health care professionals presented information on a broad range of women's health issues.

During the existence of the Alliance, its members have successfully undertaken a wide range of projects to benefit the medical profession and the community. Because of changes in national AMA membership requirements, The Memphis and Shelby County Medical Alliance now operates as a local organization with four vice-presidents, each responsible for its major programs: the Memphis Medical Foundation, Health Promotion, Legislation, and Membership. The names of Alliance presidents who have led the organization are listed in the appendix. Many also have held leadership positions at the state and national levels, and Mrs. Rex (Johnnie) Amonette served as president of the American Medical Association Alliance.

R. FRANKLIN ADAMS, M.D.

Descended from three generations of Mississippi physicians, a heritage that has produced four notable Memphis plastic surgeons, Dr. Frank Adams was drawn to the puzzle-solving of internal medicine and to its new subspecialty of rheumatology. A founding member of The Arthritis Group of Memphis, he provides diagnostic services and medical care for patients suffering from acute and chronic rheumatic diseases.

Named among the "Best Physicians in Memphis" by *The Commercial Appeal* and *Memphis Magazine*, Dr. Adams also received Methodist Hospital's "Living Award." Named an Outstanding Volunteer by the National Arthritis Foundation, he has also been chairman of the Memphis Arthritis Foundation.

Dr. Adams was president of Methodist Hospital's medical staff and was the initial director of its Internal Medicine Training Program, as well as its long-term rheumatology teaching head, training numerous internal medicine residents and orthopedic trainees in clinical rheumatology. He is a rheumatology consultant to the Methodist and Baptist hospital systems.

As an officer in the U.S. Navy Medical Corps stationed at Camp Pendleton, California, he was medical director for naval medical personnel embarking for Vietnam.

Dr. Adams graduated from Washington and Lee University and received his M.D. degree from the University of Tennessee. He completed his internship at the University of Pennsylvania, his internal medicine residency at the University of Michigan, and his rheumatology fellowship at the University of Tennessee. He has been board certified in both internal medicine and rheumatology and is an associate clinical professor of medicine at the University of Tennessee Center for Health Sciences.

ADRIAN BLOTNER, M.D.

New Orleans native Dr. Adrian Blotner opened his practice there in 1988, and planned to stay. But after Hurricane Katrina, the devastation and the questionable timetable for rebuilding prompted him to relocate to Memphis in March, 2006.

Dr. Blotner trained at Louisiana State University and Harvard Medical Schools, as well as the Ochsner Clinic. His areas of study were psychiatry, internal medicine, and neurology. Dr. Blotner's board certifications in pain medicine and psychiatry reflect his interest in the connections between medical, physical, and emotional aspects of complex and related illnesses.

His innovative treatment model combines non-addictive medication, physical therapy, and stress management, and led to his founding the Stress Management for Chronic Pain Program, for which he serves as medical director.

His faculty appointment at the University of Tennessee Health Science Center provides a unique opportunity for Dr. Blotner to share his expertise. His lectures to medical staff, nurses, mental health professionals, and consumers provide them with a better understanding of chronic pain, anxiety, mood, and sleep disorders.

Dr. Blotner's innovative approach to treatment continues to enhance quality of life for his patients. The devastation of Hurricane Katrina has brought to Memphis new hope for those who suffer from these illnesses.

BAPTIST MEMORIAL HEALTH CARE

From a Tiny Seed

In the early 1900s, Memphis was in dire need of a new hospital. This sentiment was strong in the Baptist church, and numerous Baptist groups across the region began to suggest the establishment of a Baptist hospital in Memphis. Funding came from a number of sources and, in 1912, the seven-story Baptist Memorial Hospital opened with 150 beds.

It was a humble beginning for a healthcare giant. Three years into service, the little hospital was still losing money and Baptist's leaders had begun to contemplate selling it. Mississippi planter A.E. Jennings stepped forward with an offer to underwrite the hospital's debt personally and serve as its director, with no pay. Jennings and the Southern Baptist Convention reenergized support for the hospital, and the Southern Baptist conventions of Arkansas, Mississippi and Tennessee allocated funds. The hospital soon began to grow and prosper and, by 1922, Baptist Memorial Hospital was the largest privately owned hospital in the South.

Care through Innovation

In the late 1920s, Baptist Memorial's leadership realized that access to the best doctors could be enhanced by building an adjoining office building just for doctors. In 1927, Baptist Memorial Hospital opened the Baptist Physicians and Surgeons Building, the first hospital-owned physicians' office building in the nation. The facility connected with every floor of the hospital, allowing for close clinical supervision of patients. The building changed the character of Baptist Memorial Hospital from a community institution to a regional referral center and helped attract world-renowned physicians.

Dr. Raphael Eustace Semmes, one of the most influential neurosurgeons in the profession, came to Memphis in the early 1900s, and centered his practice at the new hospital. Dr. Francis Murphey came to Memphis to be Semmes's first trainee, and the two eventually started one of the most well-respected neurology and neurosurgical practices in the country, the Semmes-Murphey Neurologic & Spine Institute.

National leadership in orthopedics also came to Baptist Memorial Hospital in its early days through Dr. Willis C. Campbell, founder of Campbell Clinic and cofounder and first president of the American Academy of Orthopaedic Surgeons. Campbell Clinic and Baptist Memorial Hospital became nationally known for expertise in orthopedics.

Leadership in cardiology has also been a longstanding tradition since the early days of Baptist Memorial Hospital. While practicing at Baptist in the early 1900s, Dr. Neuton Stern founded the cardiology practice that has since become the well-known Stern Cardiovascular Center.

New Leadership, Same Success

In 1946, Frank S. Groner became the hospital administrator and began a new era for Baptist Memorial Hospital. During his 34 years of leadership, Baptist grew to become the largest privately owned hospital in the country and achieved numerous medical and industry "firsts." A year into Groner's tenure, Baptist became one of the first hospitals in the country to install automatic elevators and, in 1959, it was among the first 10 hospitals in the nation to establish a satellite rehabilitation unit. The following year, it made history again by becoming the first hospital in the country to install a computer for accounts billing.

Two large expansions, a 13-story Madison East addition and the 18-story Union East addition, added hundreds of beds to the hospital. By 1975, the hospital had grown to 2,068 beds. The 1980s brought about great change for Baptist Memorial. Joseph Powell succeeded Groner as president in 1980, and the following year, the Baptist Memorial Health Care System was established. Powell oversaw the growth of the organization into a multihospital system that spanned three states—Tennessee, Mississippi, and Arkansas. Baptist also maintained its reputation for providing leading-edge

Opposite page: *When Baptist Memorial Hospital opened on July 22, 1912, its 150 beds were vital to a booming city in desperate need of more hospital space. Built at a cost of $235,000, it included a reception room, parlor, offices, dining room, kitchen, operating room, even an elevator.*
Above: *By the mid-1900s, Baptist Memorial Hospital was treating a remarkable number of patients, a record 23,223 in 1950 alone, with 30 percent receiving free treatment.*

care by becoming the site of the Mid-South's first adult heart transplant and adult bone marrow transplant center.

A New Mission for a Century-Old Vision

In 1994, Stephen C. Reynolds became only the fourth person to lead Baptist Memorial Health Care when he was named president and chief executive officer of the organization. His tenure is highlighted by continued growth and medical excellence.

Under his leadership, many of the hospitals that became part of the Baptist Memorial Health Care family in the 1980s and early 1990s attracted a tremendous influx of patients who wanted health care closer to their homes. Baptist responded by increasing capacity in several of its hospitals. Multistory bed towers were added onto Baptist Memorial hospitals from Memphis to Union City, Tennessee, and Columbus, Mississippi. Baptist equipped its facilities with leading-edge technology, bringing big city medicine to smaller cities and rural areas.

Baptist Memorial hospitals also continued to win accolades for delivering high-quality care. Baptist has been listed among *U.S. News and World Report*'s top 50 hospitals in the nation for orthopedics as well as neurology and neurosurgery.

Each year, the Baptist system has added more advanced services to its communities, including the Baptist Heart Institute, a freestanding women's hospital, a remarkable rehabilitation hospital, and advanced new hospitals in Memphis, Collierville, and Southaven. As new facilities and care delivery models grew in popularity, it became time to say goodbye to the building where Baptist's mission began. In 2000, the original Baptist Memorial Hospital closed. The following year, the property was donated to the University of Tennessee and the Memphis Bioworks Foundation, an organization that leads a collaboration of public, private, academic, and government organizations in an effort to build upon the city's bioscience industry. The gift is valued conservatively at $80 million.

Donations like this and Baptist's other community benefit efforts have grown as rapidly as its facilities. During the past decade, the organization's community benefit— everything from unpaid patient accounts and charity care to cash contributions—increased more than sixty percent. Through its Baptist Operation Outreach healthcare van for the homeless, countless free health fairs and sponsorships, uncompensated care, and many other community activities, Baptist provided more than half a billion dollars in community benefits in 2009.

CAMPBELL CLINIC ORTHOPAEDICS

Before Dr. Willis Campbell became a giant in the field of orthopaedics, he was simply a giant of a man who practiced pediatric medicine. And it was as a young pediatrician that he made a lengthy round of house calls one sultry summer day nearly a century ago. Late in the afternoon, he stopped in to see a bed-ridden boy and to speak with the child's rather nervous and overprotective mother. Upon entering the room, the good doctor strode mightily toward the boy's tiny bed, sat down in a modest wooden chair, and removed the stethoscope from his black leather bag.

Then suddenly and with a loud snap, Dr. Campbell's long limbs were sent flailing as the chair he had just occupied splintered under the weight of his sizable frame, setting in motion a series of startled overreactions. The little boy quickly recoiled in fear. The mother shrieked and dived onto the bed to shield her child. The doctor sat stunned at the panic he had caused. And thus the revolution began.

After deciding then and there that pediatrics would not be his calling, Dr. Campbell instead made his way to Vienna to learn the emerging field of orthopaedics from the world's foremost orthopaedic practitioners.

Upon his return to Memphis, he advanced the science by huge leaps and bounds as if the entire field of orthopaedic medicine was being carried along by his own lengthy strides.

First he established his own clinic. Then he organized the Department of Orthopaedic Surgery at UT Memphis, which included establishing the first orthopaedic residency program. Dr. Campbell ultimately cofounded the American Academy of Orthopaedic Surgeons and also cofounded the American Board of Orthopaedic Surgery—the very board that to this day still accredits, certifies, and oversees all doctors in this special area of medicine.

Now fast forward almost 100 years... Campbell Clinic is now 42 physicians strong. The clinic has four locations in the Memphis area: Germantown, Medical Center, Collierville, and Southaven. Campbell Clinic is a full service orthopaedic practice and has its own surgery center and MRI units, and offers physical therapy services at all four of its locations.

The physicians at Campbell Clinic specialize in sports medicine, shoulder, spine, total joint replacement, foot and ankle, hand, pediatric orthopaedics, knee, elbow, general orthopaedics, total joint replacement, orthopaedic oncology, and physical medicine and rehabilitation. Campbell

Clinic physicians are proud to be the sports medicine providers for the Memphis Grizzlies, the Memphis Redbirds, the University of Memphis, Rhodes College, Christian Brothers University, and many area high schools.

And one can't talk about Dr. Willis Campbell and Campbell Clinic without mentioning *Campbell's Operative Orthopaedics*. First authored by Dr. Campbell in 1939, this textbook, which is published by the Campbell Foundation, is used by nearly all medical students and referenced by all orthopaedic physicians. The book is revised and updated every five to seven years by Campbell Clinic staff members. *Campbell's Operative Orthopaedics* is now in its eleventh edition, is a four-volume textbook, and has been translated into seven languages (Chinese, Greek, Italian, Japanese,

OPPOSITE PAGE, ABOVE: *Dr. Willis Campbell, founder of Campbell Clinic, is one of the pioneers of orthopaedic medicine.* OPPOSITE PAGE, BELOW: *Dr. Campbell (center) with early members of the Campbell Clinic staff.* ABOVE: *Campbell Clinic has four locations in the Memphis area as well as its own surgery center on Brierbrook Road.*

Korean, Portuguese, and Spanish). The book contains over 4,000 pages with 1,800 descriptions of orthopaedic procedures and over 9,000 illustrations. *Campbell's Operative Orthopaedics* has been called 'the definitive orthopaedic surgery reference' and 'the bible of orthopaedic surgery' for surgeons all over the world. It is, without a doubt, the bible of orthopaedics.

Campbell Clinic physicians serve as the chair and faculty for the Department of Orthopaedics at the University of Tennessee Health Science Center and work hand in hand with the department's researchers and scientists conducting pioneering research of new treatments and preventive medicine.

The physicians are active in local and national professional societies and organizations and have held leadership positions within these organizations. In fact, six Campbell Clinic physicians, including Dr. Willis Campbell, have served as president of the American Academy of Orthopaedics.

Campbell Clinic offers numerous diagnostic services including MRI, electromyography (EMG), and nerve conduction studies. Other services offered are physical therapy, orthotics, and prosthetics.

Campbell Clinic physicians perform many surgeries at the clinic's own surgery center, The Campbell Surgery Center, which is located at 1410 Brierbrook Road, directly behind the Clinic's Germantown location. Outpatient orthopaedic surgical procedures are performed at the 11,285-square-foot, freestanding facility.

Whether it is by developing new treatment methods, inventing new joint implants, or treating patients, Campbell Clinic physicians are committed to providing the most advanced orthopaedic care through their private practice and their active programs in research and education. The orthopaedic specialists at Campbell Clinic continue to carry on the legacy of Dr. Willis Campbell's leadership while adhering to his primary concern—unsurpassed patient care.

CHARLES RETINA INSTITUTE

THE CHARLES RETINA INSTITUTE IS BUILT SOLIDLY ON the reputation and expertise of Dr. Steve Charles. After 40 years of innovative ocular techniques and surgical procedures, a 108-page curriculum vitae highlights what continues to be a stellar career.

Dr. Charles's educational training began as an engineer and expanded to include pre-med. After finishing at the University of Oklahoma in 1965, he entered medical school at the University of Miami, receiving his medical degree in 1969. Charles completed his internship at Jackson Memorial Hospital, Miami, Florida, in conjunction with a research fellowship in ophthalmology and a residency at Miami's Bascom Palmer Eye Institute. Charles continued with a fellowship at the National Eye Institute in Bethesda, Maryland and was also a retinal consultant at the National Naval Medical Center.

In 1975, Charles entered private practice gaining medical licensure in multiple states, keeping his main practice in Memphis, and holding membership in many medical and ophthalmology societies. His career includes hospital and university appointments, medical and technology board positions, publication of medical papers, and the receipt of many honors.

In 1984, Dr. Charles opened the Charles Retina Institute in Memphis. Named in *Ocular Surgery News's* top ten innovators during the past 25 years, Charles has gained national and international recognition through 26,000 vitreoretinal surgeries and the treatment of patients in every U.S state as well as 42 countries around the world. This recognition made the institute well-known worldwide and influenced doctors such as Dr. Mohammad Rafieertary and Dr. Jorge I. Calzada to later join the Institute.

Dr. Charles has 105 patents issued or pending for surgical devices. His is founder of MicroDexterity Systems, Inc. which builds medical robots to improve the precision of hip and knee replacements, in addition to spine and neurosurgery.

Authoring textbooks, chapters in medical books, articles in medical publications, and speaking in 16 named lectures would be in keeping with sharing his vast knowledge with others. Dr. Charles has traveled to 42 countries not only to lecture and attend clinics, but to teach world-class eye care to people who cannot access care at this level. Patients may not grasp the magnitude of his expertise, but like all patients, they will appreciate the life-changing results.

Patients worldwide are referred to the Charles Retina Institute for treatment. Many doctors also visit the institute for advanced training in treatment of retinal detachment, proliferative vitreoretinopathy, macular holes, diabetic retinopathy, age-related macular degeneration, epimacular membranes, retinopathy of prematurity, as well as other macular and vitreous problems. Charles is listed many times in *Best Doctors in America* and in *Guide to America's Top Ophthalmologists.* He has also earned the prestigious Wacker Medal from the Club Jules Gonin and was the first to receive the Founders Medal from the Vitreous Society.

As a clinical professor at the University of Tennessee and an adjunct professor at Columbia College of Physicians and Surgeons, Charles is actively training future retinal surgeons. Among his greatest accomplishments are three daughters, two of whom are physicians and one who is working with children at the nonprofit organization Bridges in Memphis.

Dr. Steve Charles opened the Charles Retina Institute in Memphis in 1984.

NANCY A. CHASE, M.D.

"I RECEIVE AMAZING MESSAGES," SAYS DR. NANCY A. Chase, reflecting on her career as a pediatric cardiologist. Entering the profession in the early 1970s, she often had to tell new parents, "I am sorry; there's nothing we can do." Now she can say, "Let's work together; here are our options." Those amazing missives come from patients who once might have had little hope for a normal life span or normal activity, but who are now running track, working, marrying, and parenting.

Daughter of a businessman and inventor and a lyric soprano, Dr. Chase studied piano, flute, and guitar. She spent her junior year in Germany near her mother, opera singer Nancy Symonds. Playing guitar to accompany songs by Joan Baez, she developed a lifelong commitment to the "message music" of the 1960s and early 1970s. Her hero remains longtime activist Pete Seeger, whose music embodies her interests in peace, civil rights, and the environment. A former president of the National Conference of Christians and Jews in Queens, New York, she marched, with her brother in tow, on Washington, DC, in 1963 and heard Dr. Martin Luther King, Jr., give his "I Have a Dream" speech.

A math major and philosophy minor at Vassar, Dr. Chase discovered medicine through volunteer work. She was drawn to pediatric cardiology because she liked children, was interested in surgery and, with her musical background, could easily distinguish fine variations in heart sounds. She received her M.D. degree from the State University of New York College of Medicine and completed her residency in pediatrics and her fellowship in pediatric cardiology at Long Island Jewish Medical Center.

In West Virginia, she established the first pediatric cardiology subdivision in West Virginia University's Charleston Division. There she learned and liked bluegrass music, and studied cello. Coming to Memphis in 1978 to join the University of Tennessee Health Science Center faculty, she also joined the local folk, blues, and classical music scenes. In Edinburgh for a medical meeting in 1990, she became fascinated with bagpipes, and she now plays with Wolf River Pipes and Drums.

Dr. Chase is the proud mother of Richard G. Wanderman, Jr., Gregory L. Wanderman, and Rebecca Hornstein Doede, and doting grandmother of Greg's Darrah and Hanna, and Rebecca's Henry. She maintains a private practice of pediatric cardiology in East Memphis and a clinical associate professorship of pediatrics at the University of Tennessee Health Science Center, and she lives her passions for medicine and music. A member of many medical organizations, she serves on the boards of both the Memphis Medical Society and Opera Memphis. She has presented scientific papers at medical meetings in the United States and abroad, and advocates politically for proper patient care. She has also sung with local and international choruses.

With her self-designed logo representing transition to health from infancy, through childhood, into adulthood—the desired progress of her patients—Dr. Chase is at home with tiny neonates and towering athletes in her clinic. She is also at home playing bagpipes outside the Orpheum before Opera Memphis's Lucia di Lammermoor or at the Center for Southern Folklore. In music and in medicine, she hears the tones, and they create a unique life. Dr. Chase urges her patients to enrich themselves—through art, sports, acting, or music—and is renowned for distributing event tickets.

WINSTON CRAIG CLARK, M.D., PH.D., F.A.C.S.

Winston Craig Clark graduated from Georgia Tech with a B.S. degree in industrial management and from Georgia State University with a M.S. degree in hospital administration. He was an administrative resident at the Medical Center of Central Georgia in Macon, Georgia, and administrative director of an eight-county, rural clinic in Tupelo, Mississippi. In 1973, Clark was working on his doctorate in health care administration at the University of Mississippi when he moved to Memphis to open the Memphis Health Center. Dr. Jon Robertson was the medical director, and they began a lifelong friendship. In 1975, Clark received his Ph.D. degree.

In 1976, Clark moved to Nashville to become the director of primary care at Vanderbilt University Medical Center, overseeing training programs in general internal medicine, pediatrics, and occupational medicine. At the age of 26, Clark knew his achievements left him with only lateral career moves. Vanderbilt's physicians encouraged him to go to medical school and, with his wife Kathleen's support, Clark entered the University of Tennessee College of Medicine in 1978.

Dr. Jon Robertson's brother, Dr. James T. Robertson, who was chairman of the Department of Neurosurgery at the University of Tennessee, asked Clark to work in the department, and became his mentor. Dr. Clark graduated in 1982, and began his residency in neurological surgery, serving his PGY-1 general surgery year at Methodist Hospital in Memphis.

From 1983 to 1984, Dr. Clark became a medical staff fellow in the surgical neurology branch at the National Institutes of Health (NIH). He won the Resident Research Award from the Southern Neurosurgical Society for work performed at the NIH. When he completed his neurosurgery residency in 1988, he joined his friend Dr. Jon Robertson at the Semmes-Murphey Clinic. Dr. Clark served as the chief of neurosurgery at the Memphis Veterans Administration Medical Center receiving the NIH Career Development Award. He received the Clinical Oncology Career Development Award from the American Cancer Society and developed Memphis's first stereotactic radiosurgery program in 1988.

In 1992, Dr. Clark started a private neurosurgical practice in Georgia, returning to Memphis in 1995, where Dr. Jimmy Miller joined him in opening offices at Saint Francis Hospital, Collierville, and Southaven, Mississippi. After Dr. Miller moved to Greenwood, Mississippi, Dr. Clark, in 2008, merged all services into the Southaven office. "We are small enough to give people individual attention, but we still apply the latest techniques and academic rigor in the practice," he explains. He specializes in degenerative conditions of the spine. After 27 years of practice, he says that at the end of the day, "It's all about taking care of people."

Dr. Clark served as chairman of the Department of Surgery at Baptist Memorial—DeSoto and as head of the Division of Neurosurgery at Saint Francis Hospital. His interest in research has prompted numerous published journal articles and textbook chapters. As a reviewer for several journals, he is currently a member of the editorial board of *The Spine Journal* and *The Southern Medical Journal*. Dr. Clark holds multiple board certifications, and is a diplomate of the American Board of Neurological Surgery, the American Board of Pain Medicine, the Certifying Commission in Medical Management, and the American College of Healthcare Executives. He is a fellow of the American College of Surgeons, the International College of Surgeons, the American College of Healthcare Executives, and the Royal Society of Health.

Dr. Clark attributes his entry into neurosurgery to his mentor Dr. James T. Robertson who has now retired to North Carolina. Dr. Robertson's contributions to Memphis medicine range from chairman of the Department of Neurosurgery at the University of Tennessee until 1996, to practicing at the Semmes-Murphey Clinic. His career included leadership positions in the Congress of Neurosurgeons, the American Academy of Neurological Surgery, and other organizations. In 1994, Dr. Robertson was awarded the Distinguished Faculty Alumni Award from Rhodes College. In 1996, he became a medical consultant to Medtronic Sofamor Danek and served as medical director of the Spinal Division until his retirement.

GEORGE A. COORS, M.D.

BORN IN MEMPHIS, DR. GEORGE ALCORN Coors is a third-generation physician. His grandfather, Dr. George Augustus Coors, was a homeopath and surgeon trained at the University of Arkansas. His father, Dr. Giles Augustus Coors, graduated from the University of Tennessee and practiced for 43 years. Young George graduated from the University of Tennessee-Memphis in 1943 and interned in Chicago, after which he went to Marine Corps training at Camp Pendleton. If the atom bomb hadn't been dropped," he says, "I wouldn't be here, because I would probably have been deployed and killed in action." At the time the bomb was dropped, George, whose nickname was Target bacause of his size (he weighed 246 pounds), was loading ships bound for the Pacific.

Back in Memphis, he completed his surgical residency at Methodist Hospital in 1950 and worked part time at the West Tennessee Chest Hospital with Dr. F. H. Alley and Dr. Francis H. Cole until the Navy recalled him in 1953 during the Korean conflict. Qualified as a chest surgeon because of experience, he was assigned to a hospital in Jamaica, Long Island, with a large tuberculosis ward. The chest surgery unit, which took all Navy and Marine chest surgeries on the East Coast, had 60 to 70 patients. Before effective drugs were available, resections were done for tuberculosis. He became a lieutenant commander.

On his return, Dr. Coors practiced with his father, learning from his father's experience and from scrubbing with others during a time of great camaraderie in the medical profession, a time—the 1940s through the 1960s—when Memphis produced many prominent surgeons, and surgery was king.

Given his specialty in chest surgery, he often operated for lung cancer, for which no effective drugs were available. In 1952, aortic grafts began with Dr. Denton Cooley in Texas. Dr. Coors, after observing Dr. Cooley's work in Texas, performed aortic grafts and peripheral vascular surgery.

Dr. Coors and his wife, Jeanne Parham Coors (who were married from 1944 until her death in 2001), had four daughters: Jeanne, Cristy, Dabney, and Cary. His legacy in medicine continues with his grandson, Dr. Tom Beasley (son of Cristy Coors Beasley), who graduated from the University of Tennessee in 2009, having earlier overcome lymphoma. He is now a resident in trauma medicine at the university. Dr. Coors is married to Jean Lewis Coors, daughter of Dr. Philip Meriwether Lewis.

Medicine was the only career Dr. Coors ever considered. He dissected frogs at the age of ten, and he greatly admired people who were doctors. He had skills, however, to become a professional golfer. He knew and played with great golfers, including Jack Nicklaus and close friend Dr. Cary Middlecoff. Dr. Coors played twice in the Pro-Am and treated Ben Hogan with quinine for leg cramps after an injury.

Dr. Coors is a living history of Memphis medicine. He has seen great improvements in chest surgery, particularly in vascular and heart surgery, but he is disappointed that medicine seems less congenial and cooperative than it once was because of increased specialization and far-flung hospital locations. In his day, he says, "If a colleague had a problem, you helped." Of his long career, he says, "I wish everyone could have the same experiences and enjoy them as much as I did. The personal relationship is everything. You need to know patients and call them by name."

HERMAN A. CRISLER, M.D.

A MODEL OF PHYSICAL FITNESS FOR HIS young patients, Dr. Herman Crisler was born in New Madrid, Missouri, where he grew up as an athlete and outdoorsman. An Eagle Scout, he was active in high school sports and played football at Central College in Fayette, Missouri.

He received his M.D. degree from the University of Tennessee College of Medicine. After a one-year rotating internship at the John Gaston Hospital in Memphis, he spent three years as an Army flight surgeon in the Canal Zone in Panama. On his return to Memphis, he completed his residency in pediatrics at the University of Tennessee, serving as chief resident at the start of the second half of his second year.

Dr. Crisler's love of medicine and of his practice is evident when he talks about his practice, which began in 1969 and has been a solo practice for the last 28 years (before his solo practice, he had "too much free time"). For years, he worked seven days a week, though he has now cut back to six days, with half days on Wednesday and Saturday. Retirement does not interest him; he plans to work "until the wheel falls off the wagon." Dr. Crisler loves his patients and their families, and he prides himself on being cost effective in his practice, skillfully and carefully performing minor procedures in the office to save families from hospital stays.

Part of Dr. Crisler's love for medicine comes from continuing to learn. Medicine remains fascinating for him because of new discoveries, especially in pediatrics, and he looks forward to going to work every day.

Married to Frances Beloate Crisler for 48 years, Dr. Crisler is the father of a daughter, Dr. Crista Crisler, an obstetrician and gynecologist, and of a son, Andrew Crisler, III, an attorney, with whom he returned to active work in scouting. He has four grandchildren who live nearby, and he delights in playing with them.

Dr. Crisler spends his free time hunting and gardening, growing roses and vegetables. He describes himself as a "Jack of all trades" because of his skills as a builder. With his son, he has built a house and two log cabins in the last ten years, in addition to stone fences and other landscaping. He cuts the trees and strips them himself, starting another project when one is finished.

Dr. Crisler has been physically active all his life. Patients recall seeing him running regularly until, after running 70,000 miles in 32 years, knee problems led him to switch to gym workouts. Active, energetic, and deeply committed to the science and art of medicine, Dr. Herman Crisler represents the healthy balance of activity, study, and work that the Greeks called the pursuit of "sound mind in a sound body." He is a member of the Memphis and Shelby County Medical Society and a diplomate of the American College of Pediatrics.

CHARLOTTE S. DE FLUMERE, M.D.

A turning point in Memphis and Shelby County medical history occurred during the almost 35 years (1965-1999) that Dr. David Knott served as director of the Alcohol and Drug Abuse Division of the Memphis Mental Health Institute (MMHI). There, he introduced to the area's medical community the practice of treating addictive behavior as a medical disease. Dr. De Flumere quickly became a strong supporter of Dr. Knott's approach to the treatment of addiction. During her years practicing medicine at MMHI (1995-1999) and through her present position in private practice, Dr. De Flumere has continued to build on Dr. Knott's innovative early work. As a specialist in addiction medicine and the medical management of chronic pain, she also continues Dr. Knott's legacy, working with Dr. Phillip Green and Associates at the Mid-South Pain & Anesthesia Group.

A 1988 graduate of the Medical University of South Carolina, she came to Memphis as a resident in internal medicine at Baptist Memorial Hospital. Dr. De Flumere then completed a joint fellowship in addiction medicine designed and taught by Dr. Christine Kasser at Baptist Memorial Hospital and in geriatrics at the University of Tennessee-Memphis. Dr. De Flumere currently is board certified in addiction medicine, internal medicine, and geriatrics.

Dr. De Flumere has become one of the Mid-South's most sought after practitioners in her field. She focuses not only on compassionately treating the patient's disease, but also on emphasizing preventive health care and supporting her patients so they can make better choices for themselves on their journey to good health.

JOSEPH FAHHOUM, M.D., F.A.A.A.A.I, F.A.C.A.A.I.

When Joseph Fahhoum changed his major from engineering to medicine, he knew exactly why. At age 18, Joseph became seriously ill with typhoid fever but winter in Syria made moving him to a hospital impossible. The family physician came to his home every day to administer IVs and medicine, sometimes arriving at 11:00 pm or midnight. That event solidified his career choice, and the memories are still clear. "She was very dedicated and took very good care of me," he says.

At the University of Damascus, Joseph enjoyed medical training and received his M.D. degree. His residency at Methodist Hospital in Memphis included being chief resident for one year. Following his residency, Dr. Fahhoum practiced emergency medicine and was the assistant director of the Emergency Department at Methodist Germantown Hospital. While working in the emergency department, Dr. Fahhoum pursued training in allergies, asthma, and immunology with a fellowship at the University of Tennessee.

After completing the fellowship, Dr. Fahhoum opened Allergy and Asthma Specialists of Memphis, P.C., in 1999, providing care for children and adults. Dr. Fahhoum spends time getting to know his patients while providing care for their allergic diseases. Several family members may suffer with the same illness. "I believe in medical care that is centered on taking care of the patient," says Dr. Fahhoum. "I like the lasting relationships that I have with my patients."

Dr. Fahhoum is board certified in emergency medicine, internal medicine, allergy, and immunology. He has enjoyed teaching and being a clinical assistant professor at the University of Tennessee since 1999. "Research and development will always have something new down the road," he says. "I attend conferences, and read journals to keep abreast of new developments."

BANANI DIRGHANGI, M.D.

IN INDIA, BANANI DIRGHANGI'S FATHER WAS A NEUROsurgeon and the first to know his daughter would become a doctor. He treated patients in a small office in their home, and he saw she was not frightened or appalled at the sight of blood. "I would go with my father on his hospital rounds and people would ask, 'Who is this little girl?'" says Banani. "If my father saw me looking through the window of the operating room door, he would tell me to put on a gown and come inside." She observed amputations and surgeries while her father carefully explained what he was doing, so it was no surprise that Banani entered pre-medical school earlier and farther ahead of her classmates. After only one year, she entered medical school and studied for five years. It was here she met Jayanta Kumar Dirghangi; after graduation, they married and moved to Memphis.

Banani could not find a pediatric residency in Memphis so, after one year, they moved to Chicago. At Cook County Hospital, Banani saw child abuse at its worst. "Babies were born with drugs in their systems, some were raped, and some had been dipped in boiling water," she states. "I came from a poor country, but I had never seen cruelty like this; it broke my heart."

When the Dirghangis returned to Memphis in 1983, Banani told her husband that her work with the disadvantaged was not finished, and she wanted to work in the poorest area of Memphis. A pediatric position opened at the Health Department's Hollywood Health Clinic, and Banani welcomed the challenge. "I was meant for this clinic," says Banani. She found the staff and nurses were very kind and helped her understand the patients' physical and social needs.

"When I came here, families had just one chart with their last name on it; the mother, father, and children's information was in it," says Banani. "I had no desk and I did everything in one room." They carried guns and knives in diaper bags, but she was never threatened or harmed. Patients were grateful to have medical care, and Banani took the time to listen and help them beyond their medical needs. Memphis police recognized her work and saw crime near the clinic decrease.

In 1999, a new Hollywood Health Loop was built which became part of Memphis's reorganized public health services. After 27 years, Dr. Banani Dirghangi has treated second and even third generations of families, and is visibly moved when she speaks of them. Patients will travel long distances to bring their children and grandchildren to the clinic, to receive the same caring treatment, and mementos of their gratitude hang on the walls.

"I never needed recognition, and the salary is not that important," says Dr. Banani Dirghangi. "I work for my own satisfaction, and I am happy with what I do." One day, she wants to return to India and care for its poor.

JAYANTA DIRGHANGI, M.D.

Jayanta Kumar Dirghangi was born in Eastern India, two hours west of Calcutta. He wanted to be a cardiovascular surgeon at R.G. Kar Medical College at Calcutta University, after being inspired by Drs. Christian Barnard, Michael DeBakey, and Denton Cooley's pioneering work. However, an instructor encouraged him to become an obstetrician and gynecologist, which he did. After graduating as valedictorian in 1972, Dirghangi married fellow medical student, Banani. They traveled to the United States and Henry Ford Hospital for his residency.

In 1980, Dr. Dirghangi came to Memphis as Dr. Lawrence Lewis and Dr. J. Earl Baker's associate. He had heard Lewis's stories about southern traditions and the glory of the old south. "I was very fond of Dr. Lewis because he was a true southern gentleman," says Dirghangi. "He used to brag about Alabama and being from Tuscaloosa." Dirghangi became very interested in southern history and even tried to perfect a southern gentleman drawl. "The southern accent is very soothing and I wish I could talk in that dialect."

Dirghangi's wife, Banani, was studying to become a pediatrician, but she could not find a residency in Memphis, so the couple moved to Chicago. Dirghangi taught in the medical school and cared for their new baby while Banani continued her medical training at Cook County Hospital. He also studied for the oral medical board exams, which he took and passed, while in Chicago. When Banani's residency was complete, they returned to Memphis where he reconnected with Dr. Lewis and Dr. Baker.

"Dr. Lewis died suddenly and Dr. Baker was about to retire, so I started a solo practice in 1983," says Dirghangi. Many colleagues discouraged Dirghangi from practicing alone, but he believed that if he worked hard and gave his best to the patients, all would be well. Thirty years later, Dr. Dirghangi is seeing the third generation of patients within a family, and they often invite him to family gatherings and celebrations.

Dr. Dirghangi works long office hours and Saturdays. "This is my soup kitchen, and I am happy about it," says Dirghangi. "The preacher, the teacher and the healer are well respected. I am blessed that I am two of them, a healer and a preacher." Dirghangi is involved in missionary work in India and offers his support in education and health care.

Dr. J. Dirghangi's focus has never been to accumulate wealth. "It does not matter how much money we have; at the end of the day, we are happy that we have other people's love, that's all that matters."

"I take care of all walks of life from very rich to very poor patients," says Dirghangi. "I will be a doctor as long as my eyes, my brain, and my health permit."

DUCKWORTH PATHOLOGY GROUP, INC.

LEADERS ARE NOT BORN BUT ARE FORGED THROUGH LIFE experiences, and Dr. John Kelly Duckworth is no exception. He was born in Hazard, Kentucky in 1928, but lived in White House, Tennessee. Dr. Duckworth laughs as he explains that "the farther up the holler you lived, the meaner you were—and my house was the last one."

During World War II, the family moved to Tullahoma, Tennessee where Duckworth's mother filled a position vacated by men going to war. She was a strong, capable woman who raised three children alone. When the family moved to Nashville, Duckworth and his two siblings attended the Peabody Demonstration School.

The Great Depression and the war had taken a toll on families, and Duckworth describes his family as "poor as church mice." His mother, a former teacher, convinced the school to let her children work to pay their tuition. "I took care of the indoor swimming pool and my brother Gordon worked in the cafeteria," explains Dr. Duckworth. Many teachers at Peabody held Ph.D.s and set the pace for the Duckworth children, all of whom would become physicians.

In 1945, Duckworth enlisted in an 18-month engineering program at the U.S. Merchant Marine Academy. This enrollment led to a commission in the U.S. Navy Reserve and the Merchant Service. The war ended just after Duckworth arrived and enlistees were given the choice of leaving, free of their duty, or staying and most likely earning a bachelor's degree. Duckworth says that for a poor kid from White House, Tennessee, it was a "no-brainer." Jobs were scarce when he graduated in 1949, so he came to Memphis and worked with the Corps of Engineers. Duckworth knew very quickly this was not the career he wanted.

In 1950, the North Koreans crossed the 38th parallel, and war followed. Duckworth volunteered and the U.S.S. *Achernar* would take him to Pearl Harbor, Alaska, the Aleutian Islands, and Greenland before the ship was dry-docked, and many sailors were discharged as the need for war ships was reduced. Duckworth was once again a civilian.

"My older brother Gordon was a physician, so I decided I wanted to go to medical school," says Duckworth. "I was told my engineering degree wasn't good enough for medical school, and that was upsetting." Duckworth enrolled at Memphis State University as a special student. Then he met, "...the prettiest girl I ever saw," Norma Jean Glover, and they were married in 1953. Duckworth entered the University of Tennessee Medical School with the help of the G.I. Bill and later worked at the Shelby County Old Folks Home, which was located at the Shelby County Penal Farm. He graduated in 1956, second in his class.

During his internship in Greenville, South Carolina, Dr. Duckworth met Dr. Art Dreskin, the chief pathologist at Greenville General Hospital. Dr. Dreskin's scientific approach to performing biopsies appealed to Duckworth's analytical, engineering side. "After several years as a general practitioner, I came to Memphis in 1960 to study pathology, and I finished in 1964," he states. "I passed the anatomical and clinical pathology boards." He began working with Dr. Douglas Sprunt at the University of Tennessee, and, in 1965, Methodist Hospital hired Dr. Duckworth as director of laboratories. Soon, he became independent and the Duckworth Pathology Group (DPG) was founded in 1967. Dr. Duckworth's attention to detail and "no second rate work" gained respect, and soon the lab needed more physicians on staff.

Dr. Art Wilson's part-time job at Lankenau Hospital introduced him to pathology, while he attended St. Joseph's College in Philadelphia. After medical school and an internship at Temple University School of Medicine, he served two years on a nuclear submarine. Dr. Wilson was completing his residency at the University of Washington when he came to Memphis for an interview with Dr. Duckworth. In 1976, he began his career with Duckworth Pathology Group.

In addition to his routine duties, Dr. Duckworth asked Dr. Wilson to develop a lab for clinical chemistry and toxicology. Later, Dr. Wilson would become president of the medical staff in 1998, and chief of staff of Methodist Healthcare in 2000, becoming the first pathologist in Methodist history to do so.

Dr. William Fidler joined the group in 1978. After attending George Washington University, he spent four years in a pathology residency at the University of Michigan. After military service and working as a pathologist at Madigan General Hospital in Fort Lewis, Washington, he completed a fellowship in cancer pathology at Memorial-Sloan Kettering Hospital in New York City. Then, he returned to the University of Michigan for five years, specializing in cytopathology. He and his wife were originally from the South, so when the *New England Journal of Medicine* listed a job in surgical and cytology pathology in the South, he was interested. The next day, an old colleague from Michigan called about an opening at Duckworth Pathology Group. It was perfect timing.

"In the first two years at Duckworth, it became apparent to me that we had more interesting material in my field of surgical pathology and cytology than we did at the University of Michigan," says Dr. Fidler. He expanded the cytology lab to include the subspecialty of aspiration cytology, and Duckworth Pathology Group became the

Duckworth Pathology Group, left to right, Drs. Pamela Sylvestre, Royce Joyner, Edwin Raines, Holly Hilsenbeck, David Robins, Olga Lasater, Charles Handorf, Robert Bradley, Richard McLendon, Frank White, Barry Randall, Cristina Shimek, Noel Florendo, David McGregor, Neyle Sollee, Thomas O'Brien, Alan Boom, and Jeffrey Roux.

first pathology laboratory in Memphis to both perform and interpret aspirations.

Dr. Duckworth established one of the first pathology protocols for quality assessments to check the validity of clinical samples. Errors were not acceptable and this earned Dr. Duckworth the reputation for being "difficult."

In 1980, Charles Handorf was a second-year pathology resident at the University of Tennessee. Uncertainty prompted him to seek a transfer, but this meant waiting six months to enter a new program. A friend recommended training at Methodist Hospital until the transfer. "I heard about this terrible tyrant, Duckworth, and really didn't want to have anything to do with him," says Dr. Handorf. "But, I was stuck, so I met with him." Handorf asked to continue training at Methodist until his transfer and Dr. Duckworth agreed. "He was very kind to me, so I came away not knowing how he got this reputation."

Handorf realized the "tyrant" was fair but was intolerant of laziness or incompetence. After six months, Dr. Handorf recalls how Dr. Duckworth more or less told him he was hired, and then informed him that all Duckworth pathologists were board certified. "It's June and you will pass your boards in September or you will be gone," announced Dr. Duckworth. Dr. Handorf passed the board examination and later became director of laboratories, president of Duckworth Pathology Group, director of outreach services, and chairman of the Pathology Department at the University of Tennessee.

Dr. Duckworth continued as director of laboratories from 1965 to 1980. He retired from Duckworth Pathology Group in 1989, but returned to work in 1991 when Methodist Hospital asked him to organize a clinical research department. He continued working there until 2003.

Dr. John Kelly Duckworth founded the Duckworth Pathology Group in 1967.

When Dr. Handorf became chairman of the Pathology Department at the University of Tennessee, he asked Dr. Duckworth to become the director of the Pathology Residency Program. "I couldn't think of anyone who would be better in that position," says Dr. Handorf, "and, besides, it completed the circle of my coming to Duckworth in 1980, staying, and eventually deciding Dr. Duckworth was okay, and hiring him."

"I tell the residents to be nice to everyone because you never know who you will be working for," says Dr. Duckworth laughing. Dr. Duckworth is "having a lot of fun" teaching the residents and credits Dr. Handorf with reviving the pathology residency at the University of Tennessee.

Dr. Thomas O'Brien is the current president of Duckworth Pathology Group. Since he joined the group in 1991, he has seen many changes in how tissue samples are analyzed. "The need to rapidly process a biopsy, and give the diagnosis in a timely fashion, has become the standard of care," he says. "We are looking forward to the next generation of testing that will take place and change the paradigm of how we treat disease."

Methodist Le Bonheur Healthcare continues to be Duckworth Pathology Group's primary customer. "Methodist has been very good to me and DPG," says Dr. Duckworth. "I would like to think we have built one of the best pathology groups in the world."

"In the future, pathologists will likely be more integrated into clinical medicine than during the last decade," says Dr. O'Brien. "Dr. John Duckworth is an extrovert, a rarity in pathology. I think this underlies his extraordinary professional success in clinical lab medicine. Our vision for the future is to carry on the legacy for superior work and remain at the forefront of advancements in pathology."

EASTMORELAND INTERNAL MEDICINE

Dr. Thomas E. Motley received his medical degree from Howard University in 1965, followed by an internship at St. Alban's Naval Hospital with affiliation at King's County Hospital in New York. After his internship, he became medical officer on the U.S.S. *Yosemite* for one year. In 1970, he returned to St. Alban's to complete his residency with affiliation at Cornell University Medical Center and the New York Hospital. Dr. Motley also served as internist for the evening Internal Medicine Clinic at Bronx's Municipal Hospital.

During his residency, he was inspired by several physicians who served as his attending physicians. "Dr. William Noble of New York taught me the art of medicine and how to keep that in perspective with the science of medicine," Dr. Motley says. " He taught me things like sitting down when talking to patients and their families, so they know you are going to take time with them. That's the kind of physician he was."

In 1972, Dr. Thomas Motley was honorably discharged from the U.S. Navy Medical Corps at the rank of Lieutenant Commander. Dr. Motley ended his naval career as assistant chief of medicine at Millington Naval Hospital in Memphis. He subsequently went into practice with Dr. Andrew Dancy and Dr. Lawrence Seymour. They later moved to a medical office in south Memphis.

About the time of the move, when Methodist Hospital began to appeal for physicians to become more associated with the hospital, Dr. Motley responded by becoming an original tenant in the Medical Arts Center on the hospital campus. He was joined in practice by Dr. Lawrence Madlock, and later by Dr. J. O. Patterson, III. In 1984, Dr. Motley and Dr. Arthur J. Sutherland were the first physicians in the Memphis area to monitor and exercise heart patients regardless of the fact that insurance companies did not recognize or reimburse for this rehabilitation. They had great success which gained the attention of other physicians in Memphis. Dr. Motley later became medical director of the cardiac rehabilitation program for the Methodist Hospital system.

Dr. Thomas E. Motley, top, opened his private practice in 1973. Dr. Stanley Dowell, above, joined him in 1995.

Lori Kessler of Methodist Healthcare asked Dr. Motley to participate in early outpatient clinical trials and he agreed. Most of the trials were grants for health education on high cholesterol and hypertension.

In 1995, Dr. Stanley Dowell and Dr. Ralph Taylor joined Dr. Motley and Dr. Patterson to form the Eastmoreland Internal Medicine group. After graduating from Meharry Medical College in 1983, Dr. Dowell completed an internship and residency in internal medicine at Henry Ford Hospital, where he practiced until 1986. He became a staff internist at the Memphis Health Center and its director in 1988. He worked in a private practice, before joining Dr. Motley, and also served as president of the Bluff City Medical Society from 2003 to 2004. Dr. Dowell is currently serving as chief of staff for Methodist Extended Care Hospital.

Dr. Dowell's benevolent work includes trips to Africa to set up clinics for the poor. "We are a global world and when disease is out of control in Africa, it affects everyone," says Dr. Dowell. "The people there are so appreciative of everything you do for them." Dr. Dowell's meetings with government officials offer hope that policy changes will lead to better health conditions.

Drs. Dowell and Motley continue to practice at the Methodist University Campus, and their patients remain their primary focus despite the challenges from the changes in healthcare systems. Dr. Motley's advice to youngsters remains to "Find something you want and know how to do and stick with it." He was recently honored for his dedicated service at Methodist Hospital and still enjoys coming to work each day, with no immediate plans to retire.

JAMES R. FEILD, M.D.

JAMES RODNEY FEILD, M.D., WAS BORN IN Memphis and graduated with honors from the University of Tennessee, Memphis where he was elected to the Alpha Omega Alpha honorary society. After interning at John Gaston Hospital, he practiced for a year in Albany, Kentucky, and then completed four years of neurosurgical residency at Baptist Hospital and University of Tennessee-Memphis, followed by a neurology fellowship at the Mayo Clinic. Board certified in neurological surgery, he is also a diplomate of the American Board of Forensic Examiners. Having practiced neurosurgery since 1965, he serves on the active staff of Baptist and St. Francis hospitals and on the courtesy staff of Methodist Le Bonheur Healthcare. In 1975, he completed his M.S. degree in anatomy at the University of Tennessee-Memphis.

Dr. Feild belongs to many professional organizations, local and international. He has received Public Health Service and Heart Association grants and has patented a surgical lumbar shield. With 23 professional publications to his name, he has taught neuroanatomy at the University of Tennessee. In 2007, the Consumers' Research Council of America named him a "Top Surgeon."

Politically active, Dr. Feild has served locally and nationally, including on the Statutory Committee for Medicare (Senator Howard Baker's nomination) and on the Health Insurance Benefits Advisory Council (1971 to 1975). He was also licensed by the New York Stock Exchange.

Married to the late Nancy (Tanner) Feild for 45 years, he has two sons and three daughters. He has been married to Glenda (McCartney) Feild for four years. Still active and interested, Dr. Feild works "almost seven days a week." For one accustomed to difficult situations, he says, "remaining active and living with stress are the keys to longevity."

JULIAN G. FLEMING, M.D.

JULIAN FLEMING'S INSPIRATION TO become a doctor came from a distant cousin. "He was very knowledgeable and a true country doctor," says Fleming. Growing up in Middle Tennessee, he received a B.S. degree from Middle Tennessee State Teacher's College and an M.S. degree from George Peabody College. In 1958, he came to the University of Tennessee Health Science Center rather than continue his medical training at George Peabody's parent school, Vanderbilt University.

Fleming chose internal medicine because "I didn't like surgery and blood scared the hell out of me." Fleming's residency, at Veterans Administration Medical Center, was where he met the chief of cardiology, Dr. Fred Knox, whom he greatly admired. His training continued at St. Joseph Hospital in Memphis with a transitional year of training at Metro Nashville General Hospital.

Dr. Fleming returned to Memphis to begin his first practice with Dr. Fred Knox and later moved to a larger internal medicine group. After eight years, Dr. Fleming began a private practice and still works alone, with the help of nurse Becky Nelson. Some of Dr. Fleming's patients have been with him for as long as 30 to 35 years. He has seen many changes in disease and treatments, and his patients are confident in the recommendations he makes for them. His dedication to caring for them brings him to the office every day. "I have enjoyed aging with my patients," he says.

Dr. Julian Fleming is a member of the American Medical Association and The Memphis Medical Society.

Editor's note: Dr. Fleming retired from the practice of medicine after this article was written.

JACK T. HOPKINS, M.D., FACC

In 1981, I entered the practice of cardiology in Memphis, when I joined Sutherland Cardiology.

Thus, by July, 2011, I will have devoted 30 years of my life to this practice. Back in 1980, the most sophisticated work in cardiology was applying cardiac catheterization, coronary angiography, and early forms of echocardiography in triaging patients with cardiovascular problems. At that time, the "pinnacle" of practice was the accurate diagnosis of cardiac pathology for which surgery corrected symptoms of prolonged survival. Cardiovascular surgery remains an extremely valuable therapy, now integrated with a host of nonsurgical techniques. The development of "percutaneous catheter-based procedures," with advances in cardiac pharmacology, has greatly expanded the benefits available to patients.

When I began performing invasive cardiac catheterizations, coronary intervention with balloon angioplasty was in its infancy. No one foresaw that, by 2010, the majority of patients would have stents implanted during cardiac catheterization.

In 1980, the earliest clinical trials of intervention in myocardial infarction were not yet published. Decades later, extensive trials made revolutionary changes in stopping an ongoing heart attack. First, intracoronary infusions of thrombolytic (clot-dissolving) drugs were used, and then systemic administration of high doses evolved to emergency cardiac catheterization and balloon angioplasty with stent implants. Today, using emergent coronary angiography, angioplasty and stent implants are "routine." The efficacy of this therapy is well-demonstrated by the dramatic decrease in the mortality rate for first- or even second-time heart attacks and the discharge of patients in three days, rather than three weeks.

In 1980, electrophysiology was also in its infancy. There were no implantable defibrillators and no biventricular pacemakers available. Large-scale studies of these devices demonstrate great benefits in preventing sudden cardiac death and markedly improving congestive heart failure.

Also, in 1980, aspirin, beta blocker drugs, digoxin, heparin anticoagulant, and spironolactone were either not available or not widely used for heart disease. Now they are considered beneficial, but not definitively proven in the therapy of congestive heart failure, dilated cardiomyopathy, and coronary artery disease. In 1980, ACE inhibitor drugs, ARB drugs, stating drugs for cholesterol, calcium channel blocker drugs, and powerful oral and intravenous platelet inhibitor drugs had not yet been released for clinical use, but today clinical trials have definitely demonstrated the benefit of these drugs in treating or preventing cardiac pathology.

Thirty years ago, virtually all cardiovascular surgery was performed utilizing cold cardiologic cardiac arrest and circulatory bypass with pump oxygenator. In fact, that technique was probably responsible for widespread proliferation of coronary bypass surgery throughout the United States. No one envisioned that most coronary bypass procedures in 2010 would be performed on "warm beating heart," avoiding the complications of circulatory bypass.

I have enjoyed 30 years of the clinical practice of cardiology. Writing this stimulates the review of an incredible cardiovascular, technological evolution yielding more effective therapies, and I am deeply humbled. However, so much more needs to be done. Dr. Thomas Frieden, director of the Centers for Disease Control, stated that the United States spends one of every six dollars on health care, and yet is doing an extremely poor job preventing high cholesterol and blood pressure and controlling smoking.

My clinical cardiology experience does convince me of at least two things. One, that technology will continue to progress and advance in ways that I cannot even imagine, and I am sorry I will not likely see another 30 years of it, and two, Dr. Frieden is absolutely correct in noting that prevention, with emphasis on control of obesity, high blood pressure, cholesterol and high risk behaviors, is incredibly cost-effective compared to the cost of developing new technologies for cardiovascular disease at its later stages.

LESTER R. GRAVES, JR., M.D., SAM J. COX, M.D., AND JACK C. SANFORD, M.D.

John Maury, M.D. was on the faculty at the University of Tennessee Medical School as head of the Department of Obstetrics and Gynecology in earlier years. He was the uncle to William P. Maury, M.D., who was the founder of our practice group. Initially, our group consisted of Dr. Maury. His first partner was Henry Leigh Adkins. I then followed Dr. Adkins. Drs. Jack Carter Sanford and Sam J. Cox became members of our practice after our new office at 910 Madison offered us more space. I recruited Dr. Sanford after he completed his residency at John Gaston Hospital, the charity hospital linked with the University of Tennessee Medical School. Several other doctors joined us through the years but entered and left by reason of existing events.

Our destination in life is directed by many events. I will try to piece together some of those happenings that directed me into the medical field. As a young person, I admired an old family practitioner who, to me, was the epitome of compassion and good patient care. He was the model of what I felt I wanted to be. A good friend and resident in obstetrics and gynecology recognized my interest in that field and encouraged it. From that point forward, my sole desire was to become an obstetrician and gynecologist. Between my internship and residency, I met Dr. Phil Schreier. He was the head of the Department of Obstetrics and Gynecology at the University of Tennessee. He became my chief, my mentor, and my good friend for many years. He gave me the opportunity to acquire a Masters Degree in Endocrinology. It was during this time that I happened upon a few sentences of prose which influenced me all of my life and established me as a pro-feminist.

> *Woman is the warehouse*
> *Of the world's chief trade*
> *On this soft anvil*
> *All mankind was made*
> —S. Leon Israel, M.D.

A good friend converted this into a handsome framed needlepoint which has hung over my study desk for years.

After completing an internship at Baptist Memorial Hospital in Memphis, there was a slight interruption during my years in the Navy. I then returned to Memphis and completed my residency at John Gaston Hospital after which I spent a few years as a clinical instructor in a Department of Medicine teaching at the University of Tennessee, and in that capacity, advanced to a position of an associate professor. It was at this point that I joined Dr. Maury and Dr. Adkins.

When first entering practice, there was less pressure related to the care of patients, however, our responsibilities were much greater because of the diversities of problems we were called upon to address. The obstetricians and gynecologists were more like family practitioners. We treated the patients almost in their entirety. All medical problems that developed were part of the care related to our patients. If our pregnant mother was a diabetic, we treated the diabetes along with the pregnancy. If our pregnant or operative patient developed a vascular problem, such as a deep vein thrombosis, we also took care of that and any other medical problems. As residents, our chief, Dr. Phil Schreier, declared that pregnancy and gynecological problems at the time of hospital triage to our hospital established the departmental admission; any other medical problem that arose out of this was our responsibility. Should a patient present with a breast mass, we did what was necessary to diagnose and treat the problem which included surgical care and/or radiation if necessary. There were no mammograms in those days. Diagnosis was by sight and examination.

Computer tomography scans and magnetic resonance imaging were not available. X-rays of the abdomen had to suffice. Cancer of the cervix and endometrium was treated by the obstetrician/gynecologist as a joint effort usually with the radiologist and somewhat later, with the oncologist and radiation therapist. The passage of time established the sub-specialties which usurped most of the responsibilities of the obstetrician/gynecologist. I say this not in a tone of resentment nor with a feeling of replacement, for it was a better development for medicine and for medical care. The responsibilities became too great to carry out without

OPPOSITE PAGE: *William P. Maury, M.D., the founder of our practice group.* ABOVE: *From left to right, Drs. Sam J. Cox, Lester R. Graves, Jr., and Jack C. Sanford.*

the help of others. Many of the requirements would have broken the backs of those who would carry the advance of newer developing technology in our field, not to mention the implication of legal destruction directed toward physicians by risks taken to advance new concepts for better patient care.

The advance of our specialty has been staggering to witness and unbelievable to be a part of. For a period of time, which has been called "The Golden Years of Medicine" it appeared that all new developments were directed to women's care and research. It was a critical time to be involved in the developing care of women. We were, indeed, on the leading edge of this advancing specialty. It was a far cry from the days of delivering women at home or the repairing of the first vaginal fistula without the aid of suture, sterile fields, or antibiotics.

There were no air conditioned operative suites at John Gaston Hospital. Windows were opened to keep the temperature bearable. Unfortunately, insects gained access and nurses armed with fly swatters were a necessity. The duty of the circulating nurses was to be sure that all perspiration dripping from our brows was removed to prevent incisional contamination. Straws were placed beneath our masks to give us liquids and prevent overheating and dehydration.

The first laparoscopic gynecological procedure in Memphis was performed at "Doctors Hospital." This was a private, for profit hospital, which did not survive. At that time, as head of the Department of Obstetrics and Gynecology at Baptist Memorial Hospital, I established a committee to acquire the equipment and personnel necessary to develop laparoscopic surgery. The changes orchestrated by the new surgical technique have been ongoing ever since. It completely renovated the culture of surgery in our area with outpatient surgery and all of its ramifications. This new development reduced post-operative recovery time which eventually reduced the number of beds, even total floors in hospitals, forcing reduction in hospital size and reorganization of hospitals.

Laparoscopy was initiated in the early 1970s and rapidly grew in popularity. A general surgeon, Dr. Charles Olim, initiated the precursor to laparoscopy in the form of peritoneoscopy in our area.

The development of laparoscopy has been a very exciting time for the field of gynecology and advances have been extraordinary. The tissue of the abdominal cavity and its contents were better approached with a new cutting instrument, enter the development of the laser and all requirements of its use. Its use spread to all the specialties with the modifications required. The results have affected not only the lives of patients but the general structure, makeup, and development of hospitals throughout the past several years. Space required by the hospitals has been reduced, primarily because of the development of outpatient surgery, shorter hospital stays, and markedly improved care associated with better patient education.

As the obstetrical unit at Baptist Central Hospital transferred to the newly established Baptist Hospital East, the last labor patient was mine. This patient was accompanied by Ms. Patsy Davis, head of the Nursing Division of Obstetrics, to the new hospital by ambulance to the new delivery area. She was the last undelivered patient at Baptist Central and became the first to later be delivered in the new Baptist East obstetrical suite.

The changes in medicine were always challenging and interesting. The new developments created an ongoing excitement in this specialty. Dr. Marion Sims, the founder of our specialty, would have enjoyed witnessing these changes would he have been able to, during his lifetime.

Our original delivery suite at Baptist Memorial Hospital was smaller. There were four delivery rooms, which substituted also for rooms used for Cesarean sections and minor surgical procedures. There were two common labor areas with adjoining doctors' sleeping quarters. This consisted of an area that would sleep six to eight doctors in double bunks, all in one room.

Twilight Sleep was the obstetrical anesthesia, which consisted of a combination of Demerol and Scopolamine. The pain relief was not optimal but there was less infant sedation. The advantage for the mother was the loss of memory for everything that happened. Sometimes this created great difficulty for the nurses as the patients were often very unruly. On or about that same time, a Trilene Mask was used, which gave the patient some sedation and some pain relief and was self administered with contractions only for brief periods. It was an early ancillary comforter before she was offered Twilight Sleep.

There were no perinatologists, no neonatologists. The obstetrician substituted for both, doing all of the resuscitation on newborns that required it. In addition, all high-risk obstetrical deliveries were the responsibility of obstetricians at that time.

There were long stays in the hospital for the mothers. Vaginal deliveries were kept in the hospital for five to seven days. Some of these deliveries were spontaneous but the majority were low forceps deliveries. Patients that were delivered by Cesarean section usually stayed in the hospital for ten days. Breech presentations were almost always vaginal deliveries. Infrequently, they delivered by Cesarean section. A marked change in policy has occurred since then with a much higher Cesarean section rate now prevailing, prompted by many factors originating in the name of patient safety and eventually to full defensive medicine steeped in malpractice concerns.

Natural childbirth was not by choice but by the surprise of early natural onset of labor. There were few epidurals at this time. General anesthetics were rushed or circumvented because of the effect that was transmitted to the child in sleepiness and sedation. Anesthesia was only started when the field of surgery was prepared and the surgeon was ready to make his incision for section. In the case of an emergency, speed was the essential factor. Local infiltration and spinals were utilized. Epidurals came later and in a very limited manner. Initially, they were given during the daytime only, when the anesthesia department had someone to oversee and administer them. It was not until much later when they were available twenty-four hours a day at Baptist Memorial Hospital East.

The first use of fetal monitors was a time of learning-as-you-go approach. We had a direct telephone line with Vanderbilt University Department of Obstetrics, which was for consultation as needed. They were the first to use fetal monitors and were most knowledgeable. As knowledge was gained, fetal heart monitors went from being used only on the most high-risk patients to being a routine attachment to every fetus during the entirety of labor for instant review. Later, there was transmission to a central panel located in juxtaposition to the central nursing station for constant review.

When I started practicing medicine, there were only three female obstetrician/gynecologists. As a resident and teacher at the University of Tennessee, I had one of the few classes containing female students. In my residency group, we had the very first female resident. I looked on this as the beginning of a major turnaround in our field. Women now make up a major part of our specialty. Several practice groups are now made up of women only. Our specialty has changed considerably from what it was then to now. We now practice out of the Baptist Hospital for Women.

There have been changes in the way we practice medicine not only in the academic approach, but also in our personal communication with our peers and patients. The closeness I once enjoyed with my partners has become more distant; even though we are in the same office for the same period, we infrequently have the time to communicate with each other. Is this because of the change in our personality? I would think not. We are required to distribute time to learning the new techniques of our specialty. We must maintain the qualifications to stay on hospital staffs including the CME credits demanded for staff admission and the responsibilities of various committees we are required to be active in. It seems we are so busy trying to keep up with all of the activities required to maintain our position in the medical community that we are having to reallocate our most precious commodity, time.

Patients demand more knowledge as they are better informed and have a thirst for more information. We are now called upon to discuss problems not only with the patient but with the entire family. The genetic involvement of illnesses requires family participation in diagnosis and treatment. Does this conflict with our family life? I would say so. Physicians now, more than ever before, feel the strain of the loss of family ties when time with their families is so limited. Doctors are driven, I feel, by changes and pressures which have surrounded us, influenced us and changed our approach to various issues. The need for time has rearranged our relationships.

Have we become more defensive? Have our patients become more defensive by reason of expense and the head-on conflict with stress in their workplace and home? I feel the answers to these questions are Yes, Yes, Yes. We have been converted to businessmen and/or businesswomen which carries us away from the caring, compassionate relationship once enjoyed with our patients who are now redefined as clients. However, the non-flexibility and expense of medical care often is a burden many cannot carry.

To cover all the changes in medical care would be beyond the scope of this writing, but we must allow exposure to these happenings or an understanding as to their influence on medical care. That would have to include the advances in medicine, changes in medical coverage, pharmaceutical dominance, political interventions, and new developments in disease previously unrecognized with the unbelievable expense in medical care these developments have created. The conflicts that they have established have developed obstacles almost impossible to hurdle.

We are no longer an island unto ourselves. We have the availability of newly developed technologies and specialties which supply us with a wealth of information and support. This has established a parameter of care unenjoyed in the past. We might feel we have peaked in patient care, that our expertise has reached the summit of the mountain, but there will be more advances to establish even better care for our patients.

Dr. Maury was very supportive to the practice family. His strength was his ability to be a cohesive force. He was not a businessman; economics was not his forte. He was more steeped in the common sense of life and medicine and had an ability to recognize a potential problem and correct it before it developed into an absolute disruption.

When I joined the group, we had no business manager. Retirement had no definition in our practice. As with all new ideas, my development of a retirement plan met, initially, with resistance. It started out as a money purchase pension plan, then a profit sharing pension plan, and finally into a 401K plan. The medical health plan for the group health insurance developed later.

In the beginning, there was a considerable amount of comradeship among the doctors, particularly in the obstetrical and gynecological field. We enjoyed an organization that was called the "Vittles Statistic Committee." If you were an Ob/Gyn doctor, you automatically qualified to become a member. The meetings followed the format of an excellent meal accompanied by beverages and a robust card game. Sometimes, we met elsewhere, but most meetings were at Dr. Finus Taylor's home. He was the originator of the event. The entertainment was ourselves. We enjoyed talking with each other and in these conversations we transmitted the newer things that each of us had picked up at various meetings or just discussed world affairs with the occasional interruption of a good joke. The dinners were always well attended. You learned such informative things as where the hunting was best, where the fish were biting this week, and who had miscued in his life lately.

This has been replaced by dinners and lunches contributed by the pharmaceutical houses to promote their products. They are greatly appreciated but are not exactly quite the same as the "physician get togethers"; nor do they establish the same effect.

Dr. Nate Atherton was one of the original obstetricians and gynecologists to practice in Memphis, Tennessee. We were all invited to his country retreat in Arlington once a month where we enjoyed the fishing, swimming, and horseback riding offered by his farm. Many of the parties for residents were held at his farm. Lunch would be prepared by Nate or other members, or occasionally a chef of one of the local clubs. We had a member who once proclaimed that Teal were the most delicious ducks in the world to eat and that he would never eat any other kind of water fowl. Dr. Atherton (to prove a point) shot a half-a-dozen crows and prepared them like ducks. Everyone else had duck of another type, but this doctor was served crows and told they were Teal. Everyone else knew about this except that doctor. He ate them, all the time expressing how delicious they were. Everyone had a good laugh but he never forgave Dr. Atherton.

Baptist Hospital contributed to our togetherness by making available a dining room just for the doctors. The largest gathering occurred at lunch. We would get together, enjoy a little conversation away from hospital stress. Many of the doctors who hunted and fished would bring in their fish and game. Those who did not hunt or fish had the advantage of eating wild game. Dr. Frank Gromer, the administrator of Baptist Memorial Hospital, was responsible for this. He was a close friend to all physicians and he was loved and respected by all those who knew him.

This has been an enjoyable period in my life and was always challenging and interesting. The new developments have created an ongoing excitement for me in this specialty. The memories of all these years spent is my wealth, not to be surpassed. Is there anything more exciting than the birth of a child or more interesting than the wonderment of its creation? Is there anything more enduring than the love felt and expressed by the mother toward her child? I have been rewarded by the expressions of happiness and joy proclaimed by patients and their family members involved in each occasion of birth. The 50 years experiencing those events have left me with a feeling of inner happiness and a lasting admiration for women.

INTEGRITY ONCOLOGY

Dr. A. Earle Weeks has begun a new decade with a new practice, the second one he has founded since arriving in Memphis in 1989. For Dr. Weeks, experience has been a great teacher. "I managed to survive, but it was a real challenge," he says of the first practice he founded in 1997. "It was a difficult learning experience. They don't teach business in medical school—they don't teach negotiations or how much rent should cost. There were a lot of lessons to be learned."

Dr. Weeks, along with two, long-time colleagues—Dr. Suhail M. Obaji and Dr. Margaret Gore—opened Integrity Oncology in September 2009. They were determined to continue the best of what was learned from their first practice and the results are a beginning that is very impressive. "I'm excited with our progress and what we've accomplished thus far," says Dr. Weeks.

In April, Integrity Oncology celebrated a milestone by opening its new location on West Poplar in Collierville. While it was a new address for Integrity Oncology, it was a familiar location for the three physicians who founded the practice. The historic, 100-year-old residence was purchased by Dr. Weeks and Dr. Obaji in 2003, six years after founding their original practice. The office sits on 3.5 acres between Houston Levee and Bailey Station. The structure was large enough for an ultra-modern health care facility, and at the same time, offered a more comfortable, friendly atmosphere that both patients and staff appreciated.

When Drs. Weeks, Obaji, and Gore established Integrity Oncology, they hoped somehow it would include the facility on West Poplar that was so special to them. Happily, seven months after founding their new practice, the friendly old "house" was retained and it has become the practice's unofficial symbol.

While Integrity Oncology doctors know that treatment is crucial, the manner in which the care is given is equally important. The three physicians feel treatment should be delivered in a convenient, friendly atmosphere. The doctors also believe patients' families and close friends should be involved.

The West Poplar office extends hours twice a week—from 8 am to 9 pm on Tuesday and Wednesday—offering patients more flexible treatment times. "We realize patients who have full-time jobs or who depend on others for transportation and assistance will benefit from our remaining open longer," Dr. Weeks states. "We want to provide scheduling options that offer greater flexibility and convenience in the lives of patients and their families." This convenience extends to three other locations in the area: Memphis, Union City, and West Memphis.

Establishing treatment centers close to patients' homes was something Dr. Weeks began in his first practice. His idea of bringing treatment to the patients rather than having the patients travel long distances to a facility proved beneficial. "We went to where the population was rather than have them come to us," he explains. "The result is more of a neighborhood feel than an industrial corporate complex. That's been my focus since day one—more of a community feeling rather than a large business."

This strategy continues as Integrity Oncology offers the newest technology and most modern equipment. Dr. Weeks still insists on convenience for the patients and their families, as well as a stress-free treatment atmosphere. As Integrity Oncology continues to grow, this focus will not change.

Integrity Oncology moved into its new location in this 100-year-old house (top) in April, 2010. Drs. A. Earle Weeks (above), Suhail M. Obaji (center), and Margaret Gore (below) founded Integrity Oncology in 2009.

ROBERT L. JACKSON, M.D.

Dr. Robert Jackson was born in Marion, Indiana. While growing up in Wilberforce, Ohio, he became interested in flying and meteorology. His father was a flight instructor who trained the Tuskegee Airmen. At Miami (Ohio) University, where he received his B.A. degree, he majored in zoology. He received a public health scholarship, which would be repaid by service. He received his M.D. degree from Meharry Medical College in Nashville, Tennessee, and completed his internship in internal medicine at Akron (Ohio) City Hospital. He was in Memphis from 1979 to 1983 with the National Health Service Corps as a general medical officer, serving in Rossville, Tennessee. From 1983 to 1986 he was at UCLA's King-Drew Medical Center for his residency in dermatology.

With this varied background, Dr. Jackson returned to Memphis to practice medical and surgical dermatology, including laser hair and keloid removal. He is also involved in community service, participating in the Memphis Health Plan through the Church Health Center. He has been married for 29 years to Debbie (Pittman) Jackson and they have two sons, Ryan (24) and Ian (20), both college students.

Dr. Jackson is a member of Alpha Omega Alpha Honor Medical Society, the American Academy of Dermatology, the Memphis Dermatology Society, the Bluff City Medical Society, the Memphis Medical Society, the Tennessee Dermatology Society, the American Society for Dermatologic Surgery, the American Medical Association, and the National Medical Association. Although no longer involved in flying, he remains interested in meteorology. He reads widely in professional journals and in Christian publications. A dedicated, scholarly physician, Dr. Robert Jackson brings to his patients both wide experience and compassionate concern.

LAKESIDE BEHAVIORAL HEALTH SYSTEMS

Since 1976, Lakeside Behavioral Health Systems has been one of the largest freestanding behavioral health hospitals in the country. With one in four American adults suffering from a diagnosable mental disorder, Lakeside professionals develop care plans and provide separate facilities to maximize healing for all ages. This includes a facility for geriatric patients addressing their physical, mental, and emotional challenges in a secure, therapeutic environment.

Lakeside's residential programs treat physical, mental, emotional, and spiritual needs through a 12-step model and are highly effective for individuals with a history of drug and/or alcohol abuse, eating disorders, or other self-injurious behavior. Lakeside's Bipolar Spectrum Program was developed to identify mood cycles, and aids in developing treatment plans.

The Psychological Trauma and Suicide Intervention Programs assist patients who have experienced abuse, violent crime, catastrophic loss, combat, domestic violence, and suicide attempts. Lakeside also works with professional peer agencies to treat impaired professionals such as doctors, nurses, and lawyers suffering from addictive or psychiatric disorders. This treatment extends to those with high-risk jobs such as firemen, policemen, and emergency/rescue workers.

Focusing on behavioral aspects of dependency, Lakeside has developed a Partial Hospitalization Addiction Program and Young Adult Program to complete a spectrum of services. Partial hospitalization (day treatment) allows patients to return home in the evening. Specialty programs include an Asperger's Day Camp focusing on social, sensory/motor integration, learning skills, and adaptive behaviors for children affected by this specific type of autism.

Lakeside continues its excellence in the development of programs and continues to receive high marks in patient satisfaction in Memphis and the Mid-South.

MAYS & SCHNAPP PAIN CLINIC AND REHABILITATION CENTER

Dr. Kit Mays and Dr. Moacir Schnapp are the first to say their partnership has defied the odds. "Moacir and I are about as different in training and in disposition as you could imagine," says Dr. Mays. "He's a Jewish guy from Brazil, and I am a Christian born in a small Arkansas town." Dr. Mays, an anesthesiologist and Dr. Schnapp, a neurologist, would become a formidable team in relieving chronic pain.

It all began before there was a "Mays and Schnapp." In 1975, Dr. Schnapp, a few years younger than Dr. Mays, was finishing school in Brazil and beginning a residency. Dr. Mays was in Memphis working with his mentor, Dr. Robert Popper, chief of anesthesiology at the Kennedy VA Hospital. Dr. Popper taught nerve block procedures to a generation of anesthesiologists, and Dr. Mays was a resident and then assistant chief of anesthesiology. From 1976 to 1982, Dr. Mays collaborated with Dr. Bill Byrne, the chairman of biochemistry at the University of Tennessee, studying spinal fluid endorphins in patients with pain.

In 1976, Dr. J. T. Robertson, Chief of Neurosurgery, Dr. William Webb, Chief of Psychiatry, and Dr. William C. North, Chief of Anesthesiology at the University of Tennessee, asked Dr. Mays, an assistant professor, to establish a multidisciplinary pain clinic at the university. It was one of only four in the U.S. and Dr. Mays became its first medical director.

In 1979, the prospect of a fellowship brought Dr. Moacir Schnapp to Memphis. "If there had not been a fellowship for Dr. Schnapp, we would have invented one," says Dr. Mays. Dr. Schnapp began working with Dr. Mays at the University of Tennessee. "They said I had new ideas, and Moacir knew how to put it all together and make it practical for patient care." But at that time, "We were considered at the bottom of the medical field," adds Dr. Schnapp. "Patients were either told the pain was in their heads or they would have to live with the pain." Pain management was a new frontier.

"Kit was my teacher," says Dr. Schnapp, "and of the many things I learned from him, the most important is not found in books. He was an anesthesiologist and, at that time, most anesthesiologists never spoke to their patients or saw them awake. Dr. Mays spent time with patients who actually felt better after seeing him. It was remarkable to watch this

After several years of collaboration, Drs. Moacir Schnapp, left, and Kit Mays, right, began a practice together in 1985.

brilliant man break new ground in the treatment of chronic pain. This has been his mission." "We broke new ground together," Dr. Mays adds, "and it takes more than technical skill to reach patients on the scale that we have and thrive in a 30-year partnership. What impressed me about Moacir, besides his tremendous intellect and drive, was his integrity. I continue to admire and respect his unquenchable quest for new advances in the treatment of chronic pain."

Dr. Schnapp completed his fellowship in 1982 and returned to Brazil. As medical director of the Instituto de Dor pain clinic, he established the first Brazilian chapter of the International Association of the Study of Pain. Today, this chapter is the largest pain society in South America. "At that time, the best six books on cancer only had five or six pages on pain management," says Dr. Schnapp. Drs. Mays and Schnapp often talked by phone about pain management techniques and research ideas. "I would call him at 2:00 am thinking it was 7:00 am or 8:00 am there, and I could see him listening, nodding, and smiling, but he never told me there was very little time difference between Memphis and Brazil," says Dr. Mays laughing. "Dr. Schnapp smiles and adds, "When you are dealing with pain management, you really need to talk to others in the field; it is very hard

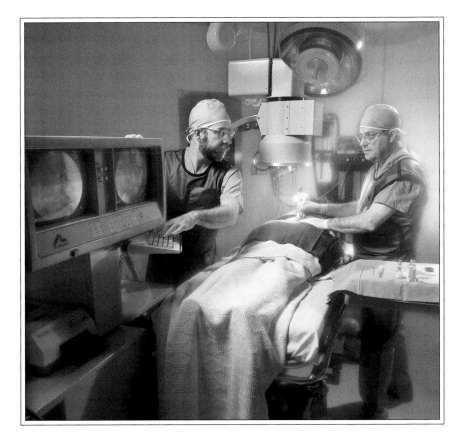

esthetics for chronic neuropathic pain are now widely practiced treatments.

In 1985, Drs. Mays and Schnapp opened a private practice downtown and quickly expanded to East Memphis. In 1993, they partnered with Baptist Memorial Hospital to open a single office as an outpatient pain clinic and rehabilitation center. "The great benefit to our patients is we could examine, diagnose, and administer treatment on the same day," Dr. Mays explains. The clinic also provided physical therapy and psychological care. "We have patients in constant pain and having a partner to discuss it with makes all the difference; you don't take the load home," explains Dr. Schnapp. "If I get tired or reach a dead end, we are working side by side, and I can consult with him. We don't have two people working; we actually have one person with two heads."

to work alone." In five years, they published 17 papers and portions of three books—5,000 miles apart. Drs. Mays and Schnapp reported on these revolutionary studies and they traveled extensively to do so.

The results of Dr. Mays's and Dr. Popper's research, on a "Review of Post Operative Morbidity in 8510 Continuous Spinal Anesthetics," was presented in 1980 to the Seventh World Conference on Anesthesiology in Hamburg, Germany. Also in 1980, Dr. Schnapp's, Dr. North's, and Dr. Mays's work on chronic pain syndromes, including a comparison of morphine and local anesthetic injections of the stellate and celiac ganglia and IV chloroprocaine administration, was reported in Germany and at the First International Post-Graduate Practical Course on Pain Therapy in Vicenza, Italy. In 1980, many of the topics that would bring hope to people suffering with pain came out of Memphis, Tennessee and are now integrated into the standard of care.

Dr. Mays began a private practice in 1984, and Dr. Schnapp returned to Memphis to become the new medical director of the University of Tennessee Pain Clinic. "The best thing that happened to Mid-South pain management was persuading Dr. Schnapp to settle in Memphis," says Dr. Mays. The two doctors were the first to report on the pioneering use of peripheral and intraspinal injections around the nerves for pain control. The techniques of injecting medication into the autonomic ganglia and using local an-

The Mays & Schnapp Pain Clinic and Rehabilitation Center was accredited by the Commission on Accreditation of Rehabilitation Facilities (CARF), in 1988, becoming the first pain management clinic in North America to be accredited by CARF and remains the only one in a 500-mile radius of Memphis. In its 2009 reaccreditation report, CARF surveyors recognized the organization's "excellent reputation locally, regionally and nationally," stating further that "the center's physician leadership has maintained a valued and long-standing presence [as well as] an open and creative atmosphere."

Dr. Mays and Dr. Schnapp's rehabilitative apparatus for treating reflex sympathetic dystrophy received a patent in 2001. A patent is also pending for Dr. Schnapp's *i-Posture Wearable Intelligent Nano-Sensor (WINS)*, featured on *The Doctors*, *Oprah*, and *Late Night with Jimmy Fallon*. This device sends a signal to correct poor posture, and wearers enjoy improved posture, balance, bone strengthening, and less back pain. The doctors also created a separate company partnering with teams in Boston and London to develop prototypes for new rehabilitation devices.

Dr. Mays and Dr. Schnapp maintain their perfect balance as personal opposites, but in pain management, they have one focus—relieving suffering patients. Retirement is not in the immediate future; after 30 years of sharing growing families, friendship, and a practice, they still enjoy working together and facing the challenges of relieving chronic pain.

MARY R. McCALLA, M.D.

Growing up in Jackson, Tennessee, Dr. Mary R. McCalla enjoyed science. After medical school at the University of Tennessee, she completed residency training in Memphis in otolaryngology—head and neck surgery.

In 1987, she began practicing in Memphis and West Memphis. Dr. McCalla has enjoyed teaching at the University of Tennessee throughout her career as a clinical associate in the Department of Otolaryngology. Her curiosity and her interest in new technologies have kept her in the forefront of changes in her field.

Dr. McCalla completed the first total joint replacement of the mandibular condyle in the Mid-South in the 1980s. She performed Memphis's first modified neck dissection with the assistance of plastic surgery resident Dr. Robert Wallace, whose California naval training proved invaluable. He later became chair of the University of Tennessee's Plastic Surgery Division. Dr. Tom Robbins, chair of otolaryngology at the university, greatly advanced this procedure, and head and neck oncologist Dr Sandeep Samant continues to refine it. Dr. McCalla was also instrumental in introducing harmonic and coblation technology to lessen the pain of tonsillectomies.

With the teaching of her former chief, Dr. James Netterville, professor of otolaryngology at Vanderbilt University, she performed Memphis's first true vocal cord medialization to restore vocalization to patients with unilateral true vocal cord paralysis. Tediously carved silastic implants were used at first, now replaced by hydroxyapatite implants.

In the late 1980s, a revolution occurred in sinus surgery. Dr. McCalla and Dr. Walter Cosby introduced the rigid nasal endoscope at the Memphis VA hospital. In 1992, South Dakota ENT specialist Dr. Reuben Setliff introduced the "hummer" at a national meeting. A shaver similar to that used in knee surgery, this instrument removes chronically diseased sinus tissue. Back home, Dr. McCalla began using the hummer, which, with new "true cut" dissection instruments, greatly improved the precision of sinus surgery. Intraoperative computers also advanced surgical techniques. Most recently, Dr. McCalla has used balloon sinuplasty in selected cases.

Dr. McCalla has greatly enjoyed her career of more than two decades continuing voluntary teaching and working to support the University of Tennessee. She has served as chair of the Department of Otolaryngology for the Methodist Le Bonheur Healthcare System and as a member of its Credentials Committee. She is a past president of the Memphis Society of Otolaryngology—Head and Neck Surgery.

With its central location, Dr. McCalla's solo practice serves a wide segment of the community. Originally drawn to medicine to provide preventive health care, she extends her friendly and open approach to her patients' ENT problems to their total care, encouraging and teaching her patients to improve their total health. She has expanded her allergy practice to include asthma therapy.

Dr. McCalla has served on the Memphis Ballet Board, Grace St. Luke's Episcopal School Board, and the Memphis Rotary Club. She is on the Memphis Opera Board and attends Grace St. Luke's Church. She and husband Judge Jon McCalla have two daughters, both grown and pursuing their own careers. Her tennis game has seen better days, but she continues to improve her piano playing. Since the straightline windstorm of 2006, she has raised a beautiful city vegetable and herb garden.

THE MEMPHIS DERMATOLOGY SOCIETY

Before the 1960s, Memphis had only a few dermatologists; one of the first was Dr. Vonnie Hall. In 1961, Dr. Elias William Rosenberg came from the University of Miami to found the Department of Dermatology at the University of Tennessee. During his 45 years at the Unicersity of Tennessee, Dr. Rosenberg trained more than 100 residents, many of whom have served worldwide. Two of the University of Tennessee dermatology graduates became presidents of the American Academy of Dermatology.

The Tennessee Dermatology Society, which became inactive during World War II, was reactivated in 1973. The Memphis Dermatology Society has always been a key supporter of The Tennessee Dermatology Society.

Founding members of the reorganized Tennessee Dermatology Society—1975 photo

Row 1: 1st from the left—Dr. Vonnie Hall; 2nd—Dr. Linda Woodbury, the first female dermatologist in Memphis; 3rd—Dr. Tom Goodman, student of electron microscopy; 4th—Dr. Dan Dunaway, 1973 president of the reorganized Tennessee Dermatology Society.

Row 2: 3rd from the left—Dr. Rex Amonette, the first Mohs Micrographic surgeon in Memphis; 4th—Dr. Don Dismukes, the first Memphis dermatologist to pursue the use of lasers in dermatology.

Row 3: 1st from the left—Dr. Ken Hashimoto, the first resident laboratory research scientist in dermatology at UT; 2nd—Dr. Jerry Whitehead, an expert diagnostician; 4th—Dr. Mark Tanenbaum, the first member of a multi-generational dermatology family; 6th—Dr. Robert Kaplan, an expert in dermatological pharmaceuticals; 8th—Dr. William Rosenberg, founding director of the UT Dermatology Residency program.

Row 4: Dr. Charles Safley, an expert clinician in the area of sebaceous gland disorders.

MEDICAL ANESTHESIA GROUP, P.A.

Medical Anesthesia Group is a prominent group of anesthesiologists who have experienced significant growth in the past 40 years. The current professional membership of the group consists of its physician owners and nurse anesthetists. The group utilizes the anesthesia care team concept to provide safe and effective anesthesia care and pain management in the many locations where its members practice.

Dr. Ray E. Curle, one of the founders of the group, notes that the father of modern anesthesiology was Dr. Ralph M. Waters, who established a residency in anesthesiology at the University of Wisconsin that became the model for later programs. Dr. Waters' program was designed to situate anesthesiology in the medical profession where it began and where it belonged.

Surgical anesthesiology in Memphis has its roots close to the foundation of the specialty. One of Dr. Waters' first two residents was Dr. Emery A Rovenstein, who went to Bellevue Hospital in New York City to start a program there. One of Dr. Rovenstein's first residents was Dr. Lee J. Vernon, the first surgical anesthesiologist in Memphis. After his training, Dr. Vernon joined the University of Tennessee faculty in 1947 and became department chair soon after. Four years later, he entered private practice. Dr. Curle came to Memphis from research and practice in Oklahoma City and in the Panama Canal Zone, where he spent two years in the Army. In Memphis, he began his career doing freelance work at Methodist and Baptist hospitals. As a practitioner, Dr. Curle became board certified and was considering returning to Oklahoma when Dr. Vernon and Dr. Robert L. Knox asked him to join them at Methodist Hospital. Dr. Knox had trained at the University of Indiana and established a solo practice in Memphis in the mid-1950s. This partnership was the forerunner of the Medical Anesthesia Group. In 1964, Dr. Leslie E. Eason joined the partnership as its first locally trained member and, soon after, the group added the first women anesthesiologists, Dr. Patsy Erwin and Dr. Ruth Dinkins. Dr. W. Heymoore Schettler joined the partnership in 1968. Dr. Schettler, the only one of the founding members of the group still practicing, notes that there were only 16 anesthesiologists in Memphis when he started practice. This was the makeup of the group in 1970.

When the rules of incorporation made incorporating professional organizations possible, Drs. Knox, Curle, Eason, Erwin, and Schettler formed the Medical Anesthesia Group, P.A., on March 31, 1970. Dr. Knox served as the corporation's first president. Soon after, the group entered into an agreement with Methodist Hospital to employ the group of nurse anesthetists who had been hospital employees, and the group's anesthesia care team was created. This group of nurse professionals continued as before, providing anesthesia services as group members in the operating rooms and obstetric suites of Methodist Hospital.

In the early 1970s, the group began expanding and has grown steadily, recruiting well-trained and highly qualified physicians and nurses. Included in the early additions was

James Heaton, D.D.S., who was trained as an oral surgeon and also as an anesthesiologist. Dr. Heaton's work ethic and character exemplified the group's spirit of service.

Throughout the 1970s, during which time Methodist Hospital began expanding, the group was the primary provider of anesthesia services at the hospital. With the opening of Methodist North Hospital in 1977, it began covering that location. The group also established cardiac anesthesia services at the new St. Francis Hospital.

New technologies in surgery and in the delivery of anesthesia care drove many changes in the practice of anesthesiology in the next decade. The use of pulse oximetry and carbon dioxide monitoring coincided with the nationwide effort to improve anesthetic outcomes and to develop standards of anesthesia care. The American Society of Anesthesiologists led the way by establishing the Anesthesia Patient Safety Foundation in 1985. The new standards and the technologies that accompanied them were incorporated into the group's practice. This important era coincided with the rapid grown of cardiovascular anesthesia services at Methodist Hospital, which also established the Memphis Neuroscience Institute in 2007, allowing for the expansion of the Medical Anesthesia Group's neuroanesthesia services.

In the years since Methodist Hospital began growing and expanding its network of hospitals throughout the greater Memphis area, the group has closely aligned its growth to meet the needs of its Methodist (now Methodist-Le Bonheur) partner and the population of the Mid-South. The group has developed comprehensive obstetric anesthesia services at several Methodist Le Bonheur facilities and developed a system of perioperative care for the growing number of outpatient surgical procedures. Indeed, it has expanded its services so that it now covers all anesthesia care at four hospitals and four ambulatory surgery centers.

Over the last decade, the Medical Anesthesia Group has dedicated significant time, energy, and resources to the development of an acute pain service. Through the use of multimodal analgesia and peripheral nerve block techniques, group physicians have seen improved patient rehabilitation and shortened hospital stays after procedures that were previously intensely painful. In 2006, when the solid organ transplant program at the University of Tennessee moved to Methodist University Hospital, the Medical Anesthesia Group was asked to participate, accepted the challenge, and is now an active participant in the Methodist University Transplant Institute.

The Medical Anesthesia Group has experienced tremendous growth and stability in the years it has been a part of the Memphis medical community. It currently employs 100 individuals, including 36 physicians and 57 nurse anesthetists. A nine-member elected board of directors serves as the governing body. The officers are elected by the board and are currently President Dr. David Leggett; Vice President Dr. Ray E. Wilson; and Secretary Dr. Edwin Cunningham. Mr. Glen Wimmer is the group's administrative officer. This core leadership, with a clearly defined mission statement, has been the foundation for all corporate and clinical decisions. Much of the group's success has been due to its governance and leadership. In the years to come, the group will continue to be challenged to adapt to the changing times, but will remain dedicated to its mission: "The Medical Anesthesia Group is to be the leading provider of a full spectrum of high quality anesthesia services to the population of the Mid-South."

MEMPHIS DERMATOLOGY CLINIC, P.A.

Memphis Dermatology Clinic, which was begun inauspiciously in 1972 with a dermatologist, his wife, and two employees, has grown into a nine-physician, 50-employee organization that is recognized as the largest center in the world for skin cancer surgery—specifically, for Mohs micrographic surgery. Dr. Rex Amonette opened his practice in a small office near the University of Tennessee Health Science Center, with the help of his wife, Johnnie, his cousin, Brenda Webb Lanier, who had just completed her business degree at the University of Texas, and one other employee.

Earlier, during his second year of residency at the University of Tennessee, Dr. Amonette's mentor and chairman, Dr. E. William Rosenberg, learned of a procedure for treating skin cancer that had been developed by Dr. Frederic Mohs at the University of Wisconsin. Dr. Rosenberg encouraged Dr. Amonette to apply for a fellowship with Dr. Perry Robins at New York University to learn this new technique, which was then called Mohs chemosurgery. He was accepted to the fellowship program and upon completion was the second fellowship-trained physician in the United States to begin a Mohs chemosurgery practice.

Advancing this procedure and gaining acceptance for it in the medical community became the driving force in Dr. Amonette's early career. As the only dermatologist with training in Mohs surgery in the South, he received referrals from all areas of the South and Midwest for treatment of skin cancers.

As the practice grew and additional Mohs surgeons were needed, Dr. Amonette began a fellowship program in partnership with the University of Tennessee Division of Dermatology. In 1975, Dr. J. Harvey Gardner was the first to complete that program, and he joined Dr. Amonette as a partner. The practice was now known as Memphis Dermatology Clinic, P.A. In 1982, Dr. Frank G. Witherspoon followed in Dr. Gardner's footsteps and became the third partner and Mohs surgeon in the practice. Dr. Lee Allen, a native Memphian, took a year from a successful practice in Portland, Oregon, to become a fellowship-trained Mohs surgeon in 1984, joining the practice in 1986.

In the last decade, a younger generation of dermatologists and Mohs surgeons has energized the practice: Dr. Gwen Beard, Dr. Robin Friedman-Musicante, Dr. John Huber, Dr. Amy Amonette Huber, and Dr. Courtney S. Woodmansee.

Standing, left to right, Dr. Lee Allen, Brenda Powell, Dr. Frank Witherspoon, Dr. Rex Amonette, Dr. Harvey Gardner, Dr. Amy Huber, Dr. John Huber, and Brenda Webb Lanier; seated, Dr. Gwen Beard and Dr. Robin Friedman-Musicante.

Over the years, 25 Mohs surgeons, who now practice across the country, have been trained in Memphis.

Mohs Micrographic Surgery requires the physician to act as both surgeon and pathologist in treating various types of skin cancer. The surgery is a staged tracing procedure performed within an office or clinic setting. Each stage involves excision of tissue, preparation of frozen tissue sections in the surgeon's on-site laboratory, and microscopic examination of the frozen tissue sections by the surgeon, who has special training in pathology. This procedure is repeated until all margins and borders of the lesion are free of tumor. One of the most valued benefits of this procedure is the conservation of normal, cancer-free skin.

The procedure has improved from "chemosurgery," which involved using a chemical fixative 24 hours before excising the tissue, to Mohs micrographic surgery, which utilizes local anesthesia and fresh tissue. One of the greatest benefits of the fresh-tissue approach is efficiency in treatment and time saving for the patient. In 1976, the physicians of Memphis Dermatology Clinic expanded the laboratory so that tissue from a biopsy could be made into a frozen section and a diagnosis can be made within several minutes. If the lesion proves to be a skin cancer, Mohs micrographic surgery can be performed on the same day as the biopsy, thereby accommodating the patient and avoiding scheduling the surgery for another time. Mohs micrographic surgery is now recognized worldwide as the most successful approach to curing skin cancer.

In addition to Mohs micrographic surgery, the Memphis Dermatology Clinic physicians practice all aspects of medical and surgical dermatology. Dr. Gwen Beard and Dr. Robin Friedman-Musicante have added a cosmetic component to their practices.

The clinic physicians actively participate in organized medicine, and Dr. Amonette was the first dermatologist in Memphis to serve as president of the American Academy of Dermatology. Clinic staff and physicians also participate in sun-safety programs and local skin cancer screenings, and the clinic has served as the title sponsor for Pale is the New Tan, a 5K race for melanoma awareness.

The goal of the practice has remained unchanged since 1972: to treat patients with skin cancer and to make the public aware of the dangers of sun exposure and tanning beds to help prevent this disease.

MEMPHIS EYE AND CATARACT ASSOCIATES

Memphis Eye and Cataract Associates, with its distinctive building on Poplar Avenue, is one of the most recognized medical clinics in Memphis. Its origins date to 1968, when Dr. Jerre Minor Freeman opened his office downtown in the Exchange Building, at the time home to many of Memphis's most prominent physicians.

MECA's reputation was built on thorough eye examinations, but also on the successive revolutions in cataract surgery that developed in the mid-1970s and still continue today, and on corneal and laser vision surgery. When Dr. Freeman began practice, cataract removal meant five days in the hospital and then wearing thick glasses that looked like the bottom of a Coke® bottle. In England, Dr. Harold Ridley began replacing the cataract with a lens inside the eye, an intraocular lens, eliminating the need for thick glasses following surgery and restoring more natural vision. In 1974, Dr. Freeman was the first in Memphis to implant an IOL and among the first to perform small-incision cataract removal using phacoemulsification, another revolutionary advance. Optometrist Dr. Edwina Campbell was an integral part of the practice from 1975 until she retired in 1985. She represented one of the first Ophthalmology/Optometry (MD/OD) practice connections in Tennessee. Dr. Thomas Gettelfinger, sharing an interest in international volunteer work with Dr. Freeman, joined the practice in 1976. Dr. Joe Weiss, a former medical missionary in Thailand, joined in 1980 and retired in 1995.

In 1985, after a number of years at Methodist Hospital, MECA moved east to its building on Poplar. With clinic space on the first floor and operating rooms on the second, it was the first freestanding ophthalmology ambulatory surgery center in Tennessee. Seeing the physician and having surgery in the same outpatient center has proven to be highly convenient for patients. In addition, millions of dollars have been saved over the years by patients, insurance companies, and Medicare because ambulatory care centers receive lower reimbursements than hospitals.

One of the most gratifying episodes in the practice's history has been work with ophthalmologists from China. In 1976 Dr. Freeman visited China, the first American ophthalmologist as China opened to the world. In 1978. Dr. Gettelfinger, invited by the Asia Pacific Academy of Ophthalmology, and Dr. Freeman, leader of a delegation from the International Intraocular Implant Society, were with the first two groups of ophthalmologists to visit. Over the years, those visits resulted in some 50 Chinese ophthalmologists and nurses coming to MECA for training in cataract removal. Prominent were Dr. Lu Dao-Yan, from Shanghai; Dr. Tang You-Zhi, from Beijing (who operated on Mao Tse Tung's cataract); and Dr. Chung Chou Lu, longtime president of the Chinese Academy of Ophthalmology, two of them studying at MECA. These physicians stayed in touch with Dr. Freeman and MECA for the rest of their lives. Dr. Meng Yongan, founder of the Ancient City Eye Hospital in Xian, among the first private eye hospitals in newly developing China has also been especially noteworthy. In 1978, Dr. Freeman started what was to become the World Cataract Foundation (WCF), providing cataract surgery and training around the world, most notably through annual surgical trips to Ometepec, Mexico. Both he and Dr. Gettelfinger have served as volunteer surgeons literally from A to Z, Afghanistan to Zaire, with many countries in between, Dr Freeman particularly in Mexico and China and Dr. Gettelfinger in the Brazilian Amazon and through ORBIS, the flying eye hospital.

Dr. Freeman holds 18 patents, including the Freeman Punctum Plug, an innovation in the treatment of dry eye, and in 1983 founded Eagle Vision, which produces a range of eye care products. He was president of the Memphis Society in 1998, president of the American Board of Eye Surgery, has been a board member of both the Outpatient Ophthalmic Surgery Society and the Society for Excellence in Eyecare, and served on the executive committee of the American Society of Cataract and Refractive Surgery. In 2006, he was inducted into the Memphis Society of Entrepreneurs.

Active in promoting staff education, Dr. Gettelfinger founded the Tennessee Ophthalmology Personnel Society in 1982, one of the first of its kind in the country. He served for years on the Communications Committee of the American Academy of Ophthalmology, producing national education materials for patients, and was a board member of both the Joint Commission for Allied Health

ABOVE: *Memphis Eye and Cataract Associates' medical staff includes, from left to right, Drs. Hal Wright, John Freeman, Thomas Gettelfinger, Jerre M. Freeman, David Irvine, and James Freeman.* OPPOSITE PAGE: *The MECA building is an architectural landmark on Poplar Avenue.*

Personnel in Ophthalmology and the Mid-South Eye Bank. On the board of The Memphis Medical Society, he helped develop its publication, the Memphis Medical Quarterly, and has written the lead editorial for many years. He was president of the Tennessee Academy of Ophthalmology in 1991. On separate occasions, both he and Dr. Freeman have received the Honor Award for teaching and service from the American Academy of Ophthalmology and the Good Samaritan Award for Volunteer Service from the Memphis Health Care News/Church Health Center. Both are clinical professors of ophthalmology at the University of Tennessee Health Sciences Center (UTHSC).

Dr. Joe Scott, known for his gracious bedside manner, joined the practice in 1985, consolidating his large solo practice; he retired in 2005. Dr. Jerre Freeman's sons—Dr. James Freeman in 1996 and Dr. John Freeman in 2004—have joined the practice, each with advanced fellowship training in corneal disease. Both have been recognized for outstanding contributions to resident instruction at UTHSC. James has served as president of the Memphis Eye Society, medical director of the Mid-South Eye Bank, and on the board of the American Board of Eye Surgery. John is a board member of the Tennessee Academy of Ophthalmology and serves on the National Board of the Aniridia Foundation. He is one of only a few ophthalmologists in the region capable of keratoprosthesis surgery, which restores sight to patients once considered hopelessly blind. Dr. Hal Wright left his position as assistant professor at the University of Mississippi Medical Center to join the group in 1996. With a postgraduate fellowship in ocular pathology, he has a broad range of ophthalmic interests, including glaucoma treatment and laser therapy. Dr. David Irvine joined the practice in 1995, adding vitreous and retinal expertise.

The physicians at MECA are particularly proud of the staff assembled over the years. There are too many to name, but Administrator Linda Fleming, office manager Suzanne Lea, and Ambulatory Surgery Center directors Debbie Booker Hull and Bernice Walters have been outstanding leaders. Linda Fleming served as president of the Memphis Medical Group Management Association and regularly delivers lectures at national meetings.

Physicians at Memphis Eye and Cataract Associates have been active in many of the enormous advances in ophthalmology in the 42 years since the group began. Improvement in the prevention and treatment of blindness and visual disability will continue, and Memphis Eye and Cataract Associates and their Memphis ophthalmologist colleagues will be part of it.

MEMPHIS LUNG PHYSICIANS, P.C.

Memphis Lung Physicians, P.C., is a unique group of practitioners serving the Memphis area since 1980. They currently provide round-the-clock, in-hospital pulmonologists at both Baptist Memphis and Baptist Desoto Hospitals (minus one 13-hour shift at Baptist Memphis). The 14-physician group is divided over three campuses, Memphis, Desoto, and Collierville.

The group was formed by Drs. William A. Potter and Michael D. Wilons in 1980 and was joined by Dr. Emmel B. Golden in 1982. The three served as directors of respiratory therapy, the pulmonary laboratory, and the School of Respiratory Therapy, and they provided consultative practice at Baptist Hospital. They expanded to Baptist East and saw patients on the first day of its opening. The group now offers both inpatient and outpatient care for all types of lung disease, including chronic obstructive pulmonary disease, asthma, pulmonary infections, lung cancer, pulmonary hypertension, and sleep disorders. The practice has grown to greater than 1,000 outpatients per month and an ever-growing hospital census including a large intensive care unit (ICU) practice.

The group reviewed its scheduling and coverage and with the Baptist Hospital administration created an intensivist program for night, weekend, and in-house coverage. The program has attracted national attention and was featured in an article in *The Wall Street Journal* in 2007. This has become a model for improved quality and safety for hospitalized, acutely ill patients. The group has gone beyond the intensivist model to a rotational system to maximize efficiency and coverage of outpatients and inpatients both in the ICU and on the floor, and being able to have a physician in the hospital on night shift. This system allows an experienced, highly qualified, critical care, pulmonary specialist to be available in the hospital for emergencies that occur through the night. "This unique intensivist model, with one specialist available at all times, has worked well," says Dr. Golden. "The physicians like the schedule and the rotation, and the night doctor responds to all calls freeing others from beepers and pagers." The rotation thus creates "the greatest win-win situation for patients, doctors, and hospitals," says Dr. Wilons.

Further innovation is evident as clinical information is "handed off" when rotations change. Night doctors have a formal hand-off: the physicians meet, list patient issues, discuss immediate problems, and suggest what might happen. When the incoming physician can respond immediately, a better outcome is likely.

The innovative doctors at Memphis Lung Physicians continue to respond to the needs of the community, with sleep centers being opened in Collierville and in Desoto County. According to Dr. Golden, "The aim is to make our services as timely and as available to our patients as possible." Dr. Wilons concludes, "Our patients are breathing easier in Memphis no matter what time of the day."

MEMPHIS NEUROLOGY

Memphis Neurology is a multipractitioner group unique in the Mid-South in that patients are cared for from infancy through adulthood. The practice opened in the early 1970s as Pediatric Neurology at Le Bonheur Hospital on Adams Avenue which periodically over the ensuing decades was the only pediatric neurology provider in the tri-state area.

Pediatric Neurology was the first practice in the area to provide care in the outlying communities, opening satellite offices in Paragould, West Memphis, Tupelo, and Jackson, Tennessee, as far back as the mid-70s. Prior to that, Dr. J. T. Jabbour and Dr. Robert M. Boehm, two of the founding partners, traveled as far as the tricity area of East Tennessee (Johnson City, Bristol, and Kingsport), Greenville, and New Albany, MS to treat children in need of neurologic care. While the practice still maintains an office at the Physicians Office Building at Le Bonheur, the main office is now located in Germantown, with a third facility in Southaven. Satellite offices are still located in Jackson, West Memphis, and Tupelo with a new Tipton County site in Brighton.

As in the past, children are still treated for conditions such as headaches, seizures, developmental delay, autism, learning disorders, and attention deficit disorder. In the past decade, an adult division has been added to the practice treating those suffering from dementia, Alzheimer's, Parkinson's, strokes, seizures, and migraine headaches.

The practice is equipped with a neurodiagnostic lab that offers services benefiting both patients and other healthcare professionals. EEGs and EMGs are performed at most locations and EEG video monitoring is provided at the Germantown office with plans to open additional EMU rooms at the Southaven office in late 2010. Botox injections for muscle spasticity and cervical dystonia are available at the Germantown location. For patients with seizures that are difficult to control through medication alone, the option of a vagal nerve stimulator (VNS), placed by a surgeon, is available. The physicians at Memphis Neurology then monitor and adjust the VNS as necessary as part of an overall comprehensive seizure monitoring and care program.

Drs. Jabbour, Eastmead, Lynn, and Natarajan and nurse practitioner Laura Porch provide pediatric patient care at all locations. To assure continuity of care, Drs. Natarajan, Sonone, and Laura Porch see adult patients at the Germantown and Southaven offices as well as the Tupelo and Brighton satellites. With multiple offices and a wide array of patient care services, Memphis Neurology strives to provide both convenient and comprehensive neurologic care for patients of all ages.

From left to right, Dr. Natarajan, nurse practitioner Laura Porch, Drs. Lynn, Eastmead (seated), Jabbour, and Sonone.

MEMPHIS SURGERY ASSOCIATES, PC

Memphis Surgery Associates, PC, is a group whose members are certified by the American Board of Surgery to practice the proud and specific specialty of general surgery. As a discipline, general surgery comprises operations of the abdomen and its contents, the endocrine glands, the breast, soft tissue tumors, and arteries and veins that lie outside the chest and skull. Care of injured and critically ill patients is particularly included.

The surgeons of Memphis Surgery Associates strive to provide each patient with the most precise and comprehensive surgical care while maintaining an atmosphere of dignity and respect. Use of the latest, safest, and least invasive surgical techniques is emphasized. Operating with a remarkable culture of congeniality and equality, the group provides its surgeons both professional gratification and family stability as practitioners of a tumultuous specialty that can threaten both.

The group was founded in 1966 by Dr. G. Randolph Turner. His vision of an efficient, highly skilled surgical team resulted in the early association of Dr. Leonard H. Hines, Dr. W. Scott King, Jr., and Dr. Carter E. McDaniel, III. The reputation of this core group within the Memphis surgical community cannot be overstated.

As the necessity of a larger group practice became clearer, Memphis's most distinguished surgical groups contributed surgeons to Memphis Surgery Associates. Doctors Malone and Francis, PA, provided Dr. Hugh Francis, Jr., and Dr. Hugh Francis, III. Coors and Laughlin, MDs, PC, provided Dr. A. E. (Ned) Laughlin, Jr. Mid-South Oncology Group, PC, provided Dr. Irving D. Fleming and Dr. Martin D. Fleming. Dr. Melvin P. Payne, III, arrived from The Surgery Group, PC.

Surgeons who have joined the group more recently include Dr. Justin Monroe, Dr. Virginia McGrath Weaver, Dr. D. Benjamin Gibson, IV, Dr. William C. Gibson, Dr. D. Alan Hammond, and Dr. Norma M. Edwards.

Each surgeon participates in the group's rigorous night and weekend general surgery call schedule, but subspecialty interests are encouraged by everyone. Fellowship training after residency has been done by Drs. Turner, Hines, and Francis in vascular surgery, Drs. Fleming and Fleming in surgical oncology, and Dr. Monroe in colon and rectal surgery. Following her advanced laparoscopy fellowship, Dr. Virginia McGrath Weaver has popularized laparoscopic gastric bypass as a preferred procedure for surgical weight loss.

The group's president, Dr. W. Scott King, Jr., leads the members in an unusually high interest in and experience with surgery of the pancreas and endocrine system. Advanced laparoscopic skills are always emphasized, but the tried and true traditional operative skills are not to be abandoned.

The group's emphasis is on the team, and teamwork includes the office staff. Practice manager Ms. Sherri Hardwick oversees a group of nurses and office staff members who share the highest commitment to patient care. Engaged and appreciated, they help to keep loyalty high and turnover rates low.

In a troubled healthcare world, Memphis Surgery Associates strives to be a proud oasis of outstanding accomplishments. Its mission is noble, and its values are clear. Memphis Surgery Associates looks to the future with an unparalleled level of thoughtfulness and concern.

MOTLEY INTERNAL MEDICINE GROUP

Drs. Ann Marie and Todd Motley began practicing in Memphis with one focus—community health. They held health fairs each month at churches, grocery stores, and businesses sites. Dr. Todd Motley created the Motley Medical Minute hoping to reach the public through radio.

Dedication to community health continued a Motley family tradition. Todd's father, Dr. Thomas Motley, was the first African-American internist in Memphis. The family moved from Rhode Island, and his father's strong medical ethic influenced young Todd to become a physician.

While at the University of Tennessee College of Medicine, Todd attended a seminar in Kansas City and met Ann Marie who was attending Creighton University Medical School in Omaha. The connection was very strong and, after graduation, she came to Memphis for a transitional year of medical training while Todd finished at the University of Tennessee. After their residencies in internal medicine at Berkshire Medical Center in Massachusetts, they returned to Memphis. Methodist Hospital was helping new medical practices in the city and, in 2000, the Motleys opened their office in Methodist South Medical Center.

Drs. Ann Marie and Todd Motley have expanded their office and remained at Methodist South Medical Center. In addition to the health fairs, Dr. Todd Motley hosts a monthly, two-hour radio question and answer show on WLOK. "The concern we have is the lack of education about medical problems and, ultimately, the damage this can cause," he says. "We know this is the closest some people will come to a doctor's appointment."

The Motleys' fulfilling hospital practice in the Methodist system includes Dr. Ann Motley's invaluable service on multiple committees, and Dr. Todd Motley's appointment as chief of staff at Methodist South Hospital for four years.

AUTRY J. PARKER, M.D.

In 1988, Yale University School of Epidemiology and School of Medicine simultaneously awarded Dr. Autry Parker a Masters degree in Public Health and an M.D. degree. After an internship at the Hospital of Saint Raphael in 1989, he received a residency in anesthesiology and critical care at Johns Hopkins University Hospital. Dr. Parker's fellowship was also at Johns Hopkins.

In 1992, Dr. Parker founded and was the director of the Center for Pain Management in Johnson City, Tennessee, in cooperation with the Anesthesia Partners of Johnson City. In 1995, he became the interim director of the Pain Care Center at Saint Francis Hospital in Memphis which became the Parker Pain and Rehab Center in 1996. In 2010, Dr. Parker joined the Semmes-Murphey Neurologic & Spine Institute.

Pain management is critical to patient wellness, and Dr. Parker's diagnostic and therapeutic services enhance patient treatment plans. Radiofrequency rhizotomy, trigger point therapy, epidural steroid injections, behavior modification, and psychiatric services are available to support a clinical diagnosis. Through the use of provocative discograms, facet blocks, anesthetic injections, and appropriate testing, Dr. Parker and his staff provide state-of-the-art care for patients.

Dr. Parker and the clinic have dedicated themselves to finding the most effective care for each patient. The clinic's mission is, "Helping people in pain gain control of their lives." Dr. Parker's goal is always to eliminate the pain when he can, and if he cannot completely eliminate the pain, to reduce the pain to a level that allows the person to be able to live a full, productive life.

MERCK/SCHERING-PLOUGH

Schering-Plough began as two companies, on two continents, created by two men who never met. Abe Plough and Ernst Schering's small companies were built on the confidence of rapid development in the pharmaceutical field. Medical breakthroughs were creating drugs to battle diseases that had plagued the world for centuries. These men were dedicated to making medical advancements available to the general public, creating new products through research and development, and diversifying product lines. This plan formed two very successful businesses that would later become a global company.

Plough Chemical Company

In Memphis, Tennessee, 16-year-old Abe Plough asked his father for a $125 dollar loan to open a drug company in 1908. Moses Plough was a Memphis merchant who had watched Abe work at a local drug store, without pay, in order to learn more about the drug industry. With his father's loan, Abe opened Plough Chemical Company; he was the only employee in a small room above his father's store. Here, he hand-mixed Plough's Antiseptic Healing Oil in dishpans and bottled it. In a bold marketing move, Abe borrowed his father's horse and buggy, and "peddled" his product to local drug and general stores. His slogan, "Sure cure for any ill of man or beast," quickly caught the attention of customers and, within two years, Plough had doubled the size of his company.

In 1920, Plough bought the Gerstle Medicine Company in Chattanooga, Tennessee, which manufactured St. Joseph Aspirin. The owner, Leopold Gerstle, was a German immigrant who came to the United States in 1865 and began forming *acetylsalicylic acid,* possibly from willow bark, into chewable, tiny orange pills. *Acetylsalicylic acid* was commonly known as aspirin, and its ability to reduce fever and relieve pain made it one of the fastest growing drugs in the United States. The St. Joseph purchase was Plough's

"first on the road to the big time." In 1950, Plough believed the increasing demand for St. Joseph Aspirin was possibly due to children liking the taste and taking too many of the orange flavored pill. He had the first child-proof cap designed for the product to prevent overdoses, and he later received the first individual award for product safety.

Plough's explosive growth in the late 1920s, was followed by the Great Depression, which would slow drug sales worldwide. Plough's employees expected to be out of work, like millions of other Americans but, instead, their salaries were increased and Abe added 100 new employees.

By 1951, Plough Incorporated had outgrown its plant and was making its move to a new, $2 million dollar facility. Abe Plough was on his way to acquiring 27 companies and sales were netting $24.5 million by 1954; this number would double by 1962. Plough met Willibald Hermann Cozen, the chief executive officer of Schering Corporation, and developed a friendship. Seeing his approaching retirement, Plough began to encourage a merger of the two companies to assure the future leadership at Plough Inc.

Schering

In 1851, Ernst Schering opened a pharmacy in Berlin. He then built one of the first German pharmaceutical and chemical companies. Schering had developed a vaccine for diphtheria and, in 1876, the company opened a division in the United States to distribute the vaccine. The venture was a successful one.

Ernst Schering died in 1889, but the company continued to expand and prosper through the early 1900s. In 1917, World War I prompted the United States to take control of all German-based businesses within the United States. In accordance with the U.S. Enemy Act, Schering was included.

By 1920, Schering was concentrating its research on female hormones, leading to the production of synthetic steroids and, eventually, birth control pills. The company's

Schering-Plough, (above and right) has grown from the two companies formed by Ernst Schering in 1851 in Berlin, (opposite page, above) and Abe Plough in 1908 in Memphis, (opposite page, below) into one of the largest pharmaceutical companies in the world with global sales of $18.5 billion in 2008.

growth and expansion prompted the name change to Schering AG in 1937.

Prosperity was interrupted again in 1939 by World War II. The Enemy Act was revived, and this time Schering's U.S. holdings were nationalized. German ownership made it the focus of an investigation into suspected development of hormone products to aid Nazi pilots. This was later proven false. The German Schering AG plant was destroyed during military battles and, after the war ended, the company literally rebuilt its new factory from metal salvaged from the rubble. Schering's nationalized U.S. company was sold to Merrill Lynch in 1952 for $29 million, and ceased to be part of Schering AG. The name was changed to Schering Corporation.

In the late 1950s, a line of antibiotics and two corticosteroids, Meticorten (prednisone) and Meticortelone (prednisolone), dominated the industry with $80 million in sales. In the 1960s, Schering's antifungals, decongestants, and the antibiotic, Garamycin, were introduced. Garamycin's use as a treatment for urinary tract infections was a huge success. The nasal decongestant, Afrin, gave Schering another product to impact the market in the late 1960s.

Schering-Plough

In 1971, Schering and Plough merged to create one of the largest pharmaceutical companies in the world. Plough's consumer-based products, such as St. Joseph Aspirin, Di-Gel antacid, and Maybelline cosmetics, were a perfect balance with Schering's antifungal and antiviral products. In 1972, Schering-Plough, in its first year in partnership, earned $500 million in sales. Although, sales slowed but Garamycin continued as the top grossing product.

Abe Plough served as chairman of the board at Schering-Plough until he retired in 1976 to pursue philanthropy. At that time, the majority of the company's sales were based on a few products, so the leadership tried to refocus the company through its acquisitions of companies in the animal health products and contact lens care industries.

During the 1980s, Schering-Plough continued Abe Plough's commitment to dedicating twenty-five percent of the company profits to research. CEO Robert Luciano

Memphis Medical Histories **275**

focused on the health industry and Schering-Plough boldly invested $100 million in biotechnology, an amount far above its competitors.

A $12 million investment with Biogen gave Schering-Plough the worldwide rights to synthesize human interferon. This alliance had far-reaching influence as interferon was studied as a treatment for diseases from cancer, to hepatitus C, and even common fevers. The Federal Drug Administration approved Intron A for sale in the United States in 1986 for a rare cancer and, by 1994, sales reached $426 million.

Expansion through company acquisitions also placed Schering-Plough as a leader in asthma and allergy treatment. In 1993, Claritin was approved for the U.S. market and, within one year, Claritin had sold $200 million in products, soon becoming the best-selling antihistamine.

During the 1990s, growing concerns about skin protection and skin cancer prompted Schering-Plough to develop Shade UVA Guard and a complete line of sunless tanning products. Coppertone took the market share with its adult and children's line of sunscreen products. Schering-Plough would also purchase the marketing rights from Pfizer to sell Bain de Soleil tanning products in the United States and Puerto Rico.

The purchases of DNAX Research Institute in Palo Alto, California, and Canji, Inc., a gene-therapy company, expanded research into the biotech area. Nasonex allergy spray, and new drug treatments for hepatitis C, Crohn's disease, and brain tumors continued Schering-Plough's worldwide presence. Claritin and Claritin-D continued to lead overall sales. Alliances to combine certain Schering-Plough products with Merck & Co., Inc. opened the door for the development of even more effective treatments for high cholesterol.

When Fred Hassan became chief executive officer and chairman in 2003, Schering-Plough was ready for new direction. Hassan's Action Agenda set a six- to eight-year prospectus expand the company worldwide to internally and externally . A renewed focus on employee relations earned the title of "100 Best Companies for Working Mothers" in 2004 and 2005 from *Working Mothers* magazine.

In 2007, the purchase of Organon and Intervet from their parent company, Akzo Nobel, expanded Schering-Plough's human and animal healthcare divisions. Organon, whose first product was insulin, had, through the years, expanded into estrogens, contraceptives, and cortisone, not unlike Schering's early products. Intervet mirrored Schering-Plough's development of the animal health division. The merger's latest developments include PreveNile vaccination for horses in the prevention of the West Nile virus.

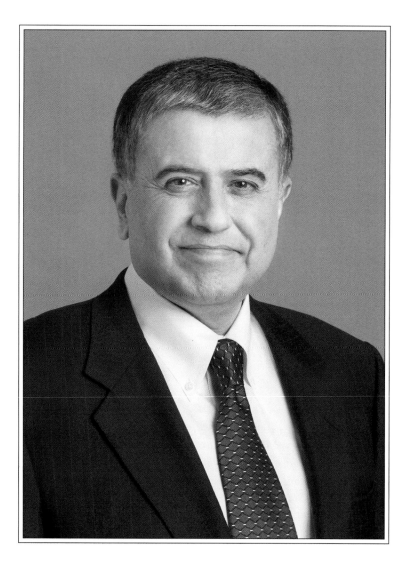

Fred Hassan became chief executive officer and chairman of Schering-Plough in 2003.

Research and development were still foundations in the company's growth with almost $3.5 billion spent in 2008 with reported net sales of $18.5 billion . The most popular products were Vytorin, Zetia, Remicade (sold outside the U.S.), and Nasonex. Vytorin and Zetia continued to be part of a collaboration with Merck & Company, Inc.

In 2009, Merck & Co., Inc. merged with Schering-Plough to create the second largest global pharmaceutical company in the world. Although the name has changed, the Memphis location remains and continues to be a major contributor in the medical and consumer health industry. Merck & Co. Inc. has strategically blended with Schering-Plough's research, science, and product excellence, much of which Abe Plough made possible.

MID-SOUTH OB/GYN PLLC

The founding father of Mid-South Obstetrics and Gynecology was Dr. Henry B. Turner, a native Memphian and World War II veteran. After the completion of his residency training in Ob/Gyn and gynecology at the University of Tennessee, Dr. Turner, in 1936, established one of the first clinics in the Memphis area dedicated specifically to women. He was not only an excellent clinician, but he had an investigative component of his personality that led him, with the help of the UT Department of Anesthesia, to develop caudal anesthesia for the pain relief of patients in labor. Over time, caudal anesthesia evolved into epidural anesthesia, the gold standard for obstetrical pain relief.

From left to right, Drs. Rye Estepp, Tom Greenwell, Judi Carney, Paul Neblett, and Herbert Taylor.

Dr. Prentiss A. Turman, who joined Dr. Turner in private practice in 1958, received his M.D. degree from the University of Tennessee College of Medicine, interned at Texas at the Scott-White Clinic in Texas, and then completed his residency in Ob/Gyn in Memphis.

The third member of this original practice was Dr. Louis Henry, who began his medical education at the Mississippi Medical College, at that time only a two-year program. He was then accepted for the final two years of his training at the renowned Washington University of St. Louis, where he graduated with honors. After an Air Force tour of duty, Dr. Henry returned to Memphis and completed his residency in Ob/Gyn, and then joined Drs. Turners and Turman in the early 1960s.

These three doctors were the pioneers in the evolution of group Ob/Gyn practices and were the first group to establish a satellite office, the first group to educate patients and establish a rotational call schedule, and the first group to incorporate a female partner, Dr. Mary Ellen (Mel) Bouldien. Dr. Bouldien, who joined the practice part time in 1966, was a story in her own right. A graduate of University of Tennessee College of Medicine, where she also completed her internship and residency, she was an accomplished aviator and would commute from her home in Clarksdale, Mississippi, land at the old Mud Island airport in Memphis, work her two-day work schedule, and then fly home to her husband and sons.

From the late 1960s to the present, Mid-South OB/GYN has been a leader in quality patient care, and a strong advocate in medical student and resident teaching. Many members of the practice have received the outstanding clinical staff teaching award, named in honor Dr. Louis C. Henry.

Mid-South OB/GYN has been a training ground for many physicians who have learned the principles of quality care and educational responsibilities and have established successful and respected practices. Strong leadership in not only the quality of patient care and education, but also in hospital governance has always been a part of the Mid-South OB/GYN heritage, in the form of department chairmanships, presidencies of hospital staffs and board of trustees, and presidency of the Memphis Shelby County OB/GYN Society.

Today, our tradition continues with Dr. Herbert A. Taylor, senior partner for 43 years; Dr. Tom Greenwell, managing partner and past president of the Baptist Metro Hospital staff; Dr. Judy Carney, current chair of Ob/Gyn at Baptist Women's Hospital; and Dr. Paul Neblett, the most recent recipient of the Baptist Women's Hospital Physician Champion Award. The newest addition to the practice is Dr. Rye Estepp, who completed her Ob/Gyn residency in June 2010 and brings with her the most up-to-date knowledge and skill in the rapidly advancing world of medical technology.

Fifty-four years or tradition, 33,000 deliveries, and many talented physicians who are also teachers and leaders have helped to build the respect and trust that defines Mid-South OB/GYN.

METHODIST LE BONHEUR HEALTHCARE

Officially titled Methodist Le Bonheur Healthcare, with Methodist University Hospital in the medical center as its flagship facility, the Methodist Le Bonheur system has a complex and distinguished history. In addition to hospitals, the system includes an array of outpatient services including home care and hospice, surgery centers, minor medical centers, diagnostic centers, and sleep centers.

Through the efforts of Mississippi planter John H. Sherard, Sr., who envisioned a hospital in Memphis where patients of all faiths would receive quality health care in privacy, Methodist Hospital began with the 1914 purchase of Captain W. B. Mallory's home at 1025 Lamar Avenue. The hospital was chartered in 1918, when the 65-bed Lucy Brinkley Women's and Children's Hospital at 885 Union Avenue was given to the board of trustees. G. T. Fitzhugh, the charter's preparer, was sent to Washington to offer the Lamar Avenue property and buildings to the U.S. government to care for World War I wounded. Scarce labor, high costs, and a government moratorium on building delayed construction until the war ended. During this time, the Lucy Brinkley Hospital building operated as Methodist Hospital.

Throughout its history, Methodist has continuously grown. Today Methodist Le Bonheur Healthcare comprises almost 1,689 beds and employs approximately 10,000 people.

Medical Achievements

As illustrious as Methodist's history of physical expansion is, it represents the profound growth in medicine itself in Memphis, the increasing medical sophistication of the hospital, and the top physicians Methodist attracts. Memphis's first open heart surgery using a heart-lung machine was performed at Le Bonheur (1959), supported by University of Tennessee physicians and the University's heart-lung machine. Today, the flagship Methodist University Hospital emphasizes neuroscience and abdominal transplantation, and maintains a strong presence in oncology.

In 2005, Dr. Osama Gaber performed Memphis's first living-donor liver transplant and the first adult cord blood

BELOW AND OPPOSITE PAGE: *Methodist University Hospital is the flagship facility of Methodist Le Bonheur Healthcare.*

transplant in the tristate area was performed at Methodist University Hospital in 2006.

Methodist is designated a Minimally Invasive Surgical Center of Excellence. With the Da Vinci-S robotic surgery system at Methodist University Hospital, Methodist Le Bonheur Germantown Hospital, and Methodist North, surgeries that once required days or weeks for recovery can be performed with only small incisions for lights and instruments. Dr. Todd Tillmanns, gynecologic oncologist at the West Clinic and professor at the University of Tennessee Health Science Center (UTHSC), routinely performs hysterectomies with the Da Vinci system. Dr. Tillmanns emphasizes the excellent visualization offered by the system, as well as the system's reduction in size of incisions, loss of blood, and recovery time compared with conventional surgical procedures.

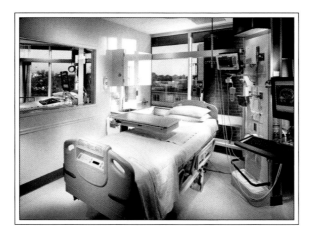

Methodist physicians also performed the first robotic colorectal surgery in the Southeast at Methodist University Hospital in 2009. Robotic surgery actually extends laparoscopic surgery, giving the surgeon a clear view of the area and precise control of the instruments. Dr. Mohammad Ismail, gastroenterology chief at the hospital and at the University,, has worked with Methodist physicians to use the minimally invasive procedure to remove large polyps. Dr. Madison Michael, medical director of the Neuroscience Institute at Methodist University Hospital, and Dr. Sandeep Samant, head of otolaryngology, head and neck surgery at UTHSC, worked together in 2009 to remove a tumor growing from the nasal cavity into the brain by cutting a hole above the hairline rather than breaking facial bones to reach the tumor. The approach not only avoids facial scarring, but affords surgeons better visualization and an excellent change of removing the whole tumor, which is treated with chemotherapy and radiation before surgery. Dr. Jon Robertson of the Semmes-Murphey Neurosurgery and Spine Institute was a mentor to Dr. Michael, and the two receive referrals from around the world.

Drs. Kevin Foley and Maurice Smith performed the first minimally invasive herniated disk repair in Memphis at Methodist University Hospital. The procedure is now the gold standard for minimally invasive back surgery.

Dr. Mathew Wilson, an ocular surgeon at the hospital and the Hamilton Eye Institute, has successfully treated melanoma of the choroid, the pigmented vascular coat of the eye, between the outer covering and the retina, using a disc-shaped applicator implanted in the eye for a few days to deliver radiation to the affected area. Methodist University Hospital is one of five centers in the United States for treatment of eye tumors, especially retinoblastoma, with Dr. Wilson a prominent figure in research and treatment.

Drs. Robert A. Sanford and Frederick Boop at Le Bonheur see neurosurgical patients from across the United States. Dr. Raymond Osarogiagbon specializes in esophageal and lung cancer and also works with sickle cell patients. Dr. Tulio Bertorini specializes in neurology and neuromuscular diseases, especially multiple sclerosis.

Ministries

Praised at its 1922 dedication as a "triumphant layman's movement," the hospital, from its inception, embraced the Methodist Church's mission of ministering to the body as well as the soul. Charity, dedication, and outreach are woven into the hospital's story. Methodist's first facility had no charity ward, and only admissions personnel knew which third of the beds represented charity patients. During the Depression, Dr. Hedden announced a group hospitalization plan for individuals to pay $1 a month to ensure hospital service in case of illness (physicians' bills not included). Methodist Hospital Auxiliary began in the 1930s with Mrs. Casa Collier's concern for a charity patient whose triplets died at birth. After arranging with friends for the babies' burial, she organized the Auxiliary to help needy patients.

During World War II, Methodist employees worked six days a week, with no pension plan, but received hospital care when needed. A few pediatric charity patients essentially lived at the hospital during the 1940s, clothed by the Auxiliary and considered family by staff. Methodist's Volunteer Personal Hostess Service ("Pink Ladies") began in 1958, when the cost of a private room with bath was $19 per day. In 1968, high-school-age "Candy Striper" volunteers began working in the summers.

Today, with its burgeoning Transplant Institute, Methodist University Hospital holds "transplant gatherings" for transplant veterans, physicians, and staff, plus a golf tournament called "Transplant Fore Life." Methodist Le Bonheur Healthcare is the largest TennCare provider in the state.

MIDSOUTH ORTHOPEDIC ASSOCIATES, P.C.

Dr. Lawrence Schrader, founder of MidSouth Orthopedic Associates, first introduced partial or unicompartmental knee replacement (PKR) to the Southeast 15 years ago. Due to its success, many orthopedic surgeons followed his lead and are now qualified to perform this procedure. In addition, Dr. Schrader is the first surgeon in the area to use the custom fit knee replacement, which offers less tissue damage and blood loss, as well as a more rapid and painless recovery.

In 1996, Dr. Schrader discovered Dr. John Repicci's work and trained with him in New York. Originally a dentist, Dr. Repicci brought knowledge of small tools and incisions to his studies in orthopedics, using a two- to three-inch incision to implant modified partial-knee components and cementing them in place to give the knee immediate full strength. Avoiding taking apart the knee (especially the knee cap), the procedure causes minimal damage to ligaments.

The key to this procedure is a combination of a custom-fit replacement and precise incision techniques. These principles have been successfully applied to the partial knee replacement and are now being applied to total knee replacement (TKR), which usually involves a much larger incision, and dislocation of the kneecap. Typically, long metal guide rods are set to "average" alignment. The ends of each bone are then "shaved" to fit the new pieces. Even if, after the new knee is applied to the bones well, it does not match a patient's normal knee alignment, ligaments are cut to fit the new knee, which causes pain and limited mobility.

A computer is now used to model a knee and build a patient-specific guide that can be used to accurately resurface a knee to its normal position based on each patient's ligaments. Using the Otis knee technique, a custom-fit knee can be made for each patient.

In order to do this, an MRI is taken of the arthritic knee, and from this scan, a computer makes a three-dimensional model of the knee, rebuilding it to each knee's "normal" shape by removing bone spurs and calcium deposits and filling in bone and cartilage defects to restore the knee with ligaments intact. Then the computer builds a custom guide that fits on the arthritic knee, allowing the surgeon to shave off the precise amount of bone to fit the new knee.

After surgery, patients can generally walk bearing full weight within a few hours, and pain is usually minimal. Physical therapy will teach protective walking for safety and demonstrate simple motion exercises and muscle strengthening. Patients can be discharged and return home within two days following surgery in most cases with follow-up from a visiting nurse and physical therapist. Generally, there is less swelling and stiffness with this procedure, and only after the wound has healed and swelling reduced would further therapy be considered.

Dr. Schrader's practice focuses on hip, knee, and shoulder joint reconstruction and sports medicine. He is board certified, a member of the Academy of Orthopedic Surgeons, and a graduate of the University of San Francisco Medical School. He is a member of the Christian Medical and Dental Society and the Fellowship of Christian Athletes. He was an All-American in Track and Field (NCAA and USA). Dr. Schrader was a trauma surgeon in Desert Storm, receiving an Army Commendation Medal. He and his wife, Kathy, live in Memphis and have five sons.

FEREIDOON PARSIOON, M.D.

Dr. Fereidoon Parsioon brings to Memphis medicine a rich history of struggle and accomplishment. Born in Iran, he bought books with his weekly allowance and at 15, read about studies of the brain in the U.S. Immediately, he determined that he would become a neurosurgeon in the U.S. He spent 1972 at Michigan Technical University, but financial problems drew him home; he received his M.D. from Pahlavi University School of Medicine, in a program based on American medicine. Through Dr. J. T. Robertson of UTHSC, he participated in a neurosurgery exchange program in 1978. When the revolution started in Iran, his father, a two-star general under the Shah, was jailed, and he had to return. During the Iran-Iraq war, he was sent for two years to a tank unit, in the trenches—an experience, he says, that "made me who I am." His daughter was born during the war.

Afterward, Dr. Parsioon worked for the government until, when his daughter was 15 months old, his young wife had an illness requiring additional treatment unavailable in Iran. Arranging for papers, hiding $200 in his camera, and carrying two suitcases, he took his family to Copenhagen, where his wife received treatment. His wife and daughter received visas, but his was delayed, and he remembers a priest of a different faith praying with him. In Memphis, friends arranged further treatment for his wife and a tiny apartment, as well as permanent U.S. residence. Waiting for a residency, he worked in a bingo parlor, and he became a surgical assistant to Dr. David Meyer. The family lived simply, with three chairs and a bed, but no car.

Although he had completed an internship in Iran, Dr. Parsioon took another internship at UTM, in surgery, and then completed residencies in surgery and neurosurgery, finishing in 1993. He was affiliated with the Canale Clinic and with Semmes-Murphey Clinic (they merged in 1998). However, as business encroached on medicine, he felt that had not escaped the Ayatollah and studied to become a neurosurgeon for money. With friends' help, he established his own practice, Phoenix Neurosurgery, PLLC, the symbolic name based on the ancient myth of a bird that rises from its ashes to new life. Having succeeded despite a half-percent chance of becoming a neurosurgeon, he now learned that establishing a solo practice could take months. With a dedicated nurse and an office administrator, it took ten days. Today, nurse Pennie Cross says the practice is a "second home."

Phoenix Neurosurgery, then, exemplifies its name. Further, it is based on an old-fashioned concept of medicine—"love, integrity, honesty, and patient care." Dr. Parsioon feels that patients lose faith when business interferes with medicine. He observes that the U.S. was built by risk-takers, and he loves this country because "you can do impossible things."

A part-time police officer in Earle, Arkansas, Dr. Parsioon works with a man whose arm he restored to normal functioning. And in another triumphant note, his wife, Dr. Faranak Motaghian, is also thriving, having completed her residency in pediatrics and affiliated with Memphis Children's Clinic.

GENARO PALMIERI, M.D.

Genaro Palmieri was born in Córdoba, Argentina where he earned a medical degree from Universidad Nacional de Cordoba in 1956. He completed an internship and a residency there before accepting a fellowship in endocrinology and metabolism at Karolinska Hospital in Stockholm, Sweden. His fellowship was supported by the Argentina Research Council with the provision that Dr. Palmieri would return to Argentina to practice medicine.

Dr. Palmieri and Swedish physician Dr. Rolf Luft conducted the first study on humans with non-thyroid hypermetabolism. In 1962, they discovered a new disease in the muscle mitochondria, and the results were published in *The Journal of Clinical Investigation*. Using their study as a foundation, other researchers discovered more than 200 mitochondrial diseases. When the fellowship ended, Dr. Palmieri was offered a position in Sweden, but he honored his promise to return to Argentina. Leaving behind excellent research facilities for clinical investigation, he used frogs and rats to continue his work in endocrinology and adrenal functions. Despite limitations, his original findings were published.

In 1964, the Oklahoma Medical Research Foundation offered Dr. Palmieri a clinical fellowship, and he moved his family to the United States. He became a senior clinical fellow in 1965, a clinical assistant in research medicine, an instructor in the Department of Medicine, and senior investigator at the Oklahoma Medical Research Foundation.

In 1969, Dr. Palmieri took a one-year sabbatical to work with Dr. Rosalyn Yalow at the Veterans Hospital in the Bronx. Dr. Yalow later became a Nobel laureate when she won the 1977 Nobel Prize in medicine. Dr. Palmieri also accepted a visiting professorship at Mount Sinai Hospital while in New York. In 1970, Dr. Palmieri returned to the University of Oklahoma and the Medical Research Foundation to become the clinical investigator at the Veterans Administration Hospital. It was there Dr. Palmieri treated the dean's secretary who suffered from Paget's disease and walked with a cane. She was given an effective new treatment, Calcitonin, but her condition deteriorated and she had to use a wheelchair. "I thought they were going to fire me," says Dr. Palmieri. After running additional tests, he found the secretary was deficient in Vitamin D, which was added to the Calcitonin and she improved rapidly. A Veterans Affairs military officer was also confined to a wheelchair and asked to be treated. After taking Calcitonin and Vitamin D, he recovered and later retired to his farm in Arkansas.

Dr. Palmieri had begun what would be a lifelong study of Vitamin D and its effects on multiple diseases. Early in his career, he met Dick Massey, the inventor of the bone densitometry table, and worked closely with him, and in the 1970s, he was the first physician in Tennessee to use bone densitometry for the evaluation of osteoporosis.

In 1974, Dr. Palmieri moved to Memphis and became an associate professor of medicine and the associate director of the Clinical Research Center at the University of Tennessee. Dr. Albert Farmer, chancellor of the University of Tennessee Center for Health Sciences, was very interested in Dr. Palmieri's Vitamin D studies and encouraged him to work with Drs. Alvin Ingram, Rocco Calandruccio, Robert Tooms, Andrew Crenshaw, and Lee Milford at Campbell Clinic.

Dr. James Pitcock realized bone biopsy analysis was not possible without a proper laboratory, and he insisted that Baptist Memorial Hospital develop a biopsy lab. One month later, Baptist opened one of only a few bone biopsy labs in the United States. The lab conducted histomorphometric analysis of bone biopsies performed by Dr. Palmieri. In 1979, Dr. Palmieri became professor of medicine at the University of Tennessee Center for Health Sciences and, in 1987, he became medical director of the Memphis Metabolic Bone Center. He served in both positions until 1998.

Diane McCommon became Dr. Palmieri's assistant in the 1980s. "It is an honor and pleasure to work with him," says McCommon. "Everyday is a learning experience. Often, he looks at lab results and knows they are not right; then he goes down to the lab and explains to the technicians why they need to run them again. He is a walking scientist who can still perform chemistry profiles manually."

Dr. Palmieri has published 85 papers, 80 abstracts, and contributed to six books. His research with Dr. James Pitcock, Dr. Tulio Bertorini, Dr. David Nutting, Dr. Raed Imseis, and many others produced publications about the effects of calcium and vitamin D deficiencies in osteoporosis, multiple myeloma, Duchenne muscular dystrophy, and Sheehan's syndrome. His publications also address mineral metabolism disorders in patients with calcinosis, cancer, AIDS, and hypoparathyroidism. "Low Vitamin D levels are harmful to the heart because calcium and Vitamin D improve function in muscles like the heart," explains Dr. Palmieri. His expertise in Vitamin D has gained him worldwide recognition and invitations to speak at some of the most prestigious conferences in the world.

Dr. Palmieri trains fellows and graduate students and doctors in basic endocrinology methods often overshadowed by new technology. He finds some of the older methods create a better patient response, and new treatments are often just different names for old procedures. In 1998, Dr. Palmieri joined The West Clinic as an endocrinologist specializing in metabolic bone diseases. "The West Clinic was interested in Vitamin D deficiencies, which was once only treated after a patient developed Rickets," says Dr. Palmieri. "More than 70 percent of African-American children in the U.S. have Vitamin D deficiencies." Most of Dr. Palmieri's patients are exhibiting irregularities in bone metabolism, calcium levels, or parathyroid gland function. "You can't get normalization of the bones unless you have adequate calcium and Vitamin D," he says. Research has shown that individuals exposed to ample sunshine are less likely to be Vitamin D deficient.

Dr. Palmieri is concerned about his patients' soaring medical costs. "He is looking for the best benefit for the least cost," explains Diane McCommon, "and he asks pharmaceutical companies to lower the costs to the patients." When Dr. Palmieri needs specific medications, which are not readily available, "There are pharmacies, such as the The Medicine Shoppe and People's Pharmacy in Memphis, that can create compounds for him," says McCommon. These are often less expensive than the manufacturers' name brands. Dr. Palmieri believes the future cost of health care could be greatly reduced by addressing Vitamin D deficiencies because Vitamin D is an immunoregulator. "This would change the concept of aging and the outcome of care needed in the future," says Dr. Palmieri. Promoting an inexpensive Vitamin D product is not of great interest to pharmaceutical companies because of the low profit ratios. This fact deeply concerns Dr. Palmieri.

The seriousness of Dr. Palmieri's research is always balanced with a good sense of humor for patients and staff. His license plate once read, "Calcium." Dr. Palmieri gives his patients another valuable commodity—time. He always has a smile and time for an unhurried explanation of their condition. Dr. Palmieri also dictates a letter reviewing each office visit, and patients receive a copy by mail. It is not surprising that people travel from the Dominican Republic, England, Germany, Chile, and several U.S. states to see Dr. Palmieri. Dr. Palmieri was voted one of *The Best Doctors in America* (1994-1995), and one of *The Best Doctors in the Midwest* (1995-1996).

Dr. Palmieri continues to enjoy the practice of medicine and, clearly, retirement is not in his near future.

PEDIATRICS EAST, INC.

Pediatrics East, front row, left to right, Drs. Shazia Hussain, Robert Higginbotham, Robert Guinter, Ja Hayes, Hugh Scott, and Bill Fesmire; back row, left to right, Grant Newman, Susan Aguillard, Charles "Tucker" Larkin, "Bubba" Edwards, Bill Threlkeld, Mel Senter, and Jim Montgomery.

Pediatrics East, which began as the practice of Drs. William Mason, Price Stepp, and William Threlkeld in the 1970s, now operates in Cordova, Germantown, Collierville, and Bartlett with 19 physicians, two nurse practitioners, and approximately 90 staff to provide service to a wide array of pediatric patients. Upon the retirement of Drs. Mason and Stepp, and with the addition of Drs. Charles Larkin and Robert Walker working with Dr. William Threlkeld, the practice at Baptist East and Germantown changed its name to Pediatrics East. Growth of the practice continued with the opening of a Cordova location in 1992, the Bartlett office in 1998, and moving the Collierville office into a new facility in 2001.

Hospital affiliations now include Le Bonheur Children's Hospital, Methodist Le Bonheur Germantown Hospital, Baptist Memorial Hospital—Memphis, Baptist Memorial Hospital for Women, and Saint Francis Hospital—Bartlett. Pediatric East's physicians include both international and "homegrown" doctors. Two physicians at Pediatrics East are actually former patients of the practice.

Pediatric East's physicians treasure their long-term relationships with patients' families, with many of the children being second and third generation patients. "Peds East" offers a full array of pediatric services during their regular office hours as well as extended office hours in the early mornings, evenings, and weekends. The extended hours of operation ensure continuity of care for patients as well as the convenience and cost savings of avoiding visits to hospital emergency departments when possible.

The physicians and staff at Pediatrics East are involved in many aspects of the community. Dr. Susan Aguillard serves on the board of directors of St. Jude Children's Hospital, and Dr. Melissa Adams is currently chairman of the Department of Pediatrics at Baptist Hospital. Physicians from Pediatrics East also volunteer at the Church Health Center, offering care to the uninsured and underserved of the community. The practice's physicians and employees actively participate in the Special Kids program, which offers support to children with Down's syndrome. "Is There a Doctor in the House?" is a program developed by Pediatrics East physicians to assist in educating schoolchildren on health issues.

At Pediatrics East, excellence in patient care means staying "state of the art" with the prevention of illnesses through vaccines, healthy living through proper exercise and diet, management of ADHD and other learning difficulties, and encouraging good infant nutrition with an in-house lactation consultant. All patient information is coordinated through electronic medical records, which Pediatrics East physicians helped pioneer in the Memphis region.

JOHN V. PENDER, JR., M.D., F.A.A.P.

WHEN JOHN PENDER LOOKS back at his medical career, he explains that his greatest accomplishments are not listed in titles and awards, but in his relationships. "I made at least one great decision in my life—marrying my wife, Kathleen Fowlkes Pender, 46 years ago," says Dr. Pender. "She is the love of my life and the mother of my three children, Mary Helen, Kate, and John, III. She has worked side by side with me as my nurse for these many years."

Self-described as "home grown," Dr. Pender was born, educated, and built his practice in Memphis. After graduating from the University of Tennessee in 1960, he was assigned to a general rotating internship at John Gaston Hospital, which is now the MED.

The desire to experience all areas of general practice led Dr. Pender to work with Dr. Bob Hollingsworth, who was a general practitioner in Shelby, Mississippi. With an enormous population and only a few doctors, they became a two-man medical center. We did everything from surgery to home obstetrical deliveries," says Dr. Pender. "In Shelby, I felt like a real doctor."

In 1963, military duty was required, and Dr. Pender served two years as a captain in the Strategic Air Command with the U.S. Air Force. When he returned home, he completed a pediatrics residency program at Le Bonheur Children's Hospital. In 1965, he opened his own practice with Kathleen by his side. The practice expanded to include more doctors and many more little patients.

When Dr Pender recalls the great influences in his career, he begins with his parents. "I thank my parents, John Vincent Pender, Sr. and Helen Gavin Pender, and thank my Aunt Mary Elizabeth Gavin for the financial sacrifices they made," says Dr. Pender. "They gave me constant encouragement and love as I pursued my dreams of a medical career."

Influential colleagues in the field of pediatrics during Dr. Pender's career include Drs. Harry Jacobson, Gilbert Levy, James Etteldorf, Jimmy Hughes, Bob Allen, Earl Wren, Barton Etter, and Clyde Croswell. Influences in the field of obstetrics were Drs. Mike Roach and Rushton Patterson, Sr.

During the time most physicians chose specializations, Dr. Pender wanted to utilize his background in general medicine, surgery, pediatrics, psychiatry, and obstetrics. By focusing on children, he would have an impact on the health of an entire family.

"I have never been and never will be one of the shining lights in Memphis Medicine," says Dr. Pender. "I have simply been one of the front line, foot soldiers." Thousands of Mid-South parents who have watched Dr. Pender care for their newborns, lower fevers, soothe strep throat, and been comforted by his voice in the middle of the night, know Dr. John Pender's light shines very bright in Memphis.

PEDIATRIC SURGICAL GROUP

ABOVE: *Pediatric Surgical Group, from left to right, Drs. S. Shochat, M. Blakely, R. Hollabaugh, E. Wrenn, Jr., S. Hixson, B. Rao, J. Eubanks, E. Huang, and M. Langham.* BELOW: *The late Robert G. Allen, M.D.*

UNTIL 1952, PEDIATRIC SURGICAL PATIENTS IN MEMPHIS were treated in general hospitals that lacked dedicated children's wards. Only the City of Memphis Hospital (now The MED) and the John Gaston Hospital with its Tobey Hospital had such facilities. Tobey was the pediatric teaching unit for the University of Tennessee Medical School, with pediatric surgery performed by the University of Tennessee faculty and residents. In private hospitals, several general surgeons and some specialty surgeons provided pediatric care. Prominent among them were Dr. Russell Patterson, Jr., Dr. Alphonse Meyer, Dr. Ed Stevenson, Dr. Ed French, Dr. Gwin Robbins, Sr., and Dr. Breen Bland.

Le Bonheur Children's Hospital opened in 1952, and Dr. Earle L. Wrenn, Jr., opened a full-time office there for pediatric general surgery in 1956. He was also a volunteer faculty member in the University of Tennessee Surgery Department, which provided rotating general surgical residents to Le Bonheur. In 1957, Dr. Robert G. Allen came to Le Bonheur for pediatric general and cardiothoracic surgery and established an open-heart surgery program. Trained at Boston Children's Hospital under Dr. Robert Gross, Drs. Wrenn and Allen were the first formally trained pediatric surgeons in Memphis at a time when few cities had more than one such trained pediatric general surgeon.

Needing full-time assistants for the open heart surgery program, Dr. Allen recruited residents, beginning the program of full-time Le Bonheur surgical residents in addition to the general surgery residents from the University of Tennessee. The special needs of open heart surgery patients led to founding an intensive care unit at Le Bonheur and recruiting trained pediatric intensivists. This collaboration improved the care of desperately ill children with a broad assortment of surgical and medical problems.

In 1968, when Dr. Sheldon Korones developed the Newborn Center at the City of Memphis Hospital to provide intensive care for fragile newborns, infants began to survive with a broad spectrum of surgical problems. Drs. Wrenn and

Allen made several important and original contributions to the care of these babies. As they advanced to become clinical professors at the University of Tennessee, they improved the training of residents at both Le Bonheur and Tobey. Their contributions were recognized nationally: Dr. Wrenn was elected chairman of the Surgical Section of the American Academy of Pediatrics, and Dr. Allen became president of the American Pediatric Surgical Association.

In 1972, Dr. Robert S. Hollabaugh, Sr., joined Drs. Wrenn and Allen in the practice of general pediatric surgery, followed by Dr. S. Douglas Hixson (1981) and Dr. Michael G. Carr (1983). The practice name became Pediatric Surgical Group, Inc. (PSG), in 1981. Dr. Allen, who was known as an innovative cardiac surgeon and a pioneer pediatric surgeon in Memphis, died tragically in an automobile accident in 1981.

When St. Jude Children's Research Hospital was established in 1962, Dr. Donald Pinkel of St. Jude consulted Dr. Wrenn about providing surgical services for pediatric cancer patients, with operations performed at nearby St. Joseph Hospital and at Le Bonheur. Founding surgeons were Dr. Irving Fleming, an oncology surgeon, and Dr. Wrenn, with other specialty surgeons called on as needed.

The new Le Bonheur Children's Hospital opened in the summer of 2010.

As St. Jude's patient volume increased, Dr. Mahesh Kumar was recruited as the first full-time pediatric oncology surgeon in Memphis. Dr. Bhaskar Rao joined St. Jude in 1980, providing innovative work on limb-sparing surgery for osteosarcoma that prevented many amputations and bringing national and international acclaim to Memphis.

In 1989, the University of Tennessee recruited Dr. Thom Lobe as the first full-time professor and chief of the Division of Pediatric Surgery at the university. He was widely recognized as an innovative surgeon who helped develop minimally invasive surgery. He founded the accredited Pediatric Surgical Residency Program that expanded and improved surgical education at Le Bonheur. Dr. Kurt P. Schropp joined the faculty in 1991. Until he left in 2001, he and Dr. Lobe worked closely with the Pediatric Surgical Group and St. Jude, although the three practices remained separate.

In 1996, St. Jude established a department of surgery with Dr. Steve Shochat as the first surgeon-in-chief. One year later, Dr. Andrew Davidoff joined him and established a basic scientific program within the Department of Surgery. Gene therapy techniques applied to treating solid tumors through anti-angiogenesis have been particularly noteworthy in Dr. Davidoff's lab.

Dr. Wrenn retired in 1999 after a long and distinguished career. Earle and Lynette still visit Memphis, always regaling the faculty with stories. Dr. Wrenn's leadership and vision of collaboration across all areas of pediatric surgery are gratefully recognized by those who carry on his legacy.

Pediatric surgery in the Mid-South has continued to evolve. Dr. James W. Eubanks, III, joined the group in 2002, followed by Drs. Martin L. Blakely (2004), Max R. Langham, Jr. (2005), and Eunice Y. Huang (2007). When Dr. Lobe left Memphis in 2004, the University of Tennessee recruited Dr. Langham as professor and chief of the Division of Pediatric Surgery after 16 years at the University of Florida. Under his leadership, the Division of Pediatric Surgery and the Pediatric Surgical Group merged into a single practice supporting Le Bonheur.

From its beginning in general surgery—including cardiothoracic, urologic, plastic, otolaryngologic, and orthopedic practices—pediatric surgery in Memphis has attracted specialists in these fields who have established practices, assuming the care of many patients. This drive to specialization reflects international trends and has allowed the group to focus on treating children with trauma, congenital anomalies, and cancer, plus the more common appendicitis and hernias.

For severely ill children or those who require complex multispecialty care, Le Bonheur remains the primary children's surgical and medical teaching hospital in the Mid-South and the primary pediatric teaching facility for the University of Tennessee. As the premier international children's cancer research hospital, St. Jude attracts children from around the world. The care in both hospitals is provided by formally trained specialists in pediatric surgery and anesthesiology.

With this distinguished past, the future appears uniquely exciting for pediatric surgery in the Mid-South. New construction at Le Bonheur is creating a $325 million, state-of-the-art hospital with the newest operating rooms, emergency rooms, and intensive care units. Recruitment of prominent surgeons, intensivists, and pediatric specialists continues, driven by the ever-improving collaboration of St. Jude, Le Bonheur, and the University of Tennessee. Working long hours, many dedicated men and women have made this success possible. Their dream continues as we strive to develop one of the best pediatric healthcare systems in the United States.

SAINT FRANCIS HEALTHCARE

In 1974, the opening of St. Joseph Hospital-East forever changed the medical landscape of Memphis. Prior to this historic event, full-service hospitals were located in the downtown medical center. The inspiration for an East Memphis location grew from the conviction that as the city expanded eastward, medical services should also move to be close to the population they served.

Sister Rita Schroeder, leader of the original St. Joseph Hospital, was an avid supporter of the proposed new suburban hospital. When she was chosen to serve as chief executive officer of both the old and new hospitals, she accepted the responsibility of raising construction funds for the latter. Sister Rita appealed to her order to finance the new hospital's proposed 20 million dollar budget. The order offered a loan of five million dollars, then lowered the amount to two million dolalrs, which only covered the project's down payment.

On August 23, 1969, Sister Rita was authorized to purchase the required land, but 18 million dollars was still needed for construction. To raise this money, she formed a plan to offer bonds, and with her indomitable conviction, persuaded influential Memphis businessmen to support her effort. Groundbreaking for the 15-story hospital was held on March 13, 1972. Her success in creating the publicly owned, not-for-profit St. Joseph Hospital-East was used as a teaching model at Harvard Business School.

Catholic Bishop Carroll T. Dozier dedicated St. Joseph Hospital-East on December 8, 1974. Sister Rita assumed her role as administrator and the not-for-profit facility, sanctioned by the Catholic Diocese of Memphis, began caring for patients according to *The Directives of Catholic Healthcare*. The hospital's basic mission is the same now as it was then: "To heal, support, and comfort all whom we serve in the tradition of Catholic healthcare."

By the time St. Joseph Hospital-East opened, St. Joseph's downtown had withdrawn its involvement and the financial burden of operating the hospital was left in Sister Rita's hands. With sheer determination, she convinced influential Memphians to support the hospital by continuing the sale of bonds. Although the two hospitals were now truly independent facilities, their similar names caused public confusion over services and events. A name change was needed and the new hospital officially became Saint Francis in 1981.

In that same year, Sister Rita's order told her she would have to rotate assignments, thereby giving up her role at Saint Francis. She chose to leave her order, rather than leave her hospital. For the next 12 years, she led the hospital as Rita Schroeder, retiring in 1993.

In 1994, Saint Francis Hospital became a for-profit facility after being sold to American Medical International (AMI), a for-profit corporation that owned hospitals nationwide. The proceeds from the sale were used to create The Assisi Foundation of Memphis, a philanthropic non-profit organization. Even as a for-profit hospital, Saint Francis retained its name and signature dedication to the ideals and guidelines of Catholic health care. In 1996, AMI merged with NME, another for-profit hospital chain, to create Tenet Health System, the current owner of Saint Francis. In 1997, David L. Archer became chief executive officer of the

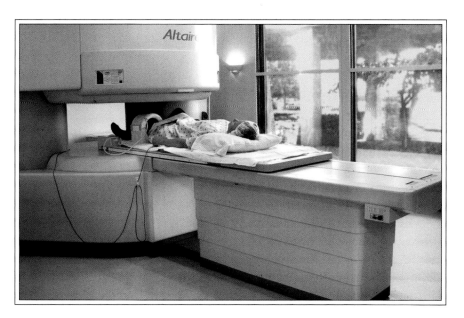

600-bed facility and under his leadership, Saint Francis has continued its innovations in medical services and expanded its involvement in the Memphis community.

During the 1980s, Saint Francis joined with the University of Tennessee to establish the Saint Francis/University of Tennessee Family Practice Residency Program on the hospital's campus. Today, the program has expanded to

Under the ownership of Tenet Health System, Saint Francis Hospital-Memphis has become the flagship of the many facilities that comprise Saint Francis Healthcare.

include residencies in psychiatry and surgery. The Saint Francis medical staff includes over 800 members, many of them serving in important hospital leadership roles. Saint Francis is one of only a few hospitals where half the members of its board of directors are physicians.

Saint Francis has brought many firsts to the Memphis healthcare market, including the first outpatient department, the first chest pain emergency center, and the first stroke emergency center. Other service lines include the Family Birthing Center; Inpatient, Outpatient and Diagnostic Oncology; the Cardiac Care Center; the Center for Surgical Weight Loss; the Outpatient Surgery Center; the Women's Center; Senior Care Services; and the TotalCare Outpatient Diagnostic Center.

Saint Francis Hospital has been recognized for its quality of care by the American Academy of Sleep Medicine; the American Diabetes Association; and the Society of Chest Pain Centers. It has also been recognized by the American Society of Bariatric Surgery as a Center of Excellence; as a Cigna Certified Hospital for Bariatric Surgery; as a Cancer Program Approved With Commendation for Inpatient and Outpatient Cancer Care; as a Center of Excellence in Diabetes Education; and by UnitedHealth as a Premium Cardiac Specialty Center, a Premium Surgical Spine Center, and as a Premium Total Joint Replacement Center and Orthopedic Center.

Tenet's focus on community involvement has allowed Saint Francis to take an active role in such Memphis organizations as United Way, the American Heart Association, the American Cancer Society, the Parish Nurse Program, the Church Health Center, and many others.

In 2004, Tenet and Saint Francis continued the hospital's bold medical tradition by opening Saint Francis Hospital-Bartlett, the first full-service hospital in Bartlett, Tennessee. The new hospital's success put it on track for a major expansion in 2010 to include 83,000 new square feet, an additional 96 beds, and expansion of a number of its ancillary services. The opening of the Bartlett facility led to the creation of Saint Francis Healthcare as a single Memphis market identity for all Saint Francis service providers.

Saint Francis Hospital-Memphis began as a Catholic-inspired vision built at the edge of an expanding city. Today, it is a modern, successful, full-service healthcare facility resting near the center of Memphis and serving a diverse and ever-growing population. "Saint Francis Hospital-Memphis succeeds because of its high quality of care and the dedication and excellence of its employees, physicians, and volunteers," Archer says, "yet our job is made easier by the high degree of trust our predecessors earned for the Saint Francis name in this community." With state-of-the-art technology; skilled and dedicated employees, respected, patient-focused physicians, and a firm commitment to continuous quality improvement and excellence in customer service, Saint Francis Hospital-Memphis will continue to meet the present and future healthcare needs of Memphis and the Mid-South.

SAINT FRANCIS HOSPITAL-BARTLETT

On May 21, 2004, the Memphis Area welcomed its newest hospital. Hundreds of people gathered for Saint Francis Hospital-Bartlett's ribbon-cutting ceremony. Dignitaries attending the ceremony included Shelby County Mayor A. C. Wharton, Bartlett Mayor Keith McDonald, and Memphis Grizzlies basketball star Shane Battier. Visitors toured the new facility and received information about the 24-hour Emergency Care Center, Women's Center, and Advanced Digital Imaging Center.

The Bartlett community was the driving force behind the creation of this hospital. Until 2004, Bartlett was the largest city in Tennessee without a hospital. Scores of elected officials and Bartlett residents drove to Nashville to show their support for the state of Tennessee's approval to build a hospital. Their efforts were successful. Led by David Wilson, chief executive officer, Saint Francis Hospital-Bartlett opened its doors to the citizens of Northeast Shelby County on June 1, 2004.

The Saint Francis-Bartlett campus encompasses 23.6 acres of land. The hospital is 204,000 square feet and cost approximately $57 million to build. In addition, a 60,000-square-foot medical office building was constructed on the hospital campus to offer physicians the opportunity to establish their offices close to the hospital. This allows patients to receive comprehensive care in one location.

"This new, state-of-the-art hospital represents a significant investment in the future health and well-being of the Bartlett community," said Trevor Fetter, chief executive officer of Tenet Healthcare Corporation. "Saint Francis Hospital-Bartlett will provide access for a fast growing community that has, up to now, been without those services."

Since the opening of the hospital, all Saint Francis-Bartlett employees have been introduced to an initiative called Target 100. The goal of this valuable, service-oriented program is to provide total patient satisfaction. Saint Francis Hospital-Bartlett was twice named a J. D. Power and Associates Distinguished Hospital for providing an "Outstanding Inpatient Experience." In addition, the hospital has been named an American Alliance of Healthcare Providers Hospital of Choice. Saint Francis-Bartlett has been honored four times with the Tenet Healthcare Five Star Service Award and was named to Tenet Healthcare's Circle of Excellence for hospitals that have achieved the highest levels of quality, service, and operational performance.

Since its opening, Saint Francis Hospital-Bartlett has undergone change and growth, which has included a change in leadership. In 2006, Kem Mullins was named to replace David Wilson as chief executive officer. Mullins was promoted to this position after serving three years as chief operating officer at Tenet's Atlanta Medical Center, where he was responsible for operations oversight of the 460-bed tertiary care facility.

Saint Francis Hospital-Bartlett has also expanded its services to include a cardiac catheterization lab featuring diagnostic cardiac catheterization as well as diagnostic and interventional peripheral studies and procedures. Digital mammography, which provides significantly better imagining than film mammography for women who are under 50 or who have very dense breast tissue, was introduced to the hospital's patients in 2008. Studies show that digital mammography uses less radiation than standard film mammography and that digital imaging generally takes less than a minute compared to 10 to 15 minutes for film mammography.

As demand for healthcare services in Northeast Shelby County continues to grow, Saint Francis Hospital-Bartlett is looking to the future. Plans include an outpatient diagnostic imaging center and a 96-bed expansion of the hospital. "The future is bright for Saint Francis Hospital-Bartlett," says Mullins. "We look forward to continuing to grow with the community while maintaining our high standards of patient care."

SALMAN SAEED, M.D., F.A.A.P.M.

"Neurology has changed more in the last 10 to 15 years, than the past 100 years," says Dr. Salman Saeed. "The period between 1990 to 2000 was called the Decade of the Brain, and new testing and medications for epilepsy, narcolepsy, Parkinson's disease, and stroke were being developed."

A residency at New York Medical College led Dr. Saeed to an elective rotation in neurology and later to becoming chief resident. Generous funding for neurological research as well as rapid breakthroughs in the field, prompted Dr. Saeed to specialize in neurology. In 1995, Dr. Saeed did a fellowship in clinical neurophysiology at the University of Louisville.

In 1996, Dr. Saeed's practice began in Covington, Tennessee and expanded to Dyersburg. In 2000, he opened West Tennessee Neurology in Bartlett, which worked closely with Saint Francis Hospital. Dr. Saeed took an active role in the construction of Saint Francis-Bartlett where he is now the director of neurophysiology. "We were able to work with administration early in the planning to integrate some services that even larger hospitals do not provide," says Dr. Saeed. "For instance, if a patient arrives within three hours of a stroke, many times the effects can be reversed, and that is exciting."

Many neurological tests, including CT scans, electromyography, electroencephalograms, and infusions for multiple sclerosis are performed in Dr. Saeed's office. As the field of neurology develops at a rapid pace, Dr. Saeed knows his treatments will also change. "In the future, there will be medicines or a vaccine for Alzheimer's disease and an oral medication for multiple sclerosis," says Dr. Saeed. "There are new therapies for epilepsy and Parkinson's disease just on the horizon."

GRADY SAXTON, SR., M.D.

Dr. Grady Saxton's medical career began in pharmacy, but later, he decided to pursue medicine. After attending Meharry Medical College, he completed a residency at Akron City Hospital. A fellowship in cardiology took him to the Medical College of Georgia, and, in 1985, he completed his residency and returned to Memphis.

Dr. Saxton practiced in public health, before moving to a larger practice where he traveled to as many as seven hospitals per day. As a medical professional herself, Blanche understood the sacrifices made in a medical career and fully supported her husband. Dr. Saxton's hectic schedule continued for several years until, in 1987, he opened his own practice. In 1992, he joined the Cardiovascular Specialists Group.

In 2000, the Saxton Heart Clinic, PC, opened, and Blanche became his business manager with CaSonya Jordan as their office manager. Dr. Saxton was committed to spending time with his patients to explain health issues. "I didn't want my patients to leave the office saying they didn't know what was happening with their health," says Dr. Saxton. He is still especially aware of older patients and knows taking the time to care for them is a credit to his mother's teachings.

The Saxtons have two sons, Grady Jr., a pharmacist, who is married to Jennifer, a teacher. They have a newborn son, Gavin. Their other son Garreth, who has a bachelor's degree in business management, is finishing a degree in cardiac ultrasound.

Dr. Saxton closed the Saxton Heart Clinic in April 2010 and and now works as an emergency department physician in Memphis.

DAN H. SHELL, III, M.D., F.A.C.S

Dr. Dan Shell remembers clearly the day he received an unusual phone call at Le Bonheur Children's Hospital. "Dr. Shell, you have a call holding from Bolivia," said the hospital operator, who had clearly said Bolivia—not nearby Bolivar, Tennessee. As Dr. Shell picked up the phone, he would become the only hope for a little girl in the Amazon.

In 1986, Dr. Darr LaFon, Jr., a U.S. Air Force flight surgeon from Memphis, was on a mission in South America. As his unit was passing through a jungle village, he met a nine-year-old girl, Fatima Lurici, who had been bitten by an extremely venomous Bushmaster snake. Amazingly, she survived, only to lie on a primitive hospital cot while her leg rotted away. Antibiotics failed to stop the loss of tissue, and Fatima was scheduled for amputation, which, in Bolivia, would seal her fate as a street beggar.

Dr. Shell was chief of pediatric plastic surgery at Le Bonheur and one of the pioneers of tissue transfer in Memphis. The procedure transfers the patient's own skin, muscle, and bone to a damaged site. Tissue, with a specific nutrient artery and vein, can be transferred, immediately revascularized, and expected to heal at the reconstructive site. This method proved very effective in saving severely damaged limbs from vehicle accidents and tumor resections. Dr. LaFon remembered Dr. Shell's work in reconstructive microvascular tissue transfer to salvage limbs and called Le Bonheur.

As Dr. Shell listened, Dr. LaFon described Fatima's condition; he believed her leg could be saved. Dr. Shell and Le Bonheur volunteered their services, and the U.S. Air Force and U.S. State Department arranged transport to the United States for Fatima and her father. When she arrived, most of the tissue from her heel to the knee was gone and Dr. Shell knew a tissue transfer was the only hope for saving her leg. First, dead and infected tissue was soaked and removed during whirlpool treatments. Dr. Shell then prepared for removal of a skin-muscle unit from the latissimus dorsi to her leg where he would quickly reconnect the tiny artery and veins for tissue survival. After hours in surgery, the transfer was successful.

During weeks of intense physical therapy, Fatima never complained. As she began to regain strength, it was clear her leg would continue to heal and develop normally. Fatima became a celebrity when her story was featured in local newspapers and then on *Good Morning America*. Memphis Mayor Dick Hackett took her to the circus, and local schools donated toys. When she returned to Bolivia, she took trunks of toys and gifts for children and friends in her village.

Fatima's story increased national awareness of microvascular tissue transfer for limb salvage, which had not previously been an option. Her story brought many children and adults to Dr. Shell for tissue transfers in the years that followed.

Fatima was later adopted and moved to Florida, where she is now married and raising a family. One phone call was the beginning of an amazing journey for a courageous little girl from the Amazon jungle—a journey with a very happy ending for her and for Dr. Dan Shell.

SYED H. SHIRAZEE, M.D., F.C.C.P.

Syed Shirazee's grandmother inspired him to become a doctor at the age of nine. He attended Dow Medical College in Karachi, Pakistan where U.S.-trained physicians had returned to teach. In 1990, Dr. Shirazee applied for a residency in the United States where health care and facilities were far more advanced. During his second year at the University of Illinois, he became interested in pulmonary care, and as a fellow at West Virginia University in Morgantown, he chose to specialize in pulmonary critical care.

As Dr. Shirazee completed his fellowship, a friend from Ripley, Tennessee asked if he would visit and consider practicing in West Tennessee. Dr. Shirazee arrived to find a receptive rural medical community, and he was immediately offered a position. He worked at Baptist Memorial Hospital-Tipton for three years before opening a private practice. He was the director of Respiratory and ICU at Baptist-Tipton until 2004.

In 2005, Dr. Shirazee moved to Memphis and became Saint Francis-Bartlett Hospital's only pulmonary specialist. He designed the pulmonary unit and became one of the first physicians in Memphis to use RotoProne therapy for Acute Respiratory Distress Syndrome (ARDS). Dr. Shirazee and the hospital staff also synchronized a team approach to caring for patients in the ICU and with remarkable results. "We had no ventilator associated pneumonia in the ICU for over a year," says Dr. Shirazee.

Today, Saint Francis-Bartlett is expanding to meet the community's growing healthcare needs. Dr. Shirazee sees his practice growing and eventually adding partners to assist with the demands of being the hospital's only pulmonary specialist. He looks forward to spending more time with his wife and three sons, and pursuing his interest in business management.

A. ROY TYRER, M.D.

Roy Tyrer and classmates attended Army boot camp and were inducted their senior year at the College of Medical Evangelists, now Loma Linda University. Medical degrees were accelerated due to the World War II physician shortage. Dr. Tyrer became a neurosurgical resident at White Memorial Hospital and then chief of neurosurgery at Letterman Army Medical Center during military duty.

In 1949, renowned neurosurgeon Dr. Eustace Semmes requested Dr. Tyrer leave Boston's Lahey Clinic and become senior neurosurgical resident at Memphis City Hospital. "In the history of neurosurgery, then and now," he says, "Memphis has been an established neurosurgery center in the South."

Neurosurgeons Drs. Nicholas Gotten and C. Douglas Hawkes asked Dr. Tyrer to join them in 1950, which constituted the founding of the Neurosurgical Group of Memphis, which progressed to include 14 physicians. Under the leadership of Dr. Hawkes, a neurosurgical residency was established at Methodist Hospital, which later merged with a preexisting residency at Semmes-Murphey Clinic, to become the current University of Tennessee neurosurgical residency. From 1950 to 1965, Dr. Tyrer taught neurological diagnosis to the University of Tennessee medical students and was sponsor of the 1954 graduating class. He represented Tennessee at the American Medical Association for nearly 25 years. Also, Dr. Tyrer and Dr. Granville Davis of Rhodes College co-chaired a committee establishing Arlington Developmental Center, which was the first facility for impaired individuals in West Tennessee and one of only three in Tennessee.

In 1951, 22 neurosurgeons formed the Congress of Neurological Surgeons, and Dr. Tyrer is one of three surviving founders, representing the organization's current national and international membership. In the 1960s, Dr. Tyrer's neurosurgical treatment of Parkinson's disease resulted in more than 200 successful operations. Dr. Tyrer retired from the active practice of neurosurgery in 2000 and, since then, has operated the Tyrer Neurosurgical Consulting Clinic.

SEMMES-MURPHEY CLINIC

When Eustace Semmes pursued a career in medicine, he did not realize he would one day be a pioneer in neurosurgery.

Raphael Eustace Semmes was born in Memphis, Tennessee on August 15, 1885. His mother's death during childbirth was thought to have played a role in Semmes's intense interest in medicine. The family spent their summers in North Carolina where young Eustace met Dr. George LeFevre, professor of zoology at the University of Missouri, who taught him how to study marine life under a microscope. The Semmes and LeFevre families became very close friends and continued to spend their summers in North Carolina.

Dr. LeFevre encouraged young Eustace to consider medicine as a career and later convinced him that he was ready for college at the age of 17. Semmes skipped his senior year at Christian Brothers to enroll at the University of Missouri. There he met friends and future colleagues Thomas Grover Orr and Walter Dandy, and the three pursued medical studies. Dr. LeFevre saw the potential in the young men and directed them to apply to Johns Hopkins Medical School. In 1907, they were not only accepted, they began as second-year medical students studying under the famous surgeons, Dr. William Halstead and Dr. Harvey Cushing. Dr. Cushing would become the father of American neurosurgery.

Three years later, Semmes graduated in the top ten of his class. He interned with Drs. Halstead and Cushing due to his keen interest in studying the nervous system. Semmes welcomed Cushing's grueling schedules and was not intimidated by his gruffness. During Semmes's residency at The Women's Hospital in New York, he studied as many neurological cases as possible. He returned to Memphis in 1912 to begin a practice and became the assistant in neurosurgery at the Memphis General Hospital.

During World War I, Semmes was one of ten neurosurgeons sent to New York Neurological Institute for three months of intensive training before deploying to Toul, France. In 1919, he was promoted to captain in the Medical Corps, but, at the request of Dr. Wittenborg, chairman of the Department of Anatomy and dean of the medical

Dr. Eustace Semmes (above left) and Dr. Francis Murphey, (above right) formed a partnership in the 1930s that became today's Semmes-Murphey Clinic (opposite page).

school in Memphis, he was released from duty to teach medical students.

When Semmes returned to Memphis, his patients, assuming he would be killed, had not paid their bills. He borrowed $75 and opened an office practicing exclusively in neurosurgery. Semmes and his friend, Dr. Walter Dandy, who was in charge of neurosurgical services at Johns Hopkins, wrote papers on lumbar disc disease which were pivotal in the development of neurosurgery.

Semmes traveled to Boston as often as possible to observe Dr. Cushing's surgical techniques. Cushing showed favoritism by ordering him a gown and mask and allowing Semmes into the surgery area, whereas other doctors were asked to observe from the amphitheater. Cushing also showed his confidence in Semmes by referring patients, who in Cushing's words, "...deserve more than I can do."

In 1930 and 1931, Dr. Semmes was nominated for membership in The Society of Neurological Surgeons, but he was not elected. After 20 years of practice in neurosurgery, this exclusion was frustrating. Semmes met with four young colleagues in Washington, D.C., and they conceived the idea of a new neurosurgical society. With Dr. Cushing's blessing, 34 neurosurgeons became founding members of The Harvey Cushing Society in 1932. Dr. Cushing suggested they hold the first meeting in Boston. The society later became the American Association of Neurological Surgeons (AANS) which is still the largest neurosurgical society in the United States. Semmes was also instrumental in the formation of the American Academy of Neurological Surgery and the Congress of Neurological Surgeons. Dr. Semmes was elected to the senior society in 1933.

For many years, Dr. Semmes was the only neurosurgeon in Memphis. In 1933, Dr. Paul Bucy visited Semmes in Memphis, and Semmes confided that he would like to have an assistant who had *not* been trained in neurosurgery. Like Dr. Cushing, Semmes wanted to teach through example. Dr. Bucy picked up the phone and called Dr. Francis Murphey at the University of Chicago and told him to take the first train to Memphis. Murphey, who had attended Harvard Medical School and was in the last class trained by Dr. Cushing, became Dr. Semmes's first assistant In 1934. "It turned out that Francis was just what I wanted," said Dr. Semmes. "Having a large neurosurgical practice in his favorite city (Memphis) was also what Francis wanted."

In 1932, Dr. Semmes established the first department of neurosurgery, and was appointed a professor at the University of Tennessee. The American Board of Neurological Surgery (ABNS) was founded in 1940, and Dr. Semmes became one of 50 members honored through exemption from the neurosurgical exam.

Dr. Semmes's diagnosis of ruptured discs, without the aid of myelography, was a landmark in the treatment of herniated lumbar discs. Dr. Semmes used Novocain as local anesthesia while training with Dr. Cushing and Dr. Halstead and believed this was the best technique. He used the method for 40 years, in ninety-five percent of his surgeries. During World War II, Dr. Murphey entered the Army and became chief of neurological surgery at O'Reilly General Hospital in Missouri. While in the Army, Murphey and Semmes published a paper, *The Syndrome of Unilateral Rupture of the Sixth Cervical Intervertebral Disc,* in 1943, proving that ruptures of the sixth cervical disc caused numbness and neck and arm pain. After Murphey's return from the war, Dr. Elmer (Dutch) Schultz joined the practice in 1948 and Dr. Richard L. DeSaussure, Jr. in 1950. At Schultz's suggestion, the clinic was incorporated and named the Semmes-Murphey Clinic.

In 1956, Dr. Murphey took Dr. Semmes's position as professor of neurosurgery at the University of Tennessee, continuing their commitment to "train the next generation."

In the early 1970s, Semmes-Murphey was instrumental in the development of the first intensive care unit at the Baptist Hospital designed for their neurosurgery patients. When computerized tomography scanning was available in 1975, Semmes-Murphey convinced the hospital to purchase the first one in the city. To avoid delays in diagnosing patients, Semmes-Murphey bought x-ray equipment from Baptist and opened its own imagining department. This added service for neurological disorders developed into a state-of-the-art diagnostic facility.

Dr. Murphey had a keen interest in cerebrovascular disorders, stroke, and brain hemorrhages. He observed that patients who had experienced a subarachnoid bleed from an intracranial aneurysm frequently had an out-pouching on the aneurysm when arteriography was performed. He referred to this out-pouching as a teat or tit. This arteriographic sign came to be known as "Murphey's teat" and is a term utilized by cerebrovascular surgeons world wide to describe the site from which an aneurysm has likely bled.

During the mid-1970s, rapid advancements in the field of neurosciences prompted physicians at Semmes-Murphey to seek specialization. Dr. Morris Ray trained in Switzerland as a cerebrovascular specialist, Dr. James T. Robertson specialized in pituitary tumors, acoustic tumors, and carotid artery disease, while Dr. Richard DeSaussure's specialty was percutaneous methods for pain and trigeminal neuralgia.

In 1994, Semmes-Murphey and Methodist Hospital founded the Medical Education and Research Institute (MERI) to train surgeons in the latest neurosurgical procedures. This was another revolutionary idea that provided the latest neurosurgical instruction from leaders in the field. Today, more than 500 physicians from the United States and 26 countries attend training at MERI each year.

Semmes-Murphey along with numerous Tennessee physicians established the physician-owned State Volunteer Mutual Insurance Company, which wrote its first policy in 1976. In order for Semmes-Murphey to have the money to invest, the sale of their duck club in Arkansas was required. This was a favorite place for Semmes-Murphey physicians to hunt ducks, relax, and entertain family, friends, and referring physicians. The investment was a wise one because the company is still the primary insurer of physicians in Tennessee and surrounding states.

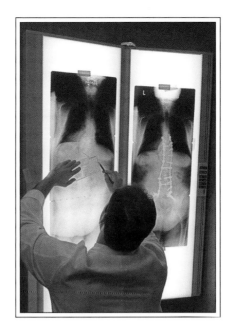

Semmes-Murphey surgeon Dr. Julius Fernandez examines a spinal x-ray. Dr. Fernandez has a special interest in surgical correction of spinal deformities.

Although Semmes-Murphey began as a practice focused on neurosurgery, it became apparent to the leaders of the group that several other specialties would be needed for the comprehensive evaluation and management of patients with neurological conditions. Neurology was the first nonneurosurgical specialty added, with neuropsychology, physical medicine, and pain management being added later.

In 2002, Semmes-Murphey opened a new facility in East Memphis which included an ambulatory surgery center where many outpatient neurosurgical procedures are performed. The clinic serves as a resource for patients to receive physician consultation, diagnostic studies, and surgical treatment.

Ongoing research with the University of Tennessee, St. Jude Children's Hospital, Methodist University Hospital, Baptist Hospital, and Le Bonheur Children's Hospital, continues to earn worldwide recognition for Semmes-Murphey's excellence in the neurosciences. Frequent publications in medical journals attest to the excellence Semmes-Murphey's physicians have achieved.

Today, Semmes-Murphey has grown to 250 employees and more than 40 physicians at three locations. The partnership of Dr. Eustace Semmes (1885-1982) and Dr. Francis Murphey (1906-1994) continues with the 100-year tradition of, "Commitment to excellent patient care, development of the field, community service, and training the next generation."

Acknowledgments to the Journal of Neurosurgery *for the information on the Semmes-Murphey clinic in their article "The history of neurosurgery in Memphis: the Semmes-Murphey Clinic and the Department of Neurosurgery at the University of Tennesee College of Medicine," by Dee J. Canale, M.D., Clarence B. Watridge, M.D., Tyler S. Fuehrer, and Jon H. Robertson, M.D., Volume 112, Number 1, pp. 189-198 (January 2010).*

SOUTHWIND MEDICAL SPECIALISTS

In 1999, two Memphis internists and a gastroenterologist left well-established practices to form their own group practice with the goal of offering comprehensive patient care in a professional and patient-driven atmosphere. Drs. Lawrence Whitlock, C. B. Daniel, and Ronald C. Michael opened Southwind Medical Specialists with only 2000 square feet of office space and a small support staff.

In 2000, Dr. Paul Dang joined the staff and additional space was added to accommodate him and his patient base. Dr. Hays Brantley then joined the group and a second location for Southwind Medical Specialists was added, where Dr. Brantley continues to practice.

Two additional internists have also been added, Dr. Tiffany Hall, who specializes in internal medicine and pediatrics, and Dr. Lamar Bailey who comes from the Veterans Administration Hospital, and a teaching position there, back to private practice.

When Southwind added Drs. Laurie Baker and Debbie O'Cain, and their practices in Family Medicine, medical services for the entire family became available. In addition to family medicine, Dr. O'Cain specializes in workmen's compensation and occupational health claims. In addition to her practice as an internist, Dr. Tiffany Hall is a pediatrician, and works closely with Drs. Baker and O'Cain.

Bridget Brady, Adult Nurse Practitioner, and Tina Dickinson, Family Nurse Practitioner, have expanded the clinic's expertise, and assist the physicians in a broad range of services including diagnosis and recommended treatment. They also focus on patient education, prevention, and wellness to encourage patients to become an active part of their own health care.

Having outgrown the original facility, Southwind Medical Specialists moved to a new location in southeast Memphis, in 2003, with 20,000 square feet. The clinic's services have expanded to include nuclear imaging, full service x-rays, ultrasounds, EKGs, bone densitometry studies, ECHOs, diabetes management, pulmonary studies, treadmills, and extensive laboratory testing. This inclusive approach provides "one stop shopping" for healthcare needs and also eliminates the need for patients to pay hospital deductibles and coinsurances for testing that would otherwise not be available in a primary care office. This provides a faster diagnosis, and allows treatment options to be expedited in a more timely manner.

ABOVE: *From left to right, C. B. Daniel, M.D., Lamar Bailey, M.D., Paul Dang, M.D., Ronald Michael, M.D., Debbie O'Cain, M.D., J. Hays Brantley, M.D., Laurie Baker, M.D., Tiffany Hall, M.D., Lawrence Whitlock, M.D.* LEFT: *Family Nurse Practitioner Tina Dickinson.* BELOW LEFT: *Adult Nurse Practitioner Bridget Brady.*

In 2004, Southwind expanded once again with the opening of Southwind Endoscopy Center, which is where Dr. Ron Michael not only acts as the medical director, but also performs his outpatient colonoscopy and endoscopy procedures. Patients may now have these procedures performed in the convenience of an office setting and under the care of the center's staff members, who report and consult with the patient's physician, on-site.

Southwind Medical Specialists continues in its commitment to provide comprehensive medical care to its patients in a professional and pleasant atmosphere.

INDURANI TEJWANI, M.D.

INDURANI TEJWANI ARRIVED IN Memphis just as a sanitation workers strike and Dr. King's death was accelerating racial tension. She had experienced unrest when India gained independence from Britain. During that struggle, Indian women's role later made their rights first priority, and their education was free. When Indurani enrolled in medical school, fifty percent of her classmates were female. During her second year of general and thoracic vascular surgery residency, she married and her husband's job brought them to Memphis in 1968.

General surgery and OB/GYN residencies were not offered to women at that time, so Indurani began a pathology residency at St. Joseph Hospital. "It was an eye-opener to a foreigner, because I expected the Western world to be more socially evolved," says Dr. Tejwani. Dr. Louis Prieto saw her surgical skills and later opened doors for an OB/GYN residency.

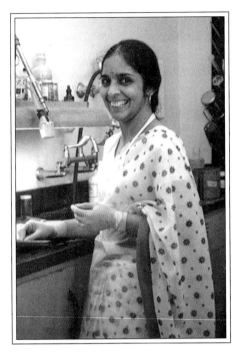

Training began at John Gaston, and of the 12 interns, only two were female, and Dr. Tejwani was the only woman of "color." One evening, another resident asked her to mop the operating room floor between surgeries. Dr. Tejwani formally established herself as a peer—and refused. When reprimanded before the entire staff, she stated she would perform any task during emergency situations, but that, that day, there were none. The residents were watching a ballgame, leaving her to operate without supervision and then mop. "No one came to my defense," she explains. "So, I told the department chair to fire me or let me get back to work."

After a year at John Gaston, Dr. Tejwani completed her residency at St. Joseph where she earned the respect and support of mentors Drs. Michael Roach, Charles Riggs, Luis

Prieto, Eugene Spiotta, Sr., Albert Grobmyer, III, Jesse Woodall, Robert Kline, and Walter and Robert Ruch. In 1974, Memphis's three female gynecologists were single, and they asked how she would "serve two masters"—a job and a husband. Dr. Tejwani, who had had a daughter during her internship and continued her education said, "I have only myself as a master."

Dr. Tejwani practiced with Dr. Charles Riggs from 1974 to 1976. After the birth of her second child, she began a private practice to be less dependent on Dr. Riggs while raising her family. Dr. Tejwani's compassionate care created a very loyal following, with so many referrals that appointments were often made three months in advance. By 1975, she had two daughters, Sujata and Natasha. Career and motherhood were balanced through the support of hospital and office staff, which became extended family.

Forty years later, Dr. Tejwani was attending her daughter's white coat ceremony at the University of Tennessee Medical School. She explained her visible emotion to the dean, "I'm happy that we are finally catching up with the Eastern world, and I'm also glad my daughter's class is fifty percent women. The class is a tapestry of color and gender, as it should be."

Dr. Tejwani is a former president of the Memphis and Shelby County OB/GYN Society and former chair of St. Francis OB/GYN Department. She has served on the boards of the Diversity Institute, the Women's Foundation for a Greater Memphis, and the India Community Foundation, and also volunteers at the Church Health Center. "This work is an added bonus in my life," she says. "Memphis and the medical community are my home and family."

UT MEDICAL GROUP, INC.

Each year, UT Medical Group, Inc. (UTMG) provides patient-centered care to nearly 250,000 patients. For more than 30 years, the Group has been the region's largest multispecialty physician group through an affiliation with The University of Tennessee Health Science Center. The Group's physicians are specialists who also teach other doctors and perform medical research.

The Group began as the Faculty Medical Practice Corporation in 1974. By 1983, the corporation became the private, tax-exempt 501(c)3 University Physicians Foundation. In 1990, the Group changed its name to UT Medical Group, Inc., emphasizing the Group's affiliation with The University of Tennessee and its dedication to quality medical care. The Group is not tax-funded nor does it accept donations.

Many of the Group's doctors have been listed in Castle Connolly's *America's Top Doctors*, and 86 of its physicians were listed in the *2009-2010 Best Doctors in America*. UT Medical Group doctors have led the Mid-South in many innovative procedures and initiatives. Among those Mid-South firsts are:

- UTMG's Dr. Sheldon Korones founded the Newborn Center at The Regional Medical Center (The MED) in 1968
- The nation's third liver transplant program (1982)
- First pediatric liver transplant (1982)
- Dr. Timothy Fabian spearheaded the area's Level 1 trauma facility at The MED (1983)
- First pediatric kidney transplant (1985)
- The MED's first Burn Center (1985)
- First pediatric heart-kidney transplant (1988)
- First kidney-pancreas and pancreas-only transplants (1989)
- Dr. Guy Voeller introduced laparoscopic surgery (late 1980s). Later, UTMG surgeons completed the area's first minimally invasive gallbladder removal, gastric bypass, and kidney removal. Surgeons now use robotic techniques for hysterectomies, ureteral reconstruction, lung lobectomy, and colon resection for cancer.
- Establishment of The MED Wound Center (1992)
- The opening of The Firefighters Regional Burn Center led by Dr. Bill Hickerson (1993). In 2009, the center remains the only full-service burn center within 150 miles.
- The Group's transplant program became one of six NIH Islet Cell Resource Centers (2001).
- The Hamilton Eye Institute opened under the leadership of Drs. Barrett Haik and Chris Fleming (2005), the only university eye center providing advanced vision care within 200 miles.
- The UTMG transplant program, affiliated with Methodist University Hospital, became one of the ten largest liver transplant centers in the U.S. (2009).
- By 2009, UTMG, with UT, The MED, Le Bonheur, and Baptist Women's Hospital, established several advanced centers to treat babies and women with high-risk pregnancies. Under Dr. Giancarlo Mari's leadership, the Group, with some of its partners, established the Maternal Fetal Institute, which provides fetal therapy, in-utero surgery and prenatal diagnosis.

Today, Group physicians provide care in offices and hospitals throughout the Mid-South. The Group has offices in the Memphis Medical Center, East Memphis, and a flagship multispecialty center in Germantown, Tennessee. Group providers also serve patients in surrounding towns including Tupelo and Corinth, Mississippi, Paragould and West Memphis, Arkansas, and Jackson, Tennessee.

Over the years, UT Medical Group's doctors and nurses have helped many men, women, and children. Beginning its fourth decade of service, the Group's providers will continue to spearheaded advances in virtually all areas of medical care—from prenatal genetic testing, pregnancy loss prevention, and treatment of infertility, to advancements in primary and surgical care of the elderly, and coping with chronic conditions such as heart disease, diabetes, or cancer.

THE UNIVERSITY OF TENNESSEE HEALTH SCIENCE CENTER

IN A SENSE MORE LITERAL THAN figurative, the University of Tennessee Health Science Center (UTHSC) is the lifeline of Memphis. Since its founding in 1911, the Health Science Center has served as a central healthcare engine for this river metropolis. Each year, more than 2,500 students enroll in six colleges, studying to earn positions in the next generation of competent, confident, healthcare professionals. Most of the students are right here in Memphis, although UTHSC has smaller campuses in Knoxville and Chattanooga. In addition, statewide the center supervises training for more than 1,000 residents, and fellows in 84 specialty programs certified by the Accreditation Council for Graduate Medical Education.

Founded in 1911, UTHSC consists of six colleges.

The Health Science Center is the lifeline of Memphis because of what it gives to the community—access to some of the best-trained, most intellectually active, most technically proficient, and most caring healthcare professionals in the world. Roughly 350 members of the UTHSC College of Medicine faculty constitute the largest physicians group in the region—UT Medical Group, a private, non-profit organization. Moreover, about 200 UTHSC faculty members hold joint appointments at St. Jude Children's Research Hospital, where they also see patients, perform research, and train graduate students. The faculty, representing a wide range of specialties, provide direct care to patients and train students through hospitals and clinics, in private practice, and in other healthcare environments.

For accident victims, the Regional Medical Center at Memphis, staffed by UTHSC physicians, has a Level 1 trauma center (Level 1 requires an academic affiliation) named for one of Memphis's most famous citizens—Elvis Presley—and known across the country. UTHSC also operates Level 1 trauma centers in Knoxville and Chattanooga. Together, these three centers treat almost 12,000 cases a year and rank in the top five in the nation in numbers of cases seen. In addition, the Regional Medical Center's Newborn Center, also staffed by UTHSC physicians, is known around the world for innovative approaches in treating at-risk infants.

Through affiliation with Methodist University Hospital, which has the only abdominal transplant program in the Mid-South, UTHSC surgeons perform more than 200 kidney, liver, and pancreas transplants each year. Chancellor Steve J. Schwab, M.D., reports in 2009 that the program ranked among the top transplant institutes in the nation, and seventh in the country for the number of liver transplants (including one for Apple's Steve Jobs).

These vital contributions to the community, however, are mainly those of the College of Medicine, although UTHSC comprises six colleges: Allied Health Sciences; Dentistry; Graduate Health Sciences; Medicine; Nursing; and Pharmacy. Each in its own way helps constitute Memphis's lifeline.

The College of Allied Health Sciences offers 15 professional degree programs in a range of disciplines. Its list of achievements is impressive: it was first in the state to award a Doctor of Physical Therapy and Doctor of Science in Physical Therapy; the first cytotechnology program in the country and the only one in the state; and it has the largest master's program in dental hygiene nationally. More than 9,000 alumni serve in clinical, administrative, educational, and service roles in the United States and around the world.

The College of Dentistry, dating from 1878, is the oldest dental school in the South and graduates more than seventy-five percent of Tennessee dentists, as well as a significant number of dentists practicing in Arkansas. The college's dental clinic, staffed by students with close faculty supervision, provides more than 40,000 patient visits per year at a fraction of the cost of private dental care.

The College of Graduate Health Sciences has granted almost 1,200 graduate degrees, including Ph.D., M.S., and M.D.S. degrees. Seven highly sought degrees are offered, and five or six individuals apply for every available space in the college.

Approximately 2,000 nurses practicing in Tennessee are graduates of the College of Nursing, which has more than 5,100 alumni. As the leading producer of graduate nurses and nursing faculty for the region, the college plays a pivotal role in addressing today's nursing shortage. The college was cited as one of America's best graduate schools in the 2010 *U.S. News & World Report*.

With more than 5,600 alumni, the College of Pharmacy ranks sixteenth out of more than 100 U.S. pharmacy schools and educates approximately forty percent of the state's pharmacists. In 2011, the college expects to move into its new, more than $65 million research facility in the University of Tennessee-Baptist Research Park, a biotechnology hub that will add 1.4 million square feet of laboratory, research, educational, and business space to the Memphis Medical Center. The new pharmacy facility will enable expanded research on new drugs and drug delivery systems affecting all branches of medical knowledge.

Research and other sponsored projects at UTHSC brought in $106 million in fiscal 2010. As a center of statewide excellence and national prominence, UTHSC has researchers who were the first to discover that stem cells can come from teeth; who developed a testosterone replacement drug and an oral drug that protects the gastrointestinal tract from the damaging effects of chemotherapy or radiation; who work with the Tennessee Mouse Genome Consortium (the mouse genome is curiously similar to the human genome); and who are using addiction research to end smoking. In addition to their diverse areas of scientific study, UTHSC researchers are also tenaciously pursuing vaccines for strep and for the strain of *E. coli* that brings illness to travelers and often brings death to children in developing countries. So successful has been such research, as well as the new biomedical companies that are its offspring, that UTHSC has established a Clinical and Translational Science Institute whose purpose is to facilitate "taking research from the bench to the bedside"—bringing new vaccines, new surgical techniques, and new treatments to consumers as quickly and as efficiently as possible.

Even more research is under way at the world-class University of Tennessee Hamilton Eye Institute, which houses the Center for Vision Research. The institute is ranked in the top ten in the nation for clinical care. Further, at the $25.3 million University of Tennessee Cancer Research Building, laboratories are configured in an open manner to encourage faculty from medicine, dentistry, and pharmacy to collaborate in understanding this deadly disease.

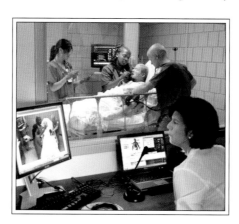

The UTHSC College of Nursing is recognized as one of the nation's best.

Opened in 2009, the UTHSC Regional Biocontainment Laboratory (RBL), Biosafety Level 3 as defined by the Centers for Disease Control, will make possible safe and effective studies of contagious materials. One of only 13 universities in the nation competitively selected as a site for an RBL, UTHSC received more than $18 million from the National Institute of Allergy and Infectious Diseases, which is part of the National Institutes of Health, to construct the facility. Infectious disease researchers will study how human and animal cells deal with some of the world's most dangerous microbes. The state-of-the-art, secure laboratory will also serve as a regional resource for emergency preparedness and emergency response in case of bioterrorism.

While the Health Science Center mission remains a four-tier commitment—education, research, clinical care, and public service—each layer is intertwined with the others. Education, always a lifelong endeavor, offers practicing physicians opportunities to share their expertise as clinical faculty and to sharpen their own skills as they work with students. Research brings together faculty and students as partners in learning, dissecting the biological, molecular, intellectual, and behavioral puzzles to forge new pathways to better health. Clinical care brings faculty and student learners into the community, sharing and enhancing expertise. Public service advances the well-being of the Health Science Center's neighbors and of our own team members by involving faculty, staff, and students in the life of the community.

With a recently developed master plan and a $130 million capital campaign, the University of Tennessee Health Science Center approaches its one hundredth anniversary in 2011 as a widely respected institution poised to move toward even greater accomplishments and recognition, while remaining a lifeline to the Memphis community.

WRIGHT MEDICAL TECHNOLOGY, INC.

Everyone is inspired by a story of success sparked by innovation. The founding and growth of Wright Medical Technology, Inc., is just such a story. From the company's humble beginnings over 60 years ago to its current success in the global orthopaedic marketplace, Wright's history reflects a continual passion for innovation.

It all began with Frank O. Wright, who was an orthopaedic sales representative in Memphis in the 1940s and '50s—and an *innovator*. In the course of his sales profession, he spent countless hours working with doctors and learning about challenges they faced when treating patients. One particular issue that doctors repeated often was concern over debilitating back pain for patients in leg casts, caused by the rigid steel heels placed in the casts. A natural problem-solver, Frank began working with one of his physician friends on an alternative to the steel heels. The solution was an all-rubber walking heel. With this simple invention, Frank Wright made great strides in helping doctors improve the standard of care for patients in leg casts—and he took the first step toward building an orthopaedic company that today serves countless healthcare professionals in over 60 countries around the world.

In 1950, Frank Wright founded Wright Manufacturing Company in Memphis to make and sell the all-rubber walking heel. But with his vast knowledge of orthopaedic products and sales expertise, it was only natural for the

ABOVE: *Wright Medical Technology, Inc., moved its headquarters and main manufacturing facility to Arlington in 1975.* OPPOSITE PAGE, ABOVE AND BELOW: *Wright has an international workforce of more than 1,300 employees.*

company's product portfolio to quickly expand to include orthopaedic hip and knee implants, as well as medical "soft goods" like arm slings. Over the next two decades, the company continued to grow and offer orthopaedic professionals the latest in implant design technology and exceptional service. During this time, success of the business necessitated moves to several locations in downtown Memphis. By the 1970s, the company began to seek a permanent home that would offer plenty of room for expansion. The small town of Arlington, just 12 miles east of Memphis, provided the perfect location. And so Wright Manufacturing moved its headquarters to the town in 1975.

While the move to Arlington in 1975 marked a significant milestone in the history of Wright Manufacturing, the year also saw the passing of the company's founder, Frank Wright. And just two years after those major developments, the business underwent a change of ownership. In 1977, Michigan-based Dow Corning Corporation purchased the company and changed its name to Dow Corning Wright. The new partnership would strengthen Wright's reach in

the medical device industry by incorporating Dow Corning's market-leading silicone technology into orthopaedic implants for the fingers and toes. The innovative new line of small joint implants allowed Dow Corning Wright to quickly gain a dominant position in this specialized segment of the orthopaedic industry.

Wright had begun a gradual retreat from the medical "soft goods" business in the late 1970s, which was complete by the early 1980s. The company continued to focus on its core orthopaedic implant product line. With Dow Corning's silicone technology, the Arlington facility also began producing a number of products for use in cosmetic surgery.

In the early 1990s, Dow Corning Corporation made the decision to exit the orthopaedic industry. A group of investors purchased the former Dow Corning Wright in 1993, changing the name to Wright Medical Technology, Inc. With the new ownership came a renewed energy for innovation and exploration in medical device technology. The mid- to late 1990s found Wright pioneering new frontiers in the burgeoning field of biologic bone repair. The exploration in the new field of biologics laid the foundation for a critical new area of growth for the company.

In 1999, Wright experienced another transition in ownership, but still remained privately held. Seeing the tremendous potential for innovation and growth that the company held, the new group of owners sought to put key leadership in place and take Wright public. Over the course of the next year, the company welcomed numerous orthopaedic industry veterans who immediately began preparing Wright for the transition from a conservative-growth private company to a high-growth, publicly held orthopaedic innovator. On July 13, 2001, the company went public as the parent entity, Wright Medical Group, Inc., offering shares on the Nasdaq Market under the trading symbol "WMGI."

Wright's commitment to the core business of hip and knee implants remained sound and the company continued to introduce new innovations within those key product lines. However, Wright also began to look more closely at underserved specialty markets, like the foot and ankle segment. The company had already developed product expertise and surgeon relationships in this area. In 2005, Wright began to invest heavily in its foot and ankle business, completing key acquisitions, establishing distribution agreements, and developing new products that would solidify the company's leadership role in this market within a few short years.

Today, Wright Medical is recognized as an innovator across numerous orthopaedic market segments, including large joint reconstruction, biologic bone and soft tissue repair, foot and ankle surgery and fixation, and upper extremity solutions. The company now employs over 1,300 people globally, including just over 1,000 at its expansive headquarters in Arlington. Although the Wright family is no longer part of the company's ownership, the company retains the name of its innovative founder, Frank O. Wright, to reflect its commitment to his pioneering spirit.

WOLF RIVER PLASTIC SURGERY

After 29 years, Dr. Ronald Johnson does not hesitate to say, "I really love what I do." At a time when he could retire, Dr. Johnson has a solo practice and frequently attends seminars; each year he earns double the required hours for membership in plastic surgery societies.

Johnson attended the University of Mississippi Medical School and, as a sophomore, he filled one of six summer positions as a scrub assistant at Mississippi Baptist Hospital. In the plastic surgery operating rooms, he liked the work he saw and decided this would be his field.

After general surgery training in Mobile, Alabama, Dr. Johnson came to Memphis for his plastic surgery residency. "When I finished in 1979, the local field seemed full, so I returned to Mobile and practiced there for seven years." He returned to Memphis in 1986 and joined a group with six plastic surgeons.

"Cosmetic surgery is a term which has been taken over by the press," says Dr. Johnson. "There are cosmetic surgeons who are not plastic surgeons, so we prefer the term aesthetic surgery."

At a recent meeting, The American Society for Aesthetic Plastic Surgery's poll found 62 percent of attendees were in solo practice. Johnson opened his solo practice in 2003 and likes the option of working when and how much he chooses

"I have board meetings in my pickup truck, and the votes are all one to nothing," says Dr. Johnson smiling. "That wouldn't suit everyone, but I like not having a lot of deliberations and talking. I make up my mind, stick to it, and I don't look back."

GEORGE R. WOODBURY, JR., M.D.

In George Woodbury's family, becoming a doctor is not a tradition, it is a continuing legacy. Woodbury's grandfather, Robert Arthur Woodbury, came to Memphis in 1946 as chairman of pharmacology at the University of Tennessee Health Science Center, where he remained for 23 years. Woodbury's father, George Woodbury, Sr., is an ophthalmologist and his mother, Linda Woodbury, is a dermatologist. Woodbury's brothers, Michael and Robert, are dermatologists and sister, Suzanne, is a pediatric rehabilitation surgeon at Baylor University. Woodbury's wife, Cathy Chapman, is a rheumatologist.

Woodbury attended the University of Chicago, and continued his education at the Pritzker School of Medicine. He interned at Baptist Hospital in Memphis, and returned to Chicago for a three-year residency at Rush-Presbyterian-St. Luke's, which he completed in 1991.

Woodbury practiced in Rochester, New York while his wife attended advanced training at Strong Memorial. The Woodburys returned to Memphis, and he was the inpatient dermatologist at the Veteran's Administration and a clinical instructor for the University of Tennessee Health Science Center. He continues as an instructor, conducting grand-rounds biyearly. Dr. Woodbury holds medical privileges at Baptist, Methodist, and St. Francis Hospitals. He teaches residents in St. Francis/University of Tennessee Family Practice program, and his office is the site of their rotations. Woodbury participates in the Dermatology Advocacy Network which works on issues for funding and safety of tanning salons.

Woodbury and wife, Cathy, opened a joint practice in 1993 with offices in East Memphis and Cordova. They have three children, and volunteer at Lausanne School and with Memphis drug prevention programs.

MEMPHIS MEDICINE
A HISTORY *of* SCIENCE AND SERVICE

IN MEMORIAM

*"Say not in grief 'they are no more,'
but live in thankfulness
that they were."*

—Hebrew Proverb

IN MEMORIAM

DAVID H. KNOTT, M.D., PH.D.
1937-2009

Iowa native David H. Knott received his B.A. in biology, *summa cum laude*, from Cornell College (Iowa). He received his M.S. degree in physiology from the University of Tennessee School of Basic Medical Sciences (UTSBMS) in 1960 and his M.D. degree, with honors, from the University of Tennessee College of Medicine in 1963. In 1965, he received his Ph.D. degree in physiology from UTSBMS. He was certified by the American Board of Family Practice and by the American Society of Addiction Medicine.

Practicing and teaching medicine, Dr. Knott both witnessed and contributed significantly to a sea change in the understanding of addiction as a disease. Ignored by the medical profession and vilified by social critics, alcohol and drug patients were once virtually denied medical care.

Planning to work for NASA studying low levels of radiation exposure experienced by astronauts, Dr. Knott began working with a friend who was studying the clinical effects of alcohol. After five years of research, he began to emphasize addiction as a disease, becoming a pioneer in the now-recognized specialty of addition medicine.

As a researcher, teacher, and practitioner, Dr. Knott was widely known in his field. He was a professor and a researcher at the University of Tennessee-Memphis, in the colleges of both medicine and pharmacy. In addition, he was a professor or a visiting professor at other institutions, including Tulane University School of Medicine, University of Utah School on Alcoholism and Other Drug Dependencies, and Duke University Summer Institute of Alcohol Studies. A Phi Beta Kappa at Cornell College and a Woodrow Wilson Fellow for graduate study, he continued to win honors. He was in the Honors Program at the University of Tennessee Medical Units and in the Alpha Omega Alpha Honor Medical Society. He was the John and Mary R. Markle Foundation Scholar in Academic Medicine from 1966 through 1971, and he received the James H. Tharp Alcohol Research Award every year from 1974 through 1995. In 1994, he received the First Annual Melville Kelly Lifetime Achievement Award in the Field of Alcohol and Drug Dependence from the Tennessee Department of Health. Also in 1994, he won the Good Samaritan Award for Volunteer Services from the University of Tennessee-Memphis, and the Crisis and Suicide Prevention Project.

During his lifetime, Dr. Knott published more than 90 articles, books, and chapters in books. He presented scientific exhibits on newer concepts in treating alcoholism in all 50 states and produced four medical/educational motion pictures. He was an invited lecturer at 252 medical schools, hospitals, and scholarly associations in the United States and abroad.

A groundbreaker in his specialty, Dr. Knott remained vitally involved in practice and research and passionate about addiction medicine throughout his life. Of retirement, he said, "If I retired, what would I do?" Although diverted from his goal of studying low-dose radiation in astronauts, Dr. David Knott left an explorer's footprints on the history of addiction medicine.

IN MEMORIAM

JOHN LUCIUS McGEHEE, M.D.
1879-1949

Known as a scholar, a gentleman, and a skilled surgeon, John Lucius McGehee, M.D., was born the son of a cotton planter in Mississippi in 1879. As a young man, he moved with his family to Memphis and attended the Memphis Military Institute, hoping to enter the U.S. Naval Academy. His naval ambition was thwarted, however, by a hunting accident that took the little finger of his right hand and paralyzed the adjacent finger. Although the injury disqualified him from Annapolis, it eventually contributed to his surgical expertise.

After graduation from Millsaps College, Lucius McGehee entered the Memphis Hospital Medical College (MHMC). He received his M.D. degree at the age of 21 and became a resident at St. Joseph Hospital. From 1902 to 1903, he took a preceptorship under Dr. W. A. Evans, a well-known pathologist, in Chicago, after which he began a general medical practice in Memphis and a long association with the medical college, which became part of the University of Tennessee in 1911. An assistant to Dr. W. B. Rogers, Memphis's first eminent professor of surgery, Dr. McGehee began a lifelong commitment to both private practice and teaching (charity patients went to the teaching hospital, Memphis General, later John Gaston Hospital). He was also a captain and assistant surgeon in the Tennessee National Guard.

In 1908, Dr. McGehee married Mary Louise Berry. Mrs. McGehee, with other physicians' wives, helped raise funds for the Baptist Hospital, which opened in 1912, providing a venue for Dr. McGehee's private patients throughout his career. From 1913 to 1917, he was chair of operative surgery at the medical school. His career was interrupted for service as a major in the Army Medical Corps in France (1917-1919), after which he studied maxillofacial surgery at Vichy with Dr. Ferdinand LeMaitre.

In 1919, Dr. McGehee returned to his practice, now limited to surgery. In the 1920s, he was chief of staff at Memphis General, and in 1925 he helped establish there one of the first postgraduate surgical residencies in the South. In 1932, he became chair of surgery at the University of Tennessee.

Occasionally intimidating, Dr. McGehee was always supportive and helpful to his students and residents. He was fond of aphorisms such as "Possess your soul with patience." His injured right hand became an asset; he could make a small incision, grasp and remove an appendix, and close all layers within seven minutes. He developed and taught the "Appendiceal Creed," emphasizing the need to diagnose and operate quickly for appendicitis, thus preventing peritonitis. He meticulously followed surgical patients, and he often received patients at odd hours from nearby small towns. Only in later years did he reserve an afternoon for golf.

Dr. McGehee was proud of his two daughters and his son, Lucius, III, who attended the Naval Academy as he had wanted to do. Active, intelligent, and dedicated, Dr. McGehee accepted retirement reluctantly, writing, "If a man has his mental faculties, even though a little unsteady on his pins, he is not old." He died in 1949 after a short retirement.

IN MEMORIAM

THOMAS C. MOSS, M.D.
1905-1995

In 1950, my father, Thomas C. Moss, M.D., left a secure job as a pathologist at St. Joseph Hospital to open a privately owned lab—Moss Pathology Lab, the first private pathology lab in Memphis. Taking this risk was not easy for one who grew up a poor farm boy in Obion County. Born in 1905, he was seven when his father died; he was reared by his mother and his grandfather. While in high school, he lost his left lower leg in a train accident. Undaunted, he became valedictorian of his high school class and worked his way through pre-med at the University of Tennessee—Knoxville and through medical school at the University of Tennessee—Memphis. He graduated from medical school in 1929, finished pathology training at Memphis General Hospital in 1932, and immediately he became the pathologist for Methodist Hospital. In 1933, he married a nurse who worked there—my mother, Ernestine Parke Moss. He stayed at Methodist until 1940, when he moved to St. Joseph Hospital.

My father's lab had humble beginnings in a rented room in an apartment house on Madison just east of Cleveland. With my mother's help, he solicited business from doctors' offices and hospitals around the Mid-South. With a typewriter, a microscope, and a tissue processor, my father did almost everything himself, including picking up specimens and delivering reports. As business grew, the lab moved around 1953 to an upstairs apartment on Eastmoreland just west of Bellevue.

In 1956, seeing the need for a clinical lab to serve local doctors more completely, my father hired medical technologist Louis Widner to start this part of the lab. With Louis's hard work, expertise, and good rapport with local doctors, the clinical lab prospered. In 1958, the lab moved to 257 Bellevue. In 1960, Dr. C. C. Farrow joined the lab as a second pathologist. When he became a full partner, the name was changed to Moss-Farrow Pathology Lab; the lab flourished and grew through the sixties.

In 1970, I finished my pathology training at Baptist Hospital under Dr. E. Muirhead and joined the lab in July 1970. When my father retired in January 1971, I replaced him, but no one could ever fill his shoes. He was a tireless worker who was loved and revered by his employees and medical associates. He continued to work in pathology and for Interstate Blood Bank until he died in 1995. He left a legacy from which we all still benefit.

Dr. Farrow directed the lab from 1970 to 1989, moving in 1979 to a building at Walnut Grove Road and Tillman and changing the name to Memphis Pathology Lab. In 1986, we sold the clinical lab to Baptist Healthcare Systems, which kept the name Memphis Pathology Lab. We changed our name to FM Pathology Lab and continued as a cytology and histology lab. In 1994, the cytology lab was sold to Baptist, and, in 1997, the histology lab was sold, thus ending a 47-year run.

Memphis Pathology Lab is still doing well. I work part time doing cytology.

Thanks, Dad.

—*In loving memory, William B. Moss, M.D.*

IN MEMORIAM

IRIS A. PEARCE, M.D.
1922-2005

Shadowing her father after her mother's early death, Iris Annette Pearce always knew what she wanted to be. The family had precedents: her father was a fourth-generation physician, and she often accompanied her father on house calls and stayed with nurses while her father made hospital rounds (the family tradition in medicine continues into the sixth generation, through cousins of Dr. Pearce). "Everything about medicine fascinated me," she said. After she studied at Southwestern (Rhodes) and graduated from Vanderbilt, Dr. Pearce's father, knowing the difficulties she would face, discouraged her from studying medicine. Iris, however, would follow her own path.

In 1943, Iris was commissioned an ensign in the United States Navy. She remained on active duty until 1946, serving in Massachusetts and in Florida, and in the reserves until 1956. Her active service completed, she enrolled in the University of Tennessee Medical Units, completing her M.D. degree in 1950. She interned at John Gaston Hospital, where she compiled a notable record of "firsts": first woman to be an assistant, associate, resident, and chief resident in medicine, then first woman to direct the rheumatic prophylaxis clinic, primary care clinic, and medicine clinic. She was the first female to be medical director of Ambulatory Services, to chair review committees and the Medical Board, and to be chief of staff.

As a professor of medicine, Dr. Pearce also compiled an impressive series of accomplishments—noteworthy for their own sake, but also because she was the first woman: Assistant Dean and Dean, Clinical Affairs; Division Chief, Ambulatory Medicine; Director, Streptococcal Disease Center; Director, Adult Nurse Practitioner Program; Chair, Commission for Equity.

Dr. Pearce received a number of awards for her service: Distinguished Service Award, City of Memphis Hospital Medical Staff, 1977; National Alumni Public Service Award from Board of Governors of the University of Tennessee National Alumni Association, 1978; L.M. Graves Memorial Health Award from the Mid-South Medical Center Council, 1981; Tennessee Hospital Association Community Service Award.

These accomplishments, however, do not overshadow Dr. Iris Pearce's concern for the poor, for those who could not afford medical care and did not understand good health practices. She once said, "I don't like for people to be sick," and continued that "people should quit putting things in their bodies that are harmful and should stop making speeding bullets of their torsos." Her wit was sharp, and her advice was often simple: laugh at yourself, not at others; seize opportunities; be active; eat a balanced diet. Her interests were wide: exploring national parks, playing organ and piano, collecting electric trains, playing pool. Her love for and dedication to medicine and teaching made her a role model for generations of young physicians.

Dr. Pearce's dedication to medicine, concern for the poor, and love for teaching came together in her estate. The $1.5 million Dr. Robert S. Pearce Chair in Internal Medicine both honors her father's career in medicine and continues to support outstanding teaching and practice at the University of Tennessee Center for the Health Sciences.

IN MEMORIAM
JESSE M. WESBERRY, SR., M.D.
1927-1998

Dr. Jess Wesberry's decision to follow his father, Dr. Jesse Wesberry, Sr., into the field of ophthalmology gave him the opportunity to learn from a man who had pursued and implemented the newest innovations in the field.

Dr. Jesse Wesberry, Sr., practiced general optometry in Batesburg, South Carolina, for 10 years. His desire to become an ophthalmologist brought him and his young family to the University of Tennessee College of Medicine. In 1962, he began a one-year internship at John Gaston and later a residency at the University of Tennessee.

Dr. Wesberry, Sr., began practicing in 1966 and opened one of the first offices in East Memphis. His concern for cataract patients led to studying abroad with innovators in the field and, in 1975, he was one of the first in Memphis to perform cataract surgery with an intraocular lens implant.

In 1982, he traveled to France to study with Dr. Aron Rosa and the use of the YAG laser. This led to adding an eye surgery center in 1985 offering patients on-site procedures. "He was concerned about controlling cost to the patients, so the surgery center was paid for when it opened," says Dr. Jess Wesberry, Jr. This was also the year Dr. Jess Wesberry, Jr., joined Wesberry Surgery Center and practiced with his father, until Dr. Wesberry, Sr.'s death in 1998. "My dad was my mentor for most of my career," says Dr. Jess Wesberry. "He was an innovator, but he was also very passionate about what he was doing."

Today, Dr. Jess Wesberry, Jr., follows his father's example in offering compassionate, innovative care in ophthalmology at the University of Tennessee Medical Group.

MEMPHIS MEDICINE

A HISTORY *of* SCIENCE AND SERVICE

APPENDICES

Baptist Memorial Hospital, founded in 1912, became
the largest private hospital in the country. Its demolition
in 2005 made space for a major biotechnology research
facility, the UT-Baptist Research Park.

APPENDIX I

PAST PRESIDENTS OF THE MEMPHIS MEDICAL SOCIETY

1876-78
SNOWDEN CRAVEN MADDOX, M.D.

1878-79
ROBERT WOOD MITCHELL, M.D.

1878-80
E. MILES WILLETT, M.D.

1880-81
ALEXANDER ERSKINE, M.D.

1881-82
SAMUEL H. BROWN, M.D.

1882-84
DUDLEY DUNN SAUNDERS, M.D.

1884-86
FRANCIS L SIM, M.D.

1886-87
SAMUEL TREAT ARMSTRONG, M.D.

1887-88
PETER R. FORD, M.D.

1888-89
HEBER JONES, M.D.

1889-90
THOMAS J. CROFFORD, M.D.

1890-92
JOHN E. BLACK, M.D.

1892-94
EUGENE PAUL SALE, M.D.

1984-95
WILLIAM WOOD TAYLOR, M.D.

1895-96
DUDLEY DUNN SAUNDERS, M.D.

1896
ANDREW B. HOLDER, M.D.

1896-97
MAXENE B. HERMAN, M.D.

1897
EUGENE E. HAYNES, M.D.

1898
THOMAS J. CROFFORD, M.D.

1899
BENJAMIN F. TURNER, M.D.

1900
EDWARD COLEMAN ELLETT, M.D.

1901
FRANK AIKMAN JONES, M.D.

1902
ALFRED MOORE, M.D.

1903
BENNET GRAVES HENNING, M.D.

1904
EDWIN MCLAREN WILLIAMS, M.D.

1905
JAMES L. ANDREWS, M.D.

1906
JAMES LUNDIE BARTON, M.D.

1907-08
ALEXANDER ERSKINE, M.D.

1909
GUSTAVUS B. THORNTON, M.D.

1910
ERNEST C. BLACKBURN, M.D.

1911
WILLIAM BATTLE MALONE, M.D.

1912
WILLIAM T. BLACK, M.D.

1913
WILLIAM KRAUSS, M.D.

1914
ALFRED B. DELOACH, M.D.

1915
JOHN CHAMBERS AYRES, M.D.

1916
JESSE JAMES CULLINGS, M.D.

1917
WILLIAM BRITT BURNS, M.D.

1918
JOHN LEMUEL JELKS, M.D.

1919
WILLIS COHOON CAMPBELL, M.D.

1920
J. LUCIUS MCGEHEE, M.D.

1921
WILLIAM GLASSEL SOMERVILLE, M.D.

1922
FRANK DAVID SMYTHE, M.D.

1923
JAMES BASSETT MCELROY, M.D.

1924
WILLIAM FRANKLIN CLARY, M.D.

1925
JOHN M. MAURY, M.D.

1926
JOSEPH AUGUSTUS CRISLER, M.D.

1927
GEORGE R. LIVERMORE, M.D.

1928
ELMER E. FRANCIS, M.D.

1929
OSWALD STUART MCCOWN, M.D.

1930
JULIAN BAKER BLUE, M.D.

1931
HIRAM B. EVERETT, M.D.

1932
PERCY WALTHAL TOOMBS, M.D.

1933
MAXIMILLIAN GOLTMAN, M.D.

1933
PERCY H. WOOD, M.D.

1934
WILSON L. WILLIAMSON, M.D.

1935
JAMES B. STANFORD, M.D.

1936
ROBIN FERGUSON MASON, M.D.

1937
OTIS S. WARR, SR., M.D.

1937
M. WILSON SEARIGHT, M.D.

1938
JOEL JONES HOBSON, M.D.

1939
JOSEPH H. FRANCIS, M.D.

1940
WILLIAM C. CHANEY, M.D.

1941
EDWARD D. MITCHELL, M.D.

1942
EUGENE M. HOLDER, M.D.

1943
WALKER L. RUCKS, M.D.

1944
ERNEST G. KELLY, M.D.

1945
WILLIAM C. COLBERT, M.D.

1946
CHARLES H. HEACOCK, M.D.

1947
ARTHUR F. COOPER, M.D.

1948
HARLEY W. QUALLS, M.D.

1949
EMMETT R. HALL, SR., M.D.

1950
CLYDE V. CROSWELL, M.D.

1951
WILLIAM D. STINSON, M.D.

1952
HENRY B. GOTTEN, M.D.

1953
E. GUY CAMPBELL, M.D.

1954
S. FRED STRAIN, SR., M.D.

1955
PHILIP M. LEWIS, M.D.

1956
SAMUEL L. RAINES, M.D.

1957
HAROLD B. BOYD, M.D.

1958
JOHN D. HUGHES, M.D.

1959
RALPH O. RYCHENER, M.D.

1960
DUANE M. CARR, M.D.

1961
BLAND W. CANNON, M.D.

1962
ALVIN J. INGRAM, M.D.

1963
GILBERT J. LEVY, M.D.

1964
WILLIAM T. SATTERFIELD, SR., M.D.

1965
ALBERT J. GROBMYER, JR., M.D.

1966
A. ROY TYRER, JR., M.D.

1967
ROCCO A. CALANDRUCCIO, M.D.

1968
BILLY G. MITCHELL, M.D.

1969
J. MALCOLM ASTE, M.D.

1970
C. DOUGLAS HAWKES, M.D.

1971
JOHN D. YOUNG, JR., M.D.

1972
FRANCIS H. COLE, SR., M.D.

1973
JOHN B. DORIAN, M.D.

1974
W. DAVID DUNAVANT, SR., M.D.

1975
WILFORD H. GRAGG, JR., M.D.

1976
MCCARTHY DEMERE, M.D.

1977
THOMAS G. DORRITY, M.D.

1978
JOHN D. PIGOTT, M.D.

1979
ALLEN S. EDMONSON, M.D.

1980
JAMES T. GALYON, M.D.

1981
DANIEL J. SCOTT, JR., M.D.

1982
HUGH FRANCIS, JR., M.D.

1983
JOHN S. BUCHIGNANI, JR., M.D.

1984
HAMEL B. EASON, M.D.

1985
DEE J. CANALE, M.D.

1986
ALBERT J. GROBMYER, III, M.D.

1987
REX A. AMONETTE, M.D.

1988
DENNIS A. HIGDON, M.D.

1989
RICHARD M. PEARSON, M.D.

1990
ARNOLD M. DRAKE, M.D.

1991
J. CHRIS FLEMING, M.D.

1992
EVELYN B. OGLE, M.D.

1993
JOE P. ANDERSON, M.D.

1994
JESSE C. WOODALL, JR., M.D.

1995
ALLEN S. BOYD, JR., M.D.

1996
JIM GIBB JOHNSON, M.D.

1997
F. HAMMOND COLE, JR., M.D.

1998
JERRE MINOR FREEMAN, M.D.

1999
CHARLES R. HANDORF, M.D.

2000
ROBERT D. KIRKPATRICK, M.D.

2001
HUGH FRANCIS, III, M.D.

2002
MARTIN D. FLEMING, M.D.

2003
MACK A. LAND, M.D.

2004
NEAL S. BECKFORD, M.D.

2005
WILEY T. ROBINSON, M.D.

2006
ROBERT A. KERLAN, M.D.

2007
VALERIE K. ARNOLD, M.D.

2008
KEITH G. ANDERSON, M.D.

2009
CLARENCE B. WATRIDGE, M.D.

2010
JAMES K. ENSOR, JR., M.D.

APPENDIX II

PAST PRESIDENTS OF THE MEMPHIS MEDICAL SOCIETY ALLIANCE

1928-1930
Mrs. Wilford H. Gragg

1930-1932
Mrs. Willis C. Campbell

1932-1934
Mrs. William T. Black

1934-1935
Mrs. Percy Toombs

1935-1936
Mrs. W. S. Lawrence

1936-1938
Mrs. W. T. Braun

1938-1939
Mrs. Edward Clay Mitchell

1939-1940
Mrs. Walter A. Ruch

1940-1941
Mrs. Calvert Chaney

1941-1942
Mrs. Clyde Croswell

1942-1943
Mrs. Jewell M. Dorris

1943-1944
Mrs. L. L. Carter

1945-1946
Mrs. M. W. Holehan

1946-1948
Mrs. Harold B. Boyd

1948-1949
Mrs. Carrol C. Turner

1949-1950
Mrs. Ernest G. Kelly

1950-1951
Mrs. George H. Burkle, Jr.

1951-1952
Mrs. Horace D. Gray

1952-1954
Mrs. Ben L. Pentecost, Jr.

1954-1955
Mrs. Roland H. Myers

1955-1956
Mrs. Harry J. Jacobson

1956-1957
Mrs. A. Lynn Herring

1957-1958
Mrs. John L. Shaw

1958-1959
Mrs. Finis A. Taylor

1959-1960
Mrs. William A. Aycock

1960-1961
Mrs. Charles W. Miller

1961-1962
Mrs. W. D. Burkhalter

1962-1963
Mrs. George Brainerd Higley

1963-1964
Mrs. Horton G. Dubard

1964-1965
Mrs. Henry B. Turner

1965-1966
Mrs. William F. Mackey

1966-1967
Mrs. Ben E. Everett

1967-1968
Mrs. Vonnie Hall

1968-1969
Mrs. Albert M. Jones

1969
Mrs. Thomas H. West

1969-1970
Mrs. Gordon L. Mathis

1970-1971
Mrs. Alvin E. Smith

1971-1972
Mrs. J. Palmer Moss

1972-1973
Mrs. John B. Hamsher

1973-1974
Mrs. J. Rodney Feild

1974-1975
Mrs. A. Hoyt Crenshaw

1975-1976
Mrs. Allen Edmonson

1976-1977
Mrs. John P. Nash

1977-1978
Mrs. Thomas L. West

1978-1979
Mrs. James W. Pate

1979-1980
Mrs. T. Kyle Creson

1980-1981
Mrs. Max W. Painter

1981-1982
Mrs. Dennis A. Higdon

1982-1983
Mrs. Gene L. Whitington

1983-1984
Mrs. George H. Burkle, III

1984-1985
Mrs. Rex Amonette

1985-1986
Mrs. Jesse C. Woodall, Jr.

1986-1987
Mrs. E. Greer Richardson

1987-1988
Mrs. Leonard Hines

1988-1989
Mrs. Herbert Taylor

1989-1990
Mrs. Kit Mays

1990-1991
Mrs. Robert Trautman, Jr.

1991-1992
Mrs. David Cunningham

1992-1993
Mrs. Sherman E. Kahn

1993-1994
Mrs. Robert E. Laster, Jr.

1994-1995
Mrs. Charles N. Larkin

1995-1996
Mrs. Joseph H. Miller

1996-1997
Mrs. Jon Robertson

1997-1998
Mrs. John V. Pender, Jr.

1998-1999
Mrs. H. David Hickey

1999-2000
Mrs. Noel Florendo

2000-2001
Mrs. Jeffrey Cole

2001-2002
Mrs. J. Cameron Hall

2002-2004
Mrs. John R. Adams, Jr.

2004-2005
Mrs. J. Chris Fleming
VP—AMA Foundation

2004-2005
Mrs. John V. Pender, Jr.
VP—Health Promotion

2004-2005
Mrs. Jesse C. Woodall, Jr.
VP—Legislative

2004-2005
Mrs. Charles N. Larkin
VP—Membership

Left to right: Former MMS presidents Jesse C. Woodall, Jr., M.D. and Wiley T. Robinson, M.D.; Former MMS and TMA President Charles R. Handorf, M.D.; Former AMA President John C. Nelson, M.D.; and George S. Flinn, Jr., M.D.

APPENDIX III

PAST PRESIDENTS OF THE TENNESSEE MEDICAL ASSOCIATION FROM MEMPHIS

1830–2009

1877
Benjamin W. Avent, M.D.

1877-1878
Benjamin W. Avent, M.D.

1881-1882
Gustavus B. Thornton, M.D.

1884-1885
D. D. Saunders, M.D.

1894-1895
Francis L. Sim, M.D.

1915-1916
Edward C. Ellett, M.D.

1918-1919
Richmond McKinney, M.D.

1921-1922
William Britt Burns, M.D.

1924-1925
Frank D. Smythe, M.D.

1927-1928
Battle Malone, M.D.

1930-1931
James Bassett McElroy, M.D.

1933-1934
H. B. Everett, M.D.

1936-1937
Wilson L. Williamson, M.D.

1942-1943
J. B. Stanford, M.D.

1945-1946
William Calvert, M.D.

1948-1949
Harley W. Qualls, M.D.

1951-1952
Ernest G. Kelly, M.D.

1960-1961
Ralph O. Rychener, M.D.

1963-1964
Bland W. Cannon, M.D.

1969-1970
Francis H. Cole, Sr., M.D.

1972-1973
William T. Satterfield, Sr., M.D.

1978-1979
John B. Dorian, M.D.

1981-1982
Allen S. Edmonson, M.D.

1987-1988
James T. Galyon, M.D.

1990-1991
Hamel B. Eason, M.D.

1996-1997
Richard M. Pearson, M.D.

1999-2000
J. Chris Fleming, M.D.

2006-2007
Charles Handorf, M.D.

2008-2009
Robert D. Kirkpatrick, M.D.

APPENDIX IV

TENNESSEE MEDICAL ASSOCIATION OUTSTANDING PHYSICIAN RECIPIENTS FROM MEMPHIS*

1963
LLOYD M. GRAVES, M.D.

1965
JAMES S. SPEED, M.D.

1967
EUSTACE SEMMES, M.D.

1971
GILBERT LEVY, M.D.

1973
HAROLD BOYD, M.D.

1979
FRANCIS H. COLE, SR., M.D.

1982
PHIL C. SCHREIER, M.D.

1988
JAMES N. ETTELDORF, M.D.

1991
HALL S. TACKET, M.D.

1997
EVELYN B. OGLE, M.D.

1999
EDWIN W. COCKE, JR., M.D.

2000
JESSE C. WOODALL, JR., M.D.

2001
THOMAS J. WHITE, JR., M.D.

2002
HUGH FRANCIS, JR., M.D.

2004
JAMES T. GAYLON, M.D.

2006
O. BREWSTER HARRINGTON, M.D.

2007
ALLEN S. EDMONSON, M.D.

2008
EDWARD W. REED, M.D.

2009
MAURY BRONSTEIN, M.D.

*Awarded since 1962

APPENDIX V

MEDICAL SPECIALTY AND SUBSPECIALTY BOARDS

To practice medicine, a physician must be licensed by the state. Tennessee has licensed physicians since 1889. As individual specialties developed, doctors formed organizations with examinations administered to ensure proper specialty competence. Memphis has had a special place in forming three specialty boards: Ophthalmology, with the first board examination in any specialty, held its first board examination in Memphis in 1916; Orthopedics, with Dr. Willis C. Campbell as a motivating influence in establishing the American Board of Orthopedic Surgery; and Radiology, with its first board examination at the Peabody in Memphis, influenced by Dr. Franklin Bogart.

To oversee the developing specialty boards, another organization, the American Board of Medical Specialties, was founded in 1933. The board works with the American Medical Association's Council on Medical Education to approve member boards, which in turn certify specialists and subspecialists. A physician who has been approved is known as a "diplomate" or a "subspecialist."

Medical boards construct and administer written and oral board examinations. After completing medical school and a year of internship, a physician can apply for a medical license. Today, however, most physicians complete additional practical training through residencies of three to five years. Then they take specialty board examination(s), becoming certified by their peers. This certification gives added credibility to the practitioner.

The number of subspecialties has grown so that now more than 124 offer certification. Internal medicine, for example, is one of 24 major specialty boards, but it also includes 19 subspecialties, such as endocrinology, gastroenterology, pulmonary diseases, and rheumatology, to name a few. Memphis has virtually all subspecialists in practice.

The following table includes the major specialty boards listed by the American Board of Medical Specialties.

MAJOR MEDICAL SPECIALTY BOARDS

- Allergy and Immunology
- Anesthesiology
- Colon and Rectal Surgery
- Dermatology
- Emergency Medicine
- Family Medicine
- Internal Medicine
- Medical Genetics
- Neurological Surgery
- Nuclear Medicine
- Obstetrics and Gynecology
- Ophthalmology
- Orthopedic Surgery
- Otolaryngology
- Pathology
- Pediatrics
- Physical Medicine and Rehabilitation
- Plastic Surgery
- Preventive Medicine
- Psychiatry and Neurology
- Radiology
- Surgery
- Thoracic Surgery
- Urology

APPENDIX VI

DATES AND EVENTS SIGNIFICANT TO THE HISTORY OF MEMPHIS AND THE MEMPHIS MEDICAL COMMUNITY

1783 Treaty of Paris, ending the American Revolutionary War. John Rice and John Ramsay enter claims for 5,000 acres on the Fourth Chickasaw Bluff.

1790 North Carolina cedes its western lands (Tennessee) to the United States.

1795 The Spanish build Fort San Fernando near the site of the present-day Pyramid.

1796 Tennessee becomes the sixteenth state on June 1.

1797 Americans erect Fort Pickering on the Fourth Chickasaw Bluff after the Spanish withdraw from Fort San Fernando.

1800 First smallpox vaccination administered in North America (in Canada).

1803 The Louisiana Purchase doubles the size of the United States.

1804 The Lewis and Clark Expedition undertakes the search for a route to the Pacific Ocean via the Missouri River. The explorers return in the summer of 1806.

1815 Andrew Jackson defeats the British at the Battle of New Orleans.

1816 Fort Pickering loses its importance as a military post and is abandoned by the army.

1818 Chickasaw Cession conveys land in West Tennessee to the United States.

1819 Surveyors lay out town of Memphis; proprietors sell first lots.

1820 Memphis population 364. George F. Graham, first doctor, comes to the new town.

1820 The first *U.S. Pharmacopeia* is published.

1827 The Shelby County seat is moved from Memphis to Raleigh.

1830 Memphis population 663. The first Memphis hospital is built on Front Street. The Medical Society of Tennessee is organized in Nashville. Gunn's *Domestic Medicine* is published in Knoxville.

1832 Cholera makes a first appearance in Memphis and the Mississippi Valley. Tennessee's first hospital for the mentally ill is built in Nashville.

1836 Land east of the town limit (now Forrest Park) is purchased as the site for a new hospital.

1840 Memphis population 1,799. *The Appeal* newspaper begins publication.

1841 The new Memphis Hospital (state-owned) is built on a ten-acre site, the present-day Forrest Park.

1846 The Tennessee legislature grants charters for the Memphis Medical College and the (Memphis) Botanico Medical College, the first medical schools in the state. The United States declares war on Mexico, and Tennessee sends volunteer regiments. Morton introduces ether anesthesia at Massachusetts General Hospital.

1847 The American Medical Association is organized.

1848 War with Mexico ends. The U.S. gains California and the Arizona Territory.

1849 Gold is discovered in California. There is a great demand for a transcontinental railroad.

1850 Memphis population 8,841. The building of the railroads brings many immigrants.

1855 The first documented yellow fever epidemic in Memphis occurs, with about 200 deaths.

1857 The Memphis & Charleston Railroad connects Memphis to the Atlantic Ocean.

1858 The explosion of the *Pennsylvania* near Memphis. Mark Twain's brother is a victim.

1860 Memphis population 22,623; 30 percent are immigrants—mostly Irish and German.

1861 Tennessee secedes from the Union. Memphians organize 1,000 hospital beds.

1862 The Naval Battle of Memphis takes place on June 6. The city surrenders to federal naval officers.

1863 Memphis becomes a federal hospital base, with more than 5,000 beds organized.

1865 The Civil War ends in April, and the last federal hospital is closed in August. The explosion of the *Sultana* on April 27, the nation's greatest maritime disaster.

1866 Tennessee transfers ownership of Memphis Hospital to the city. The hospital is now called City Hospital. Memphis experiences a serious cholera epidemic.

1867 Memphis has its second yellow fever epidemic, with 2,500 cases, and about 550 deaths.

1870 Memphis population 40,226. Chelsea, Greenlaw, and Fort Pickering areas are annexed.

1873 Memphis suffers from a severe winter, and a smallpox epidemic followed by a summer yellow fever epidemic, with 5,000 cases, and 2,000 deaths.

1876 The Memphis Medical Society is reorganized as the Shelby County Medical Society.

1878 The first meeting in Memphis of the Medical Society of Tennessee. A devastating yellow fever epidemic, with 17,600 cases and 5,150 deaths in Memphis; there were 25,000 cases in the Mississippi Valley, and a heavy death toll.

1879 Yellow fever strikes again, with 2,000 cases and 600 deaths. William H. Welch introduces the science of bacteriology in the U.S.

1880 The population of Memphis drops to 33,592. Memphis is made a taxing district of the state. The Waring sewer system is installed, and other sanitation reforms undertaken. The Memphis Hospital Medical College opens at the southwest corner of Union and Dunlap. The Tri-State Medical Association of Mississippi, Arkansas & Tennessee is formed. Louis Pasteur, French microbiologist, and Robert Koch, German physician, develop the germ theory of disease.

1882 The Poor and Insane Asylum is built on Raleigh Road (later Jackson Avenue).

1883 Drs. W. E. and W. B. Rogers open Rogers Surgical Infirmary at 171 Vance.

1884 The U.S. Marine Hospital is constructed at the former site of Fort Pickering.

1885 Drs. Mitchell and Maury open the Infirmary for Diseases of Women at 73 Court. Drs. Overall and Knox open the Electric Infirmary at 381 Main Street.

1886 Dr. T. J. Crofford opens the Infirmary for Diseases of Women at 251 Main Street.

1887 Artesian wells are sunk. The Memphis aquifer is an enormous source of pure water. Western State Hospital for Insane opens in Bolivar.

1889 St. Joseph Hospital, the first private hospital in Memphis, opens in a frame house. The Sisters of St. Francis provide the nursing staff.

1890 Memphis population 64,495. Diphtheria is treated with antitoxin.

1891 Memphis Sanatorium, with modern facilities, opens at Second and Poplar.

1892 The Great Bridge opens on May 12, the first bridge between St. Louis and New Orleans.

1895 Roentgen discovers x-rays. Marconi introduces wireless telegraphy.

1896 Vaccination for prevention of typhoid fever is introduced.

1897 Dr. Ronald Ross identifies the malaria parasite in the stomach of the anopheles mosquito.

1898 The New City of Memphis Hospital opens in July on Madison east of Dunlap. Spanish American War. American soldiers die of yellow fever and cholera.

1899 The Walter Reed Commission in Cuba proves the mosquito transmission of yellow fever. Memphis nurse Lena Angevine Warner is a member of the research group.

1900 Memphis population 102,320. Planning takes place for Overton and Riverside parks.

1901 Roentgen and von Behring receive the first Nobel Prize for work in radiology.

1902 The Medical Society of Tennessee reorganized as the Tennessee State Medical Association. The Memphis Medical Society becomes the Memphis and Shelby County Medical Society.

1906 The College of Physicians and Surgeon opens for its first session in a new building on Madison opposite the City of Memphis Hospital. The Pure Food and Drug Act is passed by Congress. The new Lucy Brinkley Hospital opens on Union near Dunlap.

1907　Myles V. Lynk moves the University of West Tennessee from Jackson to Memphis.

1909　The Jane Terrell Hospital opens, organized by the Negro Baptist Association. Collins Chapel Hospital is organized by Collins Chapel CME Church. Memphis adopts a commission form of government.

1910　Memphis population 131,105. E. H. Crump takes office as mayor. The Gartly-Ramsay Hospital opens in the McDavitt Home on Jackson Avenue. The Abraham Flexner Report brings major reforms in medical education.

1911　The University of Tennessee College of Medicine moves to Memphis.

1912　The first of four major Mississippi River floods between 1912 and 1937. Baptist Memorial Hospital opens in June. Union Railroad Station opens.

1913　The second major Mississippi River flood.

1914　The Panama Canal opens. LePrince develops controls for tropical diseases.

1915　Goldberger's research in Mississippi proves pellagra is a nutritional deficiency disease caused by the lack of B vitamin, not bacteria. The Memphis Society of Ophthalmology and Otolaryngology is organized.

1916　The American Board of Ophthalmology, first American Board to be organized, gives the first board examination in Memphis.

1917　The United States enters World War I on April 6, but few troops are deployed until 1918. The Harahan Bridge with vehicle roadways opens.

1918　The Spanish influenza pandemic begins, and lasts through 1919, with world deaths exceeding 50 million. Influenza arrives in Memphis in October. Schools, churches, and businesses close.

1919　The Crippled Children's Hospital opens, founded by Dr. Willis C. Campbell.

1920　Memphis population 162,351. Campbell Clinic opens on Madison Avenue on December 26.

1921　Methodist Hospital opens its first hospital in November. The building was sold to the Veterans Bureau six months later. The Methodist staff moved to the Lucy Brinkley Hospital. The Memphis Pediatric Society is formed. The Oakville Tuberculosis Sanitarium opens in July.

1922　The UT Doctors football team wins national recognition with an undefeated season.

1923　The first Hospital for Crippled Adults opens, organized by Dr. Willis C. Campbell. The University of Tennessee purchases the Rex Club on the corner of Dunlap and Madison for a student center.

1924　The Memphis and Shelby County Medical Society begins publishing the *Memphis Medical Journal.* The New Methodist Hospital opens at 1265 Union Avenue on September 16. The Memphis Urology Society is formed.

1925　The Peabody Hotel on Union Avenue opens. The Women's Hospital opens in the former Lucy Brinkley facility on Union.

1926　Medical teaching clinics at Baptist Hospital are conducted by Edward Ellett, ophthalmology, Eugene Johnson, surgery, and Eustace Semmes, neurosurgery. The Memphis Roentgen Society is organized. The Memphis Eye, Ear, Nose & Throat Hospital opens on August 1.

1927　The third great Mississippi River flood. Memphis houses hundreds of refugees.

1928　The Memphis Obstetrical and Gynecological Society is formed. The Baptist Memorial Hospital Physicians and Surgeons Building opens, the first doctors' office in the country owned by a hospital.

1929　The Stock Market Crash occurs on October 24, a national economic disaster, with many banks failing.

1930　Memphis population 253,143. The Tennessee Medical Association marks its centennial in Nashville.

1931　The Cotton Carnival is inaugurated in the depths of the Depression.

1932　The Tri-State Medical Association is renamed the Mid-South Post Graduate Medical Assembly.

1933　President Franklin D. Roosevelt takes office on March 4. Thirty percent of the workforce is unemployed.

1935　The Works Progress Administration (WPA) construction of Riverside Drive and riverfront cleanup is completed.

1936　The John Gaston Hospital, a new city hospital, opens in June on Madison Avenue. The Firestone Tire & Rubber Company opens a large plant in North Memphis.

1937　The greatest Mississippi River flood in history, leaving more than one million homeless. Drs. Turner and Gotten purchase Hayes Sanitorium on Raleigh LaGrange. A new U.S. Public Health Hospital replaces the old Marine Hospital.

1938　WPA slum clearance. Lauderdale Courts and Dixie

Homes built. Dr. L. W. Diggs organizes a blood bank at John Gaston, the first in the South.
1939 Florey and Chain develop penicillin for therapeutic use.
1940 Memphis population 292,942. The first peacetime draft takes place. A flu epidemic in occures in Memphis.
1941 A TB division established in the Health Department, with "TB nurses." The Health Department established a crippled children's service.
1942 The city and county health departments merge. The Thomas F. Gailor Psychiatric Hospital and Diagnostic Clinic opens, the first public psychiatric hospital in Memphis, on Dunlap adjacent to John Gaston Hospital.
1943 Kennedy Hospital is built in Memphis.
1946 A symposium on War Medicine and Surgery is held in Memphis.
1947 Baptist is the first hospital in Memphis to provide a central oxygen system.
1948 The first Radioisotope Unit in Memphis is installed. The Health Department stresses follow-up treatment for children with defects found in testing at school. There is a 25 percent decrease in the infant death rate since 1945.
1949 Les Passees establishes a treatment center for children with cerebral palsy. Polio cases increase to 155. Dr. Alys Lipscomb is the first in Memphis to prescribe 131I for hyperthyroidism.
1950 Memphis population 396,000.
1951 The Van Vleet Cancer Center is established. St. Joseph opens Memphis's first physical therapy department with a registered therapist in charge. Dr. Thomas Stern develops a cardiac catheterization laboratory.
1952 The opening of Le Bonheur Children's Hospital and the first Holiday Inn. The University of Tennessee participates in the first testing of the Pap smear on 200,000 women aged 20-50. A major polio epidemic occurs in Memphis and the United States. Isonaizid is introduced to treat tuberculosis.
1953 The University of Tennessee begins requiring three years of premedical training.
1954 The Memphis Health Department physicians participate in polio field trials. There are 174 cases, with 8 deaths from polio.
1955 The Salk vaccine is first administered to third-graders in Memphis. There are 55 cases, with 2 deaths from polio. The Gailor Annex, medical and surgical, is completed. The new Collins Chapel Hospital, at Lane and Ayers, is dedicated by Bishop J. Arthur Hamlett of Christian Methodist Episcopal Church, with Dr. W. S. Martin as chief of staff. Methodist Hospital opens its Thomas wing.
1956 Dr. John Shea's first Teflon® stapedectomy in the world takes place at Baptist. The E. H. Crump Hospital is built by the city to serve African Americans and as a teaching hospital for African-American doctors, nurses, and medical technicians.
1957 The new Les Passees Center is built. The Pediatric wing of Gaston is rebuilt. The first ICU in Memphis opens at Le Bonheur.
1958 The Sickle Cell Center is established at the University of Tennessee by Dr. Lemuel Diggs. Dr. Bill Runyan establishes the first Visiting Nurse Program for the Memphis Health Department.
1959 The Van Vleet Center purchases the first cobalt radiation unit in the region. Dr. Robert Allen performs the first open heart surgery in Memphis using a pump oxygenator.
1960 Memphis population 497,524. The Pigg-O-Stat is developed at Le Bonheur (in Radiology).
1961 The Shelby County Medical Examiner system replaces the coroner system.
1962 The first Cuban doctors immigrate to Memphis, including top Cuban physicians. St. Jude Children's Research Hospital opens as the world's first children's catastrophic disease research and treatment hospital. The Tennessee Psychiatric Hospital and Institute is established (later known as the Memphis Mental Health Institute).
1963 Drs. Bill Runyan and George Lovejoy describe the Visiting Nurse Program of the Memphis Health Department. They win a Rockefeller Award for their work.
1964 The first fully accredited alcohol and drug facility in Tennessee is established at Tennessee Psychiatric Hospital and Institute by Drs. David Knott and James Beard. Dr. H. Edward Garrett, Sr., performs the world's first successful coronary bypass in Houston, shortly before moving to Memphis.
1965 The Social Security Act of 1965 (Medicare/Medicaid) brings changes to medicine. Memphis hospitals integrate in response to Medicare/Medicaid requirements.
1966 The William F. Bowld Hospital opens in the medical center as the practice arm of the University of Tennessee Health Science Center (UTHSC).
1967 The Alcohol Rehabilitation Research Center and Children's Unit are established at Memphis Mental

Health Institute. The new Veterans Administration Hospital opens in the medical center.

1968 Martin Luther King, Jr., comes to Memphis to support striking sanitation workers and is assassinated on April 4 in Memphis. The first Newborn Center in Memphis is established by Dr. Sheldon Korones. Le Bonheur establishes the first comprehensive orthopedic service in Memphis.

1969 Methodist establishes the first Memphis Cardiovascular Diagnostic Facility. Dr. Gene Stollerman joins UTHSC's College of Medicine.

1970 Memphis population 623,530. Memphis's first human kidney transplant by Dr. Louis G. Britt.

1971 Memphis is the first city in the South to have Mohs surgery with Dr. Rex Amonette. Dr. Muirhead begins the first Memphis Center for Research in Hypertension. Dr. James Pate is the first in the U.S. to implant an artificial valve in a pediatric patient.

1973 The first Shelby County rural health department clinic opens in Eads. Methodist South, the first satellite hospital, opens. The Health Maintenance Organization Act passes, requiring HMOs to be offered as an insurance option.

1974 St. Joseph East is dedicated as a satellite of St. Joseph downtown. The first Community Blood Plan Donor Center in Memphis opens at 1715 Union Avenue.

1975 Dr. Donald Pinkel achieves the first five-year cure for acute lymphocytic leukemia at St. Jude Children's Research Hospital.

1976 State Volunteer Mutual Insurance Company (SVMIC) officially licensed.

1977 Elvis Presley's death turns the spotlight on his Memphis autopsy. The Health Care Financing Administration (HCFA) is established under the Department of Health, Education, and Welfare.

1979 Baptist East opens on Walnut Grove as a satellite of Baptist Memorial Hospital.

1980 Memphis population 646,356. St. Joseph East becomes St. Francis to emphasize the connection with the Sisters of St. Francis of Mishawaka, Indiana. The AIDS epidemic is recognized. HMOs begin expanding.

1982 Dr. Santiago Vera performs Memphis's first liver transplant at Bowld Hospital.

1983 The Elvis Presley Trauma Center opens at The MED. DRGs are imposed on hospitals nationwide as part of a prospective payment system. MetroCare at Methodist marks the beginning of alphabet soup of MCOs.

1985 The first freestanding ambulatory ophthalmic surgery center in Tennessee, Memphis Eye and Cataract Associates, opens. A burn center opens at The MED, designated in 1993 as the Firefighters Regional Burn Center.

1986 Dr. James Porterfield implants the first cardiac defibrillator in Memphis. The Church Health Center opens.

1988 The state's first skin bank is established by the Medical Examiner's office.

1989 The first kidney/pancreas transplant is performed in Memphis by Drs. Osama Gaber, Louis Britt, and Santiago Vera. The first laparoscopic gallbladder surgery in Memphis is performed by Drs. Guy Voeller and Richard Patterson. John Gaston Hospital is demolished.

1990 Memphis population 610,337; Metro area population 1,205,204. Campbell Clinic and the University of Tennessee Department of Orthopaedics enter a formal residency agreement.

1991 The first bariatric surgery journal is established by Dr. George Cowan.

1992 The Wound Center opens at The MED.

1993 The growing influence of HMOs in medicine is seen. The payment system is replaced by a fee schedule. New standards for reimbursement, relative value units (RVUs), are established. Methodist's first rural clinics, in Wiener and Trumann, Arkansas, open. Campbell Clinic moves its main facility to Germantown Road. Drs. Mark Hammond and Wiley Robinson form the first group of hospitalists.

1994 A Pancreas Islet Transplant Laboratory is established at the University of Tennessee under Dr. Osama Gaber. The Diggs-Kraus Sickle Cell Center rededicated to honor the three who had spent their lives studying and treating this disease. TennCare is established, offered through private HMOs.

1995 Methodist and Le Bonheur hospitals merge to form Methodist Le Bonheur Healthcare. The Gamma Knife facility opens at Methodist. Tenet buys St. Francis Hospital. Christ Community Health Services opens its first clinic. The Medical Education Research Institute (MERI) is established. Semmes-Murphey at Baptist and Neurosurgical Group at Methodist merge.

1996 Dr. Peter Doherty becomes Memphis's first Nobel Prize winner, and shares the prize in medicine with Dr. Rolf Zinkernagel. TennCare separates behavioral health from its program.

1997 Baptist Memorial Hospital and St. Joseph Hospital

Healthcare reform was a contentious issue at Congressman Steve Cohen's (standing left) August 8, 2009 town hall meeting, with overflow attendance. Among others, Drs. Neal Beckford, James Klemis, Laura Bishop, Autry Parker, and Frank McGrew spoke to the crowd. The Patient Protection and Affordable Care Act was signed into law on March 23, 2010. It remained a prominent issue in the fall Congressional election of that year.

begin a merger. The Balanced Budget Act of 1997 establishes a sustainable growth rate (SGR). There is a cap established on Medicare.

1998 The Memphis and Shelby County Mental Health Summit is established.

2000 Memphis population 650,100. Baptist Central closes. Its last patient is transferred to Baptist East. Baptist East becomes the flagship hospital of Baptist Memorial Hospital Memphis. Baptist Heart Institute and Baptist Women's Hospital open.

2001 HCFA is renamed the Centers for Medicare and Medicaid Services. The West Nile virus comes to Tennessee.

2003 The first Da Vinci® Surgical System in the city is installed at Baptist.

2004 St. Francis Hospital Bartlett opens, the first full-service hospital in Bartlett. The University of Tennessee Transplant Program moves to Methodist University Hospital, becoming the Methodist University Hospital Abdominal Transplant Institute. The Medical Examiner's Office is taken over by the Center for Forensic Medicine, with Dr. Karen Chancellor in charge in Memphis.

2005 The implosion of Baptist Central occurs on November 6. Baptist donates land for UT-Baptist Research Park. TennCare2 begins.

2007 The Hamilton Eye Institute opens (begun in 2004; third phase, 2007). The new Memphis Mental Health Institute opens on the former site of Bowld Hospital. The University of Tennessee's Translational Science Institute is formed.

2008 St. Jude becomes the first pediatric institution designated a Comprehensive Cancer Center by the National Cancer Institute of the National Institutes of Health.

2009 Financial woes threaten services at The MED.

2010 The population of Memphis is estimated at 676,640. St. Francis Hospital establishes a Spine Institute. The Patient Protection and Affordable Care Act passes in Congress.

MEMPHIS MEDICINE
A HISTORY *of* SCIENCE AND SERVICE

SELECTED BIBLIOGRAPHY

Part One: Chapters 1-8

Adams, George W., "Confederate Medicine," *The Journal of Southern History,* Volume 6, 151-166, 1940.

Adams, George W., *Doctors in Blue: The Medical History of the Union Army in the Civil War,* Henry Schurman, Inc., New York, New York, 1952.

Barnes, Joseph K., *The Medical and Surgical History of the War of the Rebellion 1861-1865.* Volume I, Part 1, Government Printing Office, Washington, D.C., 1870.

Biles, Roger, *Memphis in the Great Depression,* The University of Tennessee Press, Knoxville, Tennessee, 1986.

Brooks, Stewart M., *Civil War Medicine,* Charles C. Thomas Publisher, Springfield, Illinois, 1966.

Bruesch, Simon Rulin, "Early Medical History of Memphis 1819-1861," *West Tennessee Historical Society Papers,* Volume 2, 33-94, 1948.

Bruesch, Simon Rulin, "The Disasters and Epidemics of a River Town: Memphis, Tennessee, 1819-1879," *Bulletin of the Medical Library Association,* Volume 40, 288-305, July 1952.

Bruesch, Simon Rulin, "Yellow Fever In Tennessee in 1878," *Journal of the Tennessee Medical Association,* Part I, Volume 71, 887-896, December 1978; Part II, Volume 72, 91-104, February 1979; Part III, Volume 72, 193-205, March 1979.

Calandruccio, Rocco A., "The History of the Campbell Clinic," *Clinical Orthopaedics and Related Research,* Volume 374, 157-170, May 2000.

Canale, Dee J., Clarence B. Watridge, Tyler S. Fuehrer, and Jon H. Robertson, "The History of Neurosurgery in Memphis: the Semmes-Murphey Clinic and the Department of Neurosurgery at the University of Tennessee College of Medicine," *Journal of Neurosurgery,* Volume 112, 189-198, January 2010.

Carter, Henry Rose, *Yellow Fever: An Epidemiological and Historical Study of Its Place of Origin,* edited by Laura Armistead Carter and Wade Hampton Frost, Williams & Wilkins, Baltimore, Maryland, 1931.

Crosby, Molly Caldwell, *The American Plague: The Untold Story of Yellow Fever, The Epidemic That Shaped Our History,* Berkley Books, New York, New York, 2006.

Cunningham, Horace H., *Doctors in Gray: The Medical History of the Confederate Army in the Civil War,* LSU Press, Baton Rouge, Louisiana, 1958.

Daniel, Larry J., *Shiloh: The Battle That Changed the Civil War,* Simon and Schuster, New York, New York, 1997.

Dary, David, *Frontier Medicine: From the Atlantic to the Pacific 1492-1941,* Alfred A. Knopf, New York, New York, 2008.

DeCosta-Willis, Miriam, *Notable Black Memphians,* Cambria Press, Amherst, New York, 2008.

Dowdy, G. Wayne, *Mayor Crump Don't Like It, Machine Politics in Memphis,* The University Press of Mississippi, Jackson, Mississippi, 2006.

Duffy, John, "Medical Practice in the Ante Bellum South." *The Journal of Southern History* Volume 25, 53-72, February 1959.

Fakes, Turner J., Jr., "Memphis and the Mexican War," *West Tennessee Historical Society Papers,* Volume 2, 119-144, 1948.

Falsone, Ann Marie McMahon, "The Memphis Howard Association: A Study in the Growth of Social Awareness," Thesis, Memphis State University, Memphis, Tennessee, 1968.

Flexner, Abraham, *Medical Education in the United States and Canada: A Report to the Carnegie Foundation for the Advancement of Teaching,* The Carnegie Foundation, New York, New York, 1910.

Flexner, James Thomas, *Doctors on Horseback: Pioneers of American* Medicine, Viking Press, New York, New York, 1937.

Ford, Joseph H., "Administration of the American Expeditionary Forces," *The Medical Department of the United States Army in the World* War, Government Printing Office, Washington, D.C., 1927.

Forman, J. G., *The Western Sanitary Commission St. Louis*, R. P. Studley & Co., St. Louis, Missouri, 1864.

Garraty, John A., *The Great Depression*, Harcourt, Brace Jovanovich Publishers, New York, New York, 1986.

Gorgas, Surgeon General William Crawford, *Report of the Surgeon General, Part 2*, 2010, 2134, 2153, 2819-20, 2851, 2864, Government Printing Office, Washington, D.C., 1920.

Hamer, Philip M., Editor, *The Centennial History of the Tennessee State Medical Association*, Tennessee State Medical Association, Nashville, Tennessee, 1930.

Hamilton, Green Polonius, *The Bright Side of Memphis*, Inter-Collegiate Press, Memphis, Tennessee, 1908.

Hamner, James Edward, III, *The University of Tennessee, Memphis: 75th Anniversary Medical Accomplishments*, The University of Tennessee, Memphis, Tennessee, 1986.

Hansen, Arlen J., *Gentlemen Volunteers: The Story of the American Ambulance Drivers in the Great War, August 1914-September 1918*, Arcade Publishing, New York, New York, 1996.

Harkins, John E., *Metropolis of the American Nile: An Illustrated History of Memphis and Shelby County.* Woodland Hills, California: Windsor Publications, 1982.

Harkins, John E., *Memphis Chronicles*, The History Press, Charleston, South Carolina, 2009.

Harrison, Gordon, *Mosquitoes, Malaria & Man: A History of the Hostilities Since 1880*, E. P. Dutton & Company, Inc., New York, New York, 1978.

Harvey, A. McGehee, *Science at the Bedside: Clinical Research in American Medicine 1905-1945*, The Johns Hopkins University Press, Baltimore, Maryland, 1981.

Hiatt, Roger L., Sr., "History of Ophthalmology in Tennessee and the TAO," University of Tennessee Health Science Center, Memphis, Tennessee, 2003.

Hicks, Mildred, Editor, "Yellow Fever and the Board of Health, Memphis 1878," Memphis and Shelby County Health Department, Memphis, Tennessee, 1964.

Hooper, Ernest Walter, "Memphis, Tennessee: Federal Occupation and Reconstruction 1862-1870," Ph.D. Dissertation, University of North Carolina, Chapel Hill, North Carolina, 1957.

Jenkins, Earnestine Lovelle, *African Americans in Memphis*, Arcadia Publishing, Charleston, South Carolina, 2009.

Keating, John McLeod, *A History of the Yellow Fever: Yellow Fever Epidemic of 1878, in Memphis, Tenn.*, The Howard Association, Memphis, Tennessee, 1879.

Keating, John McLeod and O. P. Vedder, *History of the City of Memphis and Shelby County, Tennessee*, 2 volumes, D. Mason & Co., New York, New York, 1889.

Klass, Morris D., "The St. Jude Children's Research Hospital Collection," Memphis and Shelby County Room, Memphis Public Library, Memphis, Tennessee, 2004.

Lanier, Robert, *Memphis in the Twenties*, Zenda Press, Memphis, Tennessee, 1979.

LaPointe, Patricia M., *From Saddlebags to Science: A Century of Health Care in Memphis, 1830-1930*, Health Sciences Museum Foundation, Memphis, Tennessee, 1984.

LaPointe, Patricia M., "History of Yellow Fever in Memphis: A Collection of Documents, Drawings, and Newspaper Articles," Memphis and Shelby County Room, Memphis Public Library, Memphis, Tennessee, 2003.

LaPointe, Patricia M., "Joseph Augustine LePrince: His Battle Against Mosquitoes and Malaria," *The West Tennessee Historical Society Papers*, Volume 41, 48-61, 1987.

LaPointe, Patricia M., "Military Hospitals in Memphis, 1861-1865," *Tennessee Historical Quarterly*, Volume 42, 325-342, Winter 1983.

LaPointe, Patricia M., "The Disrupted Years: Memphis City Hospitals 1860-1867," *The West Tennessee Historical Society Papers*, Volume 37, 9-29, 1983.

LaPointe, Patricia M., "The Memphis Crippled Children's Hospital School 1917-1982," Memphis and Shelby County Room, Memphis Public Library, Memphis, Tennessee, 1986.

"Lest We Forget," *Memphis Medical Journal*, Volume 2, 244-245, October 1925.

Lynk, Miles Vandahurst, *Sixty Years of Medicine: Or the Life and Time of Dr. Miles V. Lynk*, Twentieth Century Press, Memphis, Tennessee, 1956.

Magness, Perre, *Elmwood 2002: In the Shadow of the Elms*, Elmwood Cemetery, Memphis, Tennessee, 2001.

Marks, Geoffrey and William K. Beatty, *The Story of Medicine in America*, Charles Scribner's Sons, New York, New York, 1973.

McNeill, William H., *Plagues and Peoples*, University of Chicago Press, Chicago, Illinois, 1976.

Miller, William D., *Memphis During the Progressive Era 1900-1917*, Memphis State University Press, Memphis, Tennessee, 1957.

Newberry, John Strong, *The U.S. Sanitary Commission in the Valley of the Mississippi During the War of Rebellion 1861-1866*, Fairbanks, Benedict & Co., Printers, Cleveland, Ohio, 1871.

O'Daniel, Patrick., "Mid-South Flood Collection," Memphis and Shelby County Room, Memphis Public Library, Memphis, Tennessee, 2005.

O'Daniel, Patrick, *Memphis and the Super Flood of 1937: High Water Blues*, The History Press, Charleston, South Carolina, 2010.

Oldstone, Michael B. A., *Viruses, Plagues & History: Past*

Potter, Jerry O., *The Sultana Tragedy: America's Greatest Maritime Disaster*, Pelican Publishing Co., Gretna, Louisiana, 1992.

Raichelson, Richard M., *Beale Street Talks: A Walking Tour Down the Home of the Blues*, second edition, Arcadia Records, Memphis, Tennessee, 1999.

Remini, Robert V., *Andrew Jackson and the Course of American Democracy, 1833-1845*, Harper & Row, New York, New York, 1984.

Roper, James, *The Founding of Memphis 1818-1820*, Memphis Sesquicentennial, Inc., Memphis, Tennessee, 1970.

Rosenberg, Charles E., *The Cholera Years: The United States in 1832, 1849 and 1866*, University of Chicago Press, Chicago, Illinois, 1962.

Sigafoos, Robert A., *Cotton Row to Beale Street*, Memphis State University Press, Memphis, Tennessee, 1979.

Sorrels, William W., *Memphis' Greatest Debate: A Question of Water*, Memphis State University Press, Memphis, Tennessee, 1970.

Steiner, Paul E., *Disease in the Civil War*, Charles C. Thomas Publishers, Springfield, Illinois, 1968.

Stewart, Marcus J. and William T. Black, Jr., Editors, *History of Medicine in Memphis*, McCowat-Mercer Press, Jackson, Tennessee, 1971.

Vogel, Virgil J., *American Indian Medicine*, University of Oklahoma Press, Norman, Oklahoma, 1970.

Williams, Greer, *The Plague Killers*, Charles Scribner's Sons, New York, New York, 1969.

Woodward, J. J., *The Medical and Surgical History of the War of Rebellion*, 2 volumes. Government Printing Office, Washington, D. C., Second Issue, 1875.

Wrenn, Lynette Boney, *Crisis and Commission Government in Memphis, Elite Rule in a Gilded Age City*, University of Tennessee Press, Knoxville, Tennessee, 1998.

Young, James Harvey, *The Toadstool Millionaires: A Social History of Patent Medicines in America before Federal Regulation*, Princeton University Press, Princeton, New Jersey, 1961.

Young, James Harvey, *The Medical Messiahs: A Social History of Health Quackery in Twentieth-Century America*, Princeton University Press, Princeton, New Jersey, 1967.

Local History Print Sources

The Memphis and Shelby County Room of the Memphis Public Library is an invaluable repository of newspaper clippings files, city directories, indexes, journals, books, photographs, and other material on the history of Memphis and its surrounding area, material often not available elsewhere.

Part Two: Chapters 9-15

Print Materials

"Awards, Dr. Lawrence Wruble," *Memphis Commercial Appeal*, B5, May 19, 2009.

Berryhill, Dale A., *Whatever It Takes: Le Bonheur Children's Medical Center: The First Fifty Years*, Urban Child Institute, Memphis, Tennessee, 2006.

"Brain Tumor Stem Cell Program Established at UTHSC," *The Record: The University of Tennessee Health Science Center*, Memphis, Tennessee, July 1, 2005.

Congressional Budget Office, "The sustainable growth rate formula for setting Medicare's physician payment rates," September 6, 2006.

Connolly, Daniel, "Med Might Not Spin Off Clinics," *Memphis Commercial Appeal*, B5, March 13, 2010.

Cordes, Frederick C., and C. Wilbur Bucker, "History of the American Board of Ophthalmology," *Transactions of the American Ophthalmological Society*, Volume 59, 295-328, 1961.

Fisher, Craig, Editor, *Legacy of Heroes*, American Academy of Orthopaedic Surgeons, Rosemont, Illinois, 2004.

Hall, Peter, *Cities in Civilization*, Pantheon Books, New York, New York, 1988.

Hamner, James E., *The University of Tennessee, Memphis, 75th Anniversary—Medical Accomplishments*, University of Tennessee, Memphis, Tennessee, 1986.

Hobson, J. J., "Poliosteria, A New Disease," *Memphis Medical Journal*, Volume 27, Issue 5, 66, 1952.

King, William P., *Follow the Green Line*, Xlibris, Bloomington, Indiana, 2003.

LaPointe, Patricia A., *From Saddlebags to Science: A Century of Health Care in Memphis, 1830-1930*, Memphis Pink Palace Museum, Memphis, Tennessee, 1984.

Lewis, James, "Letters to Editor," *Memphis Commercial Appeal*, September 23, 2009.

Lollar, Michael, "The Last Word," *Memphis Commercial Appeal*, A1, A7, August 10, 2008.

Methodist Health Systems, *Building a Dream: The Story of Methodist Hospitals of Memphis*, Taylor Publishing, Memphis, Tennessee, 1986.

Monthly Morbidity Report, *Memphis Medical Journal*, Volume 23, Issue 11, 5, 25, 1948.

Monthly Morbidity Report, *Memphis Medical Journal*, Volume 24, Issue 2, 11, 17, 25, 36, 50, 82, 118, 119, 151, 152, 184, 188, 219, 1949-1950.

Monthly Morbidity Report, *Memphis Medical Journal*, Volume 26, Issues 11 and 12, 15, 188, 1951.

Monthly Morbidity Report, *Memphis Medical Journal*, Volume 7, Issue 1 through Volume 27, Issue 11, 30, 46, 62, 77, 91, 104, 176, 1952.

Naik, Kalyani, and M. Sue Zaleski, "Cytotechnology: A Program on the Move—Then and Now," *Labmedicine*, Volume 39, issue 4, 201-206, 2006.

"Pastor's dream fulfilled—Serves entire Mid-South," *Memphis Press-Scimitar*, June 6, 1952.

LaPointe, Patricia M., "Joseph Augustus LePrince: His Battle Against Mosquitoes and Malaria," *West Tennessee Historical Society Papers*, Volume 41, 48-61, 1987.

Smith, Hugh, "The Legacies and the Legends of The Campbell Clinic, 1920-1941," *The Campbell Journal*, Volume 2, Issue 2, 61 pages, October 1988. Articles about early physicians of the clinic.

Stewart, Marcus J., Jr., *One More Step*, Quebecor Books, Memphis, Tennessee, 1999.

Stewart, Marcus J., Jr., William T. Black, and Mildred Hicks, Editors, *History of Memphis Medicine*, McCowat-Mercer Press, Jackson, Tennessee, 1971.

Stollerman, Gene H., "Rheumatogenic and Nephritogenic Streptococci," *Circulation*, Volume 4, 915-21, 1971.

"25 Years Ago—Jan. 14, 1962," *Memphis Commercial Appeal*, January 14,1989.

Numerous publications from the University of Tennessee Health Science Center were made available by Sheila Champlin and by other hospitals. Articles from *Memphis Business Journal, Memphis Business Quarterly*, and from *Memphis Medical Quarterly* provided background. Other publications included the *Church Health Center Newsletter* and the *Bluff City Medical Society Newsletter*. Telephone directories in the Memphis Room at the Memphis Public Library and Information Center provided information about physicians practicing at a given time.

Archival Materials

Memphis and Shelby County Public Library and Information Center, Memphis Room collection of clippings from the *Memphis Commercial Appeal* and the *Memphis Press-Scimitar*.

Memphis Medical Society, Clippings, Awards, Licensure, and Obituary files.

Memphis Medical Society, Bound volumes of minutes, 1955-2009.

University of Memphis, Local History Collection.

University of Tennessee Health Science Center Library, Simon R. Bruesch collection, plus copies of *Memphis Medical Journal* (now defunct), drawings of Dorothy Sturm, and various other materials

Typescript Materials

Adams, Frank, M.D., Clippings and photographic files of Dr. Milton Adams's career, 2 volumes.

Baskin, Reed, M.D., "History of oncology in Memphis," Typescript, 25 pages.

Chang, Cyril F., Howard P. Tuckman, and Diego Nocetti, "Economic Contributions of the Memphis Medical Community: Business Perspectives," Typescript, 23 pages.

Crenshaw, Hoyt, M.D., family and descendants material, submitted to The Memphis Medical Society.

Elliott, Maurice, "The Racial Integration of the Memphis Medical Center," Presented to the Memphis Medical Historical Society, January 17, 2007, Typescript, 22 pages.

Gotten, Nicholas, Sr., M.D., Bound typescript document, 2 volumes, submitted by family.

Lathram, M.W., Jr., M.D., Talk Presented to Memphis Methodist Hospital Auxiliary on Psychiatry, April 1, 1980, Typescript, 5 pages.

Long, Jean Morris, "The First Twelve Men," Typescript history of radiology in Memphis featuring Wilson & Long, plus predecessors, lent by family.

Memphis Chamber of Commerce, "Medical & Research," 2007, Typescript, 2 pages.

Novick, William, M.D., Personal history, Typescript, 48 pages.

Shugart, Anita, Transcriber, "The effects of the polio epidemic on Memphis: Interview with Dr. Alvin J. Ingram," October 25, 1989. Master's thesis, The University of Memphis Special Collections, Memphis, Tennessee.

Voeller, Guy, M.D., "History of minimally invasive surgery in Memphis," Typescript manuscript. 20 pages.

Personal Interviews

Adams, Frank, M.D.
Amonette, Rex A., M.D.
Barron, John M., M.D.
Britt, Louis, M.D.
Coors, George A., M.D.
Crisler, Herman, Jr., M.D.
Cunningham, David, M.D.
DeFlumere, Charlotte A., M.D.
Feild, James Rodney, M.D.
Francis, Hugh, III, M.D.
Francisco, Jerry, M.D.
Gotten, Nicholas, Jr., M.D.

Hale, Ruth Ann, and Jill Fazackerly, Methodist Le Bonheur Healthcare.
Higdon, Dennis, M.D., Medical Anesthesia Group.
Hixson, Douglas, M.D., Pediatric Surgical Group.
Jackson, Robert A., M.D.
Johnson, James Gibb, M.D.
Lovejoy, George, M.D.
McGehee, Lucius, M.D.
Mid-South OB/GYN Group, Newspaper clippings, letter.
Miller, Steven T., M.D.
Pate, James, M.D.
Parsioon, Fereidoon, M.D.
Pearce, Iris, M.D., collected printed information supplied by family.
Riggs, W. Webster, M.D.
Schrader, Lawrence F., M.D.
Stern, Thomas N., M.D. Typescript history of cardiology in Memphis submitted to The Memphis Medical Society.
Turner, Randolph, M.D.
Tyrer, A. Roy, M.D.
Wilons, Michael D., and Emmel B. Golden, Jr., Memphis Lung Physicians.
Woodard, Shirley, Coordinator, Continuing Medical Education at Methodist Le Bonheur Healthcare.

Websites

American Academy of Physical Medicine and Rehabilitation, "History of the specialty of PM&R," http://aapmr.org/academy6/historyb.htm, accessed September 15, 2009.
American Medical Association, Women Physicians Congress, "Women in Medicine: An AMA Timeline," http://www.ama-assn.org/go/wpc, accessed June 22, 2009.
Baptist Health Services Group, "Welcome," http://www.bhsgonline.org, accessed October 1, 2009.
"Baptist Memorial Hospital for Women—Key Physicians," http://bmhcc.org/media/news/archivecontent.asp, accessed October 23, 2009.
"Baptist Milestones in Heart Care," http://www.bmhcc.org/services/medical/heart/milestone.asp, accessed November 4, 2009.
Centers for Disease Control, "The History of Malaria, an Ancient Disease," http://www.bhsgoline.org, accessed August 3, 2008.
Doherty, Peter C., "Autobiography," http://wwwnobelprize.org, accessed February 14, 2009.
HealthChoice Home Page, http://www.myhealthchoice.com/layouts/internetsite/default.aspx, accessed November 4, 2009.
Haynie, Holli W., "Covering the Uninsured," http://www.memphismedicalnews.com, accessed June 9, 2009.
Lanfranci, Anthony R., et. al., "Robotic Surgery: Current Perspectives," *Annals of Surgery*, http://www.pubmed-central.nih.gov, accessed April 17, 2009.
McDade, J. F., and Bozeman, F. M, "Legionnaires' Disease Bacterium Isolated in 1947," *Annals of Internal Medicine*, http://www.ncbi.nlm.nih.gov/pubmed/373548, accessed September 19, 2009.
Maynard-Garrett, Elizabeth, "Winning the War against Strep," *Tennessee Alumnus Magazine*, Volume 85, Winter 2005, http://pr.tennessee.edu/alumnus/alumarticle, accessed August 7, 2008.
Methodist University Hospital, "Virtual brain tumor board," http://methodisthealth.orglive.com/vbtb/faculty/index.cfm, accessed November 14, 2009.
MetroCare Physicians, "Who we are, history," http://www.metrocaredocs.com/ history, accessed January 3, 2009.
Mid-South Eyebank for Sight Restoration, http://www.msebtn.org/history.php, accessed May 22, 2009.
National Institutes of Health, *Stem Cell Basics*, http://www.stemcells.nih. gov/info, accessed August 23, 2009.
Regional Medical Center at Memphis, "Elvis Presley Memorial Trauma Center," http://www.the-med.org/trauma, accessed October 21, 2009.
"Robert Webster," *Wikipedia*, http://en.wikipedia.org/wika/Robert_Webster, accessed November 3, 2009.
Sloan, F. A., P. J. Rankin, D. J. Wheland, and C. J. Grover, "Medicaid, Managed Care, and the Care of Patients Hospitalized from Acute Myocardial Infarction," *American Heart Journal*, Volume 139, 567-576, http://www.TBIonline.org, accessed March 19, 2009.
University of Tennessee Medical Group, *Timeline*, http://www.utmedicalgroup.com/pages/Anniversary/Timeline, accessed October 30, 2009.
University of Tennessee Medical Group, http://www.UTMG, information about physicians, history, and research.
UT-Baptist Research Park, http://www.utbaptistresearchpark.com, accessed July 31, 2009.
"William Hickerson, M.D., Named Chief, Burn Surgery," http://www.utmem.edu/news/ne.php?HYPER, accessed October 21, 2009.

Each of the major hospitals and most physicians' groups in Memphis maintain websites. Such online information provided details too numerous to cite. Particularly valuable were the varied websites—from history to public health to disease histories to scholarly articles—of the United States National Library of Medicine, http://www.nlm.gov. Additional information came from many articles retrieved through the database of Newsbank.com.

MEMPHIS MEDICINE
A HISTORY *of* SCIENCE AND SERVICE

INDEX

Numbers in italic indicate a picture.

Abbott, Sarah E., 162
Acchiardo, Sergio, 217
Acker, Jim, 180
Acquired Immune Deficiency Syndrome (AIDS), 178, 179, 180, 185, 196, 202, 225
Active Implants, 194
Adams Hospital, 28
Adams, John R., 193
Adams, Lorenzo, *126*, 127
Adams, Lou, 224
Adams, Louis R., 127
Adams, William Milton, 125, *126*, 127
Adams, R. Franklin, 127, 163
Adams, William M., Jr., 127, 224
Adams-Graves, Pat, 174
Adler, Justin, 102
Aivazian, Garo, 192
Alcohol Rehabilitation Research Center, 148
African-American physicians, *65*, 66, *72, 122*, 133, 155-160
Allen, Charles G., Jr., *151*
Allen, Robert, *131*, 138, 169
Alley, F. H., 125, *135*
Allissandratos, Jane, 163
ambulatory surgery centers, 191
American Association of Neurological Surgeons (AANS), 106
American Board of Radiology (ABR), 96
American Cancer Society, 134, *152*
American Red Cross, 108, 109, 119-20
American Orthopedic Society for Sports Medicine, 117
American Pharmaceutical Association, 20
American Society of Tropical Medicine, 75
Amiri, Hosein, 169
Amonette, Rex, 165
Anderson, Ernest L., 79
Anderson, Lynn, 79

Andrew, James Lindsey, 77, 96
Andrews, William "Chubby," 127, 182
Angevin, Lena, 52, *54*, 59
Anthony, D. H., 87
Archer, Everett B., 114
Archer, Myrtle, 79
Arkin, Charles, 163
Armstrong, William James, 42-43
Armstrong, Louise Hanna, 43
Army Hospital, 28
Arnold, Valerie, 171
Arrhythmia Consultants, 180
Ashworth, Mrs. William B., *169*
Assisi Foundation, 186, 195, 205
Aste, Malcolm, 127
Athey, A. Phillip, 39
Atkins, Henry Lee, 162
Atkins, Leland L., *133*, 159
Atkinson, Edward, 109
Avent, Benjamin W., 38, 39
Avergis, John, 219
Ayers, J. C., 96

Bade, Mrs. W. H., 119
Bagwell, Troy, 69
Bailey, P. W., *72*
Baker, Laurie M., 219
Ballenger, Peter, 122
Baptist East Hospital, 185, 211
Baptist Foundation, 181, 205
Baptist Health Services Group, 186
Baptist Memorial Hospital, 61, 70, 73, *73*, 76, 79, 85, 90, 92, 93, 104, 106, *129*, 138, 139, 156, 174, 180, 200, 201, 207, 209, 210-11, 213, 223, 227, 229; Comprehensive Breast Center, 212; Heart Institute, 212; School of Nursing, *177*
Baptist Memorial Health Care Foundation, 181
Baptist Memorial Health Care System, 185, 186
Baptist Memorial Health Corporation, 195

Baptist Women's Hospital, 212
Barnett, Martin K., 174
Barron, John M., 138, 142, 145
Barry, Edward, *166*
Baskin, Reed, 162, *174*, 186, 189
Battle of Memphis, 23
Battle, Allen O., *198*, 199
Beale Street, 50
Beale Street, The Spirit of, 120
Beale, Leo, 177
Beard, James D., 148, 152, 192
Beck, Otis, *69*
Beckford, Neal, 157, *157*
Behrman, Steve, 204, 221
Belcher, Aline, 136
Bell, Ann, 98, 144, 145
Bell, Emmett, Jr., 130
Belz family, 133
Bender, Charles, *79*
Berkenstock, Oran L., 219
Berry, Edward, 150
Berry, Michael, 221
Berry, Reuben, 34
Bertorini, Tulio, 144, 217
Besh, Steven, 181, *182*
Bethea, William, 96
Bicks, Richard, 189
Bierman, Paul, 221
Billings, John Shaw, 59
Birdsong, Sidney, 224
Bisson, Wheelock A., *72*, 133, 159
Black, Dennis, 224
Black, William T., 96-97
Blakney, Eric D., 219
Blassingame, C. D., *79, 81*
Blue, Julian B., 77
Bluff City Medical Society, 158, 159, 160, 171
Blumenfeld, H. P., 163
Boala, J. O., 79
Board of Health, 11, 20, 46
Boatwright, Frank, *193*

Boggs, W. E., 42
Boldley, Jim, 69
Bond, Samuel, 10
Bond, William, 10
Bond's Station, 10
Bookout, Alton C., 114
Boop, Frederick A., 132
Borland, Solon, 9
Boston, Barry, 162, 174
botanic medicine, 6
Botanico Medical College, 16-17, 30, 50
Bouldin, Mary E., 162
Bowld, William F., Hospital.
  See University of Tennessee
Bowman, Alexander H., 11
Boyd Harold B., 95, *173*
Boyd, Hugh, 80
Boyd, Robert F., 66
Brackett, Elliot G., 78, 87
brain-death law, 168
Brandon, William H., 96
Brannon-McCulloch Primary Health Care Center, 186
Braund, Ralph, 133, *134*
Brawner, Clara A., 133, 158, *159*, 170
Brawner, J., *72*
Brennan, Bill, 68
Bringle, Carey, 122
Brinkley, John Richard, 101
Brinson, S. N., *79*
Britt, Louis G., 167, 168-69, 202, 224
Britt, Louis P., *136*, 140
Brock, Joseph, 127
Brody, Dan, 139
Bronstein, Maury W., 160
Brown, R. F., 39
Brown, Thomas A., 50
Bruegge, Colin F. Vorder, 109
Bruegge, John Vorder, *112*
Buck, Kinsey Mansfield, *79*
Buford, Dr. George G., 64, 65
Burch, Lucius, *156*
Burke, F. Noel, 31
Burkett, Albert Sidney J., 50
Burns, Tina K., 219
Burrell Alexander, 50
Burt, E. E., *72*
Bucy, Paul, 103
Byas, A. D., 66
Byas, James, *72*

Cabrera, Antonio, *218*
Caffey, Shed, 114
Calandruccio, James, 194

Calandruccio, Rocco, *95*
Caldwell, Felix L., II, 219
Callison M. K., 126, 142, 175
Camp Greenleaf (Georgia), *79*
Campbell, Charlie, 68-69
Campbell Clinic, 90, 92, 95, 136, 222
Campbell Clinic Foundation, 95, 195, 205
Campbell, Edward G., 96
Campbell, Guy, 96
Campbell, Willis C., 64, 76-77, 92-93, 94, 95, 123
Campbell's Operative Orthopaedics, 94-95, 123
Canale Clinic, 204
Canale, Dee J., 204, 206
Canale, Phil M., *146*
cancer, 134, 145, 147, 151, 157, 162, 164, *165*, 166, 172, 174, 189, 190, 192, 205, 208, 210, 211, 213, 216, 221, 222, 225, 228
Cannon, Bland, 144
Cannon, Newton A., 114
Carbone, Laura, 163
cardiology, 138, 139, 164, 173, 211, 214, 217, 219
Cardiology Associates of Memphis, 138
Cardiovascular Physicians of Memphis, 138
Cardiovascular Specialists, 220
Carl Zeiss Meditec AG, 222
Carnes, E. H., 107
Carnes, W. A., *79*
Carroll, David S., 145, *152*, 153
Carter Louis, 182
Carter, Anne, 182
Carter, Grover, 88
Carter, Jacqueline, *163*
Carter, Parvin
Carter, Peter W., 221
Cartman, Eva, *134*
Cates, George, 223
Caudill, R. Paul, 156
Chancellor, Karen, 145, 172
Chandler, Walter, *134*
Chaney, William C., 142
Chang, Cyril, 227
Chapman, Lyman, 65
Chappell, Fenwick, 127
Charles Retina Institute, 191
Charles, Ravi D., 213
Charles, Steve T., 176, 185, 190, 213
Chase, Dr. Nancy, *171*
Chauban, Rair, 193
Chaum, Edward, 223

Cheek, Richard, 224
Chelsea-Watkins Health Center Clinic, 59
Chesney, P. Joan, 204, 205
Chesney, Russell, 132, *204*, 205
Children's Foundation of Memphis, 93
Children's Foundation Research Center of Memphis, 93
Children's Miracle Network, 132
Chin, Thomas, 218
Ching, Richard, *69*
Chipps, H. D., 98
Christ Community Health Services, 181
Christensen, Ronald E., 181
Church Health Center, 183
Churchwell, Luella, 172
Churchwell, Mary Ashley, 172
Civil War, 20, 22-33, 34, 38, 39, 49, 50, 52, 57, 74
Clark, Dwight W., Jr., 153, 165, 220
Clark, Glenn M., 162, 163
Clark, Mrs. Glenn, *163*
Clark, J. C., 67
Cleveland Clinic Foundation, 98
Clinical Orthopedics Society, 94
Cocke, Edwin, Sr., 126, 141
Cocke, Edwin W., Jr., *222*
Cocke, Mrs. Marian, 176
Cockroft, William B., 69
Cohen, Lawrence E., 114
Cohen, Steve, 137
Cole, F. Hammond, Jr. 221
Cole, Francis H., Sr., 115, 125, 127, 221
Coleclough, Liz, 182
Coley, Steven, 96
College of Physicians and Surgeons, 64, 65, 70
Collier, Casa, 87
Collins Chapel CME Church, 72
Collins Chapel Hospital, 72, 133, 155
Community Blood Plan, 168
Conrad, Lynn W., 193, 213
Conway, John, 173, 219
Cook, Annie, 42
Cooley, Denton, 153
Cooper, A. F., *79*
Cooperative Study for the Natural History of Sickle Cell Disease, 144
Coopervision, 194
Coors Clinic, 125
Coors, George A., 115, *125*, 126, 127, 140, 153
Coors, Giles, *112, 125*
Coppedge, Thomas, *79*
Coughlin, T. F., *79*

Covington, C. J., *72*
Cowan, George, Jr., 204
Cowley, Malcolm, 80
Cox, Clair, 205
Cox, Sam J., 162
Cox, T. E., 67
Crawford, Alvin, 159
Crawford, Lloyd, 132
Crawford, Thurman, 114
Crenshaw, Andrew Hoyt, 95
Crenshaw, Mrs. Hoyt (Ruth), *95*
Creson, T. Kyle, Jr., 219
Crews, James, 126
Crile, George, 78-79
Crippled Children's Hospital and School, 93-94
Crippled Children's Vitreoretinal Research Foundation, 190
Crisler, Gus, 125, 127
Crisler, James A., 64, 98
Crisler, Joseph Augustus, Sr., 76
Crisler, Joseph, Jr., 115
Crosby, Glenn, 153, 212
Croswell, Clyde, *69*
Crow, Robert L., 68
Crump, Edward Hull, 60, 103
Crump stadium, 108
Cullings, Dr. Jesse, 65, 96
Cummings, Alva, 189
Cummings, E. E., 80
Cunningham, David, 122, 206, 229
Cushing, Harvey, 78, 83
cytotechnology, 147

Dagogo-Jack, Samuel, 220, 224
Dakin, Henry, 81
Dale, James, 175, 224
Daughtry, C. O., 133
Da Vinci® surgical system, 213, *214*
Davis, Charles H., 30
Davis, Edward E., 2
Davis, Reuben, 7, 9
Davis, Shed, 69
Davison, Bob, 192
Dean, J. Patrick, 220
Deere, Charles J., 138, 172
Deitcher, Steven R., 174
Deitel, Mervyn, 204
Dellinger, Hubert, Jr., 160
Deloney, J. L., 66
DeMere, McCarthy, 115, *115,* 168
DePass, Matthew, 49
Derby, Nelson, 26
dermatology, 165, 172

DeSaussure, Richard, Jr., 106, 115, 173
Deutch, Alice, 170
DeWeese, Melvin, 190, 191
DeWitt, Washington J., 11
Dhanireddy, Ramasubbareddy, 161
diagnosis related groups (DRGs), 188
Diggs, Lemuel W., 98, 109, *144,* 151, *151,* 168, 202, 228
Diggs-Kraus Sickle Cell Center. See University of Tennessee
Dilawari, Raza, *189,* 192, 217
Diseases of the Army of Occupation in the Summer of 1846 (Robards), 13, *16*
Dixie Homes, 107, *107*
Doak, Maurice "Bully," 69
Dobbs Foundation, 152
Dobbs, James K., *129,* 152
Dobbs Research Institute, 152, *153,* 162
Doherty, Peter C., 203
Dominican priests, 41
Dominican Sisters, 30, 41
Donato, Robert A., 213
Donlon, Richard, 181, *182*
Dorrity, Thomas G., *115*
Dowell, Greensville, 44
Dr. Gunn's Domestic Medicine, 4
Dragutsky, Michael S., 221
Drake, John R., 80
Drewery, Richard, 177
Duckworth, John K., *162*
Duckworth, Nancy, 192
Duckworth Pathology Group, 162
Duckworth, William C., *220*
Dugdale, Marion, 171, *172, 174*
Dunavant, David, 127
Dunnavant, Benjamin, F, 79
Dupuy, Starke, 17
Durbin, I. N., 98
Dye, Mrs. Thomas, *140*

E. H. Crump Memorial Hospital, 129, 133, 156, *281*
Eagle Guards (Mexican War), 13, 16
Easley, E. T., 38
Eason, Hamel B., *142* 173
Eason, James D., 215
Eber, Paul, 193,
Edmondon, Allen, 177
Eggers, Frank, 180
Einhaus, Stephanie, 132
Elfervig, John, 177
Ellett, Edward Coleman., 53, 57, 64, 76-77, 82-83, *84,* 122
Ellett, Judge Henry T., 84

Ellett, Kathrine, 84
Elliot, Maurice, 156, 157, 168, 177, 206
Ellis, Alan, 202
Elmwood Cemetery Association, 17
Elsakr, Raif W., 221
Elvis Presley Memorial Trauma Center, 167, *187,* 188, 194, *211*
Emmett, John, 143
epidemic diseases, 11, 13
    See also: malaria, polio, yellow fever, typhoid, etc.
Episcopal Sisters of St. Mary, 41
Erickson, Cyrus C., 147, *147,* 162
Ernest, Kelly, 127
Erskine, Alexander, 38, 50, 56, *59*
Etheridge, Ken, 181
Etteldorf, James, 112, 130-131, 169
Etter, C. Barton, *113,* 130
Evens, William S., 151
Exchange Building, *18*
Expanding Orthopaedics, 194

Fabian, Timothy, 215
Fagin, Robert, 77
Family Cancer Center, 221
Farmer, Michael R., 216
Farrington, Pope M., 59, 77
Faulkner, William, 73, 128, 145
federal (Civil War) hospitals, 27-31
Federal Express, 195, 228
Fields, Kenneth, 221
Finn, Cary, 219
Firefighters Regional Burn Center, 167, 188
Firestone Children's Outpatient Center, 132
Firestone, Raymond C., 132
Fisher, Charles G., 37, 39, 41
Flaggs, R. L., 67
Fleenor, John D., 219
Fleming, Chris, 185, 223
Fleming, Irvin, 134, 174
Fleming, James Chris, 177
Fleming, Julian G., 219
Fleming, Martin D., 174
Flexner, Abraham, 67
Flexner Report, 67
Flinn, George, Jr., *165*
floods of 1912-13, 77
Flood of 1937, 90, 91, *108*
Florendo, Noel, 177
Flowers, Nancy, *139*
Foley, Kevin, 206, 207, 218
Fontaine, Bryce W., 73
Ford, Hobart "Hobie," 69
Ford, Mary Nell, 183

Ford, Robert, C. "Jitney," 69
Forrester, Eugene S., *169*
Forrest Park, 11
Fort Adams, 5
Foundation for International Education Neurological Surgery, 122
Fowlkes, Jeptha, 10
Francis, Hugh, 126-127
Francis, Hugh, III, 177
Francis, Hugh, Jr., 201, 224
Francis, Joseph H. *135*
Franciscan priests, 41
Francisco, Jerry T., 145, 177
Frankum, Charlie, 201
Frayser, John R., 9, 10, 11
Freeman, Anne, 185
Freeman, Jerre Minor, 177, 185, 190, 191, 223
Freeman, John, 185, 191
Frierson, S. E., *79*
Funderburg, W. Roger, Jr., 189
Fuste, Ricardo J., 217
Futrell, J. B., 69

Gabbert, Michael, 16, 34
Gaber, Osama, 169, 202, 214
Gailor Memorial Hospital, 102
Gailor Psychiatric Hospital, 124
Galloway, Robert, *112*
Gangrene Hospital (First Baptist Church), 28
Gardner, Cody, 165
Garrett, Edward, Sr., 126, 152, 153, 167, 212-13
Garrett, Edward., Jr., 212-13, 229
Gartly Ramsay Hospital, *73*, 102, *114*, 124, 145
Gartly, George, 73, 96
gastroenterology, 164, 189, 220
Gayden, Lynn W., 212
Gayoso Hospital, 28, 30
Gehorsam, Elizabeth, 192
Gelfand, Michael, 218
genetic research, 202
Germantown Baptist Church, 182
Gettelfinger, Thomas C., 185, 190, 191
Ghandi, Mohandas, 154
GI Pathology Partners, 220
Gibson, Ben, 201
Gibson, Francis D., 115
Gibson, Jeffrey B., 212
Gibson, Will, 201
Gieschen, Holger, 216
Gilbert, Christian, 215

Gillespie, Clarence, 127
Gilliland, Elizabeth Jordan, 128
Gilton, J. H., *72*
Givens, Preston G., 219
Gladding, T. C., 97-98
Goelz, Mrs. E. L., *119*
Golden, Alfred, 98
Goltman Clinic, 90
Goltman, Alfred M., 60
Goltman, David W., 60
Goltman, Jack S., 61
Goltman, Max, 57, 60, 64, 84
Gompertz, Mike, 189
Gooch, Jerry B., 212,
Goodman, Mrs. Joseph, *142*
Gordon, James O., 96
Gore, Margaret, 174
Gorgas, William C., 74, 79, 83
Goss, A. C., 98
Gotten, Henry B., 69, *114*, 115, 142
Gotten, Nicholas, Jr., 219
Gotten, Nicholas, Sr., 69, 101, *102*,115, 122, 144, 182
Gowling, Rachel, 57
Graevnor, Don, 221
Gragg, Wilford, 48
Graham, George Franklin, 7
Graham, George, 69
Granoff, Alan, 203
Graves, Lester R., 160
Graves, Lloyd M., 103, 121, 122, *127, 136, 137,* 147, 160, *160,* 162
Graves, L. M. Award, 157, 169
Graves, L. M. Health Building, 122
Gray, Horace, 96
Great (Frisco) Bridge, 58
Great Flood of 1927, 75
Green, C. L., 69
Green, Douglas R., 224,
Green, James, B., 127
Green, Jim, 201, 224
Griffin, Dan 189
Griffin, Daniel E., 221
Griffith, John F., 132
Grise, Jerry, 165
Grizzard, Tom, 190
Grobmeyer, Albert, 172
Grogan, Fred, 132
Groner, Frank, 126, 156, 168
Gullett, Roy, *69*
Gunn, John C., 4
Guy, Rodney K., 224

Haase, Marcus, 64, *76*

Haik, Barrett, 185, 223
Hairston, Jacob C., 50, 66-67, 72
Haiti Medical Missions of Memphis, 185
Halford, Jack, 172
Hall, Sir Peter, 128
Halle, Margaret, 170
Hamilton Eye Institute. See University of Tennessee
Hamilton, Barbara, 223
Hamilton, Catherine, 50
Hamilton, Fred, 165
Hamilton, Joe, *95*
Hamilton, Ralph F., 185
Hamilton, Ralph S., 223
Hammond, Mark, 198
Hancock, James C., Jr., *192*
Handorf, Charles, 205
Handwerker, John J., *120*
Hanna, Jeff, 69
Hardin, William G., 221
Harell, B. D., 67
Harkess, James, 173
Harrington, Brewster, 153
Harrington, O. Brewster, 212
Harris, B. T., 217
Harris, John, 192
Harris, Zeno T., 34
Harvey Cushing Society, 106
Hashimoto, Ken, *220*
Hatch, Fred, 168
Hawkes, Douglas C., 122, 173, 182
Hawkes, Jean, 173
Hay, Leon, 173
Hayes, John, 101
Hayman, O. W., *131*
Haynes, Eugene E., 56
Haynes, Leigh K., 127
Heacock, Charles Hunter, 91, *116*
Health Choice, 186
Health Loop, 188
Heart and Vascular Institute, 138
Heart Center of Memphis, 138
Heck, Robert, 192, 222
hematology/oncology, 174
Henderson, Lucius Samuel, 66
Henderson, R. D., 87
Hendrix, James, 127, *127*
Henning, Bennett Graves, 52, 53, 56
Henning, David M., (Max), 80
Henning, Donald, *150*
Henry G. Hill Orthopedic Clinic, 90
Herman, M. B., 57
Heros, Roberto C., 217
Hiatt, Roger, 185

Hickerson, Bill, 224
Hidaji, Fred, 185
Hill, Fontaine S., 192
Hill, J. F., 77
Hill, J. M., 16
Hill, Theron S., 192
Hill-Burton Act, 125
Hines, Leonard, 201, 202, 224
Hines, Nancy, *224*
Hobson, Joel J., *79*
Hoffman, Walter K., Jr., 138, 142, 219
Holbert, James M., 174
Holder, Eugene M., 64, *76*
Hollabaugh, Robert, Sr. 131
Hollabaugh, Robert, Jr., 193
Holmes, Georgia, 80
Holt, Hugh T., Jr., 163
Holt, Hugh T., Sr., 163
Holt, R. T. "Tarzan," *69*
Hooker, Rufus, W, 77
Hoover, Jeffery N., 219
Hoover, Susan, 221
Hope House Day Care Center, 185
Hopkins, Jack, 220
Horne, Arthur E., 115, 159
Horne, Howard, 175
hospice movement, 193-94
Hospice of Memphis, *193*
Hospital for Crippled Adults, 93, *95*
hospitalists, 198
Hospitals. See institutions by name.
Hotel Peabody *69*
Howard Association, 34, *40*
Howard Nursing Corps, 37
Howard Society, 37
Howard, Hector, 153
Howard, William, 77
Huffman, Claude R., 114
Hugh, Nash, 80
Hughes, Allen, 224
Hughes, Bobby, 192
Hughes, James D., 138
Hughes, James G., *114*, 115, 130
Hughes, John D., *114*, 115, 142
Hughetta, Sister, 42
Hummel, John, 128
Hunt-Phelan House, 50
Hurd, Bill, 185
Hurwitz, Julia, 182
Hyde Family Foundation, 195, 223
Hyer, William, 16
Hyman, O. W., 109

Ignatova, Tatyana, 225, *226*

Ihle, James, *203*
immunology, 132, 162, 163, 169, 173-74, 175, 180, 203
Indian remedies, 5-6
infant mortality rate, 161
infirmaries, 51-54
influenza, 78- 89
Ingram, Alvin, 95, 115, 117, 135, *136*, 141
InMotion Orthopedic Research Center, 226
Inpatient Physicians of the Mid-South, 198
internal medicine, 97, 138, 142, 160, 162, 163, 164, 172, 173, 174, 198, 218, 219, 221
Integrated Research Tower, St. Jude Hospital, 202
integration, 148, 155, 156-57, 158
International Children's Heart Foundation, 184
international medical graduates and students, 217-19
Irvine, David, 191
Irving Block, the, *23*, 24
Irwin, B. J. D., 25, 27, *28*, *29*, 33
Isolation Hospital, 135, 136

Jackson Hospital, 28
Jackson, President Andrew, 2, *3*, 4, 8
Jackson, Clay, 194
Jackson, George R., 50, 67
Jackson, Terrence L., 221
Jacobi, Abraham, 57
Jacobs, Arthur G., 64, 76
Jain, Manoj, 218, *219*
Jalfon, Isaac, 221
James K. Dobbs Medical Research Institute, 128, 153
Jane Terrell Hospital, 50
Jankov, Aleksandar, 221
Jay, Jacob P., 50
Jefferson Hospital, 28, 30
Jelks, John L., 65
Jennings, David K., 183
Jerald, Barry, 164
John D. Hughes Clinic, 172
John Gaston Hospital, 98 *107*, 109, 132, 156
Johnson, Eric, 213
Johnson, Eugene J., 49
Johnson, H. H., *72*
Johnson, J. Eugene, 75-76, 79, 87, 98
Johnson, James Gibb, 127, 175
Johnson, Joseph E., *223*
Johnson, L. A., 159
Jones and Laughlin Steel Corporation, *121*

Jones Clinic, 221
Jones, C. Michael, 174, 221
Jones, Frank A., 97
Jones, Heber, 55, 64, 70
Jones, Minerva Telitha, 101
Jones, Riley, 186
Jones, Tiberius Graccus, 101
Jones, Wesley E., 221
Jordan, Oakley, 219
Jost, Charles M., 180
Junior League of Memphis, 185

Kane, Elizabeth, 70, 90
Kang, Andrew, 163, 172
Kang, Ellen, 172
Kaplan, Stanley B., 163
Kearney, Monsignor Merlin F, *166*
Keating, John M., 36
Keith, Alice Jean, 109
Keller, J. M., 38
Keller, L. L., *79*
Kelley, Bobby, 168
Kelley, Richard, 180
Kennedy Army Hospital, 124
Kennedy General Hospital, 106, 116
Kennedy, H. H., 67
Kennedy, James N., 116, 118
Kerlan, Jeff, 220
Kerlan, Robert, 172, 189
Kerr, Natalie, 177, 223
Kersey, John, 165
Khan, Raja, 216
Khandekar, Alim, 212
Kilmer's Swamp Root Almanac, 6
King, Martin Luther, Jr., 154-55, 177; Award, 177
King, Cash, 96, 115, *116*, 126, 164
King, Ortie, 69
King, Scott, 201
Kirkpatrick, Robert, 172
Kitabchi, Abbas, 220
Kittrelle. A. N., 50
Kneeland, Francis "Fannie," 50, 66-67
Knight, Robert, 121
Knott, David H., 148, *152*
Knott-Craig, Christopher J., 218
Knox, Robert L., 54
Koleyni, A., 229
Konziolka, Douglas, *206*
Koonce, Marshall, 163
Korean War, 112, 125, 140-41
Korones, Sheldon,161
Kraus, Alfred, 144, 202
Kraus, Bernard M., 142

Kraus, David, 220
Kraus, Gordon, 185, 219
Kraus, Lorraine, 144, 202
Kraus, Robert, 219
Krauss, William, 48-49, 57, 61, 64-65, 84
Kroetz, Frank, 139, 165
Kukekov, Valery, 225, *226*
Kyle, Warren, 142

L. M. Graves Award, 157, 169
L. M. Graves Health Building, 122
Lachina, Mike, 217
Laird, R. G., *69*
Lan, David Z., 217
Land, Mack A., 179, 181
Langston, Jim, 180
Lankford, William A., 176
Lasker, Bob, 180
Lassiter, Gene M., 144
Laster, Robert, 216
Latham, Swayne, *112*
Lathram, Frank, 124, 192
Lauderdale Courts, 107
Laugheed, Joe, 127
Laughlin, Ned, 201
Lawrence, Walter Sibley, 65, 91, *96*
Lawson, Ronald D., 174, 189, 216
Le Bonheur Children's Hospital, see Methodist Le Bonheur Children's Hospital
Le Bonheur Club, 128
Leake, N. E., 123
Lee, Paul, Jr., 176
Leffingwell, Edward H., 17
Leggett, Louis, *69*
Leiberan, Phil, 175
Lemond, Mike, 219
LeMoyne College, 67
LePrince, Joseph A., 74-*75*
Leroy, Louis, 64, 65, 77, 97
Les Passees Rehabilitation Center, 132, 136, 140
Levy Gilbert, 121, 123, 135 *151*, 228
Levy, Louis, 68-69, 77, *79*, 87, *88*
Lewis, A. C., 122
Lewis, Archibald, 76-77
Lewis, James, 172
Lewis, Philip Meriwether, 122, 126
Lewis, Myron, 189, 2211
Lewis, Thomas, 87
Lichterman Nature Center, 168
Lichterman, Ira, 166
Lichterman, Lottie Loewenberg, 166
Lieberman, Gerald, 221

Lincoln Memorial University, 70
Linn, John, 177
Lipscomb, Alys, 170, *171*
Lisk, Laura, 132
Livermore, George, R., 96, *124*, 127
Lobe, Thom, 202
Loeb, William, *112*
Loewenberg, William A., 166
Loewenberg-Lichterman Foundation, 166
Lorenz, Adolph, 76
Lovejoy, George S., 136, *141*, 160
Lovejoy, John, *95*
Loving, Martha A., 162
Lown, Bernard, 175
Lucas, Petro W., 11
Luce, Edward, 224
Lucy Brinkley Hospital, 56, 70, 98
Lutton Jessie L., *166*
Lynch, Michael, 173
Lynk, Beebe Steven, 67
Lynk, Miles, 67, *72*
Lynn, T. J., 38

Mackey, William F., *151*
Madu, Ernest C., 180
Magoffin, Frank, *142*
Mahan, Kamal, 198
Maker, Howard, 163
malaria, 6, 10, 13, 24, 31, 34, 36, 49, 56, 57, 74, 75
Malcolm, Stevenson, 127
Malone, William Battle, 49, 79, 98, 126
Malone, William Battle, II, *126*, 127
malpractice, 101, 177, 186, 189, 196
managed care, 178-79, 185-86, 196, 198, 199, 201, 209
Mandell, Alan, 177
Mangiante, Eugene, 201, 202
Mann, Hiram, 79
Mann, Theresa Gaston, 107
Marcus Haase Nurses Home, 99
Marine Hospital, 28
Marsh, Homer F., *116*
Martin, Arthur T., 50, 73
Martin, B. B., 73
Martin, Eva Cartman, 73
Martin, J. B., 73
Martin, Robert G., 66
Martin, William S., 72-73, 133, *134*
Masi, Alphonse, *163*
Mason, Robin, *79*
Mastin, C. H., 24
Maur, Alvin, 151

Maury, John M., 53, 57, 64, 76, 96-97, 162
Maury, Richard B., 33, 50, 51, *52*, 53, 56, 57, 70
Maury, William P., 162
Maury, William P., Jr., 115
McArthur, A. N., *69*
McBurney, Robert, 127
McCaughan, Joseph, 118, 125
McCleave, B. F., 67 *158*
McCool, Dick Cauthen, *102*, 124, *127*
McCown, Oswald S., 64, 96
McCullers, Jon, 225
McDaniel, Carter, 201
McDonald, Mary, 172
McElroy, James Bassett *96*, 97
McEwan, Robert C., 219
McGehee, John Lucius, Jr., 79, *119*
McGrew, Frank, 226
McIntosh, John A., 80, 123
McKellar, Kenneth D., 103
McKinney, Richard, 64, 76-77
McLaughlin, C. R., *69*
McLean, J. L., 64
McNeer, Paul, 190
McSwain, H. Michael, 193, 213
MED, The, 133, 135, 144, 157, 167, *179*, 183, 187, 188, 199, 202, 208, 209, 214, 224, 227
Medical Arts Building, 90
Medical College of Ohio, 1819, 6
Medical Education Research Institute, 202, 204, 206
medical education, 5, 6, 61, 64, 119. See also names of institutions.
medical journals, 56-57
    See also names of specific journals
medical missions abroad, 181
medical practice, early regulation of, 7, 49, 51
medical societies, early, 49, 51
Medicaid, 157, 158, 159, 199, 208
Medicare, 153-54, 156, 158, 188, 191, 193, 196, 197, 198, 199, 208-10, 224
MedPlex Ambulatory Care Center, 188
Medtronic, 194-195
Medtronic Sofamor Danek, 194
Meduri, G. Umberto, 205
Meharry Medical College, 67, 72
Memphis and Shelby County Arthritis Research Program, 163
Memphis and Shelby County Medical Society, 94, *95*, 108, 112, *119*, 123, 158
Memphis and Shelby County Mental Health Summit, 200

INDEX   335

Memphis Belle, The, 120
Memphis Bioworks Foundation, 210
Memphis Blood Bank, 109
Memphis City Hospital, 10, 11, 16, 24, 61, 64
Memphis Education and Research Institute, 195
Memphis Eye and Cataract Associates, 191; Ambulatory Surgery and Laser Center, 180
Memphis Eye, Ear, Nose and Throat Hospital, 85, 90, 99
Memphis Health Center Clinic, 159
Memphis Heart Association, *131*, 138, *139*, *142*, 167
Memphis Heart Clinic, 138, 180, 217
Memphis Hospital Medical College, 16-18, 49
Memphis Journal of the Medical Sciences, 56-57
Memphis Lancet, 57
Memphis Medical Hospital, 70, *129*
Memphis Medical Journal, 56, 98, 112, 122, 136, 137, 189
Memphis medical milestones, 1980-89: new infections, 179; blood and testing, 180; ambulatory surgery centers, 180; imaging and radiology, 180
Memphis Medical Monthly, 56
Memphis medical practices: allergy, 219; anesthesiology, 189; bariatric surgery, 204; cardiology, 219; endocrinology, 220; gastroenterology, 189, 220; neurosurgery, 204, 206-207; oncology, 189, 221; ophthalmology, 190-91, 222; orthopedics, 205, 224; pathology, 205; pediatrics, 205; plastic surgery, 224; psychiatry, 192; pulmonology, 205; urology, 192, 205
Memphis Medical Recorder, 17
Memphis Medical Society, The (formerly Memphis and Shelby County Medical Society), 49, 56, 105, 107, 114, 156, 157, 170, 171
Memphis medicine, economic impact of, 226-27
Memphis Mental Health Institute, *198*
Memphis Neurological Society, 106
Memphis Pathology Laboratory, 144
Memphis Pediatric Society, 90, 91
Memphis Regional Medical Center See The MED
Memphis Roentgen Society, 91
Memphis Sanitarium, 53
Memphis Society of Pathologists, 98

Memphis, capsule medical history of, 228-29
Memphis, city of: adoption of commission government, 60; capsule medical history of, 228-29; growth of, c. 1908, 60; immigration to, 13; Mexican War and, 13, 16; Navy Yard, 12; parks system in, 61; railroads in, 13; riverboat commerce in, 11, 12; slum clearance in, 107
Memphis-Afghan Friendship Summit, 182
Memphis Urological Society, 96
Merrill, Ayres P., 17, 18, 20
Methodist Health Systems, 185
Methodist Healthcare System, 132, 186, 200, 204, 214
Methodist Hospital, 70, 98, *99*,101, *118*, 123, 126, *129*, 132, 135, 143, 156, 167, 185, 200, 201, 205, 206, 207, 220, 222, 223, 226, 227; Neurosurgery Group, 101, 122, 206
Methodist Le Bonheur Children's Hospital, 93, 113, 129-132, 138, 144, 150, 153, 157, 176, 200, 205, 214, 216, 227, 229; 'Bunny Room,' 132; Early Intervention and Development Program, 132;
Methodist Le Bonheur Healthcare, 200
Methodist North Hospital, 185, 213
Methodist University Hospital (MUH), 213, 214, 215, 216, 222, 225, 229; Cancer Center, 216
Methodist South Hospital, 185
Methodists, in early Memphis medical care, 5
MetroCare, 186
Metz, John, *69*
Meyer Eye Group, 191
Meyer Orthopedic Hospital, 87
Meyer, Alphonse, H., Jr., *86*
Meyer, Alphonse, H., Sr., *86*, 87
Meyer, David, 176, 190
Meyer, L. L., *79*
Meyers, Roland, 122, 130
Michael, L. Madison, II, 216
Michelson, I. D., 98
MicroDexterity Systems, 191, 213
Mid-South Cancer Center, 221
Mid-South Eye Bank, 222
Mid-South Regional Blood Center, 168
Mid-South Transplant Foundation, 167
Mid-South Urology Group, 193
Miles, Robert, 127
Milford, Lee, W., 95
Miller, Antonius, 174

Miller, Charles, 124, 168
Miller, Joseph, 122
Miller, Karen, 181, *182*
Miller, Mark, 201, 202
Miller, Richard, 122
Miller, Stephen T., 171, 217
Milligan, Kerry, 219
Millington Naval Air Technical Training Center, 120
Milnor, J. Pervis, Jr., *115*, 126, 153, 172
Minor, H. F., 77
Minor, James L., 77
Mississippi Valley Medical Monthly, 56
Mitchell and Maury Sanitarium, 39
Mitchell Maury Training School for Nurses, 59
Mitchell, Carol J., 219
Mitchell, David C., *146*
Mitchell, Edward Clay, 64, *79*, 80, 90-91
Mitchell, Edward D., Jr., 114
Mitchell, Frank Thomas, 130, *130*
Mitchell, Joseph, *95*
Mitchum, Bill, 165
Moening, Francis, 55
Monroe, Justin, 201
Montgomery, T. R., 65
Moore, Bruce, F., 77
Moore, David, 192
Moore, F. A., 67
Moore, Fontaine B., Jr., 193
Moore, Admiral Moore, 116
Moore, Thomas D., 124, 145, 176
Moore, Y. S., 51
Morgan, Allen, *131*
Morgan, Carroll H., 96
Morgan, Jerome L., 96
Morgan, Logan, 80
Morehead, A. B., *69*
Morris, Albert W., Jr., 160
Morris, Scott, 183
Morris, Tom, 173
Moss, Thomas C., 97, 98, 99, 123, 144, 162
Mroz, Christine T., 221
Muhlbauer, Michael, 132, *206*
Muirhead, E. Eric, 125 168, 177
Mullins, Brent, 174, 220
Munn, Chuck, 219
Munn, E. K., *69*
Murphey, E. D., 98
Murphey, Francis, 103, 104, 105-106, 122, 173
Murphy, Garnett, 224
Murrah, William, 122
Murrman, Susan, 172

Nair, Satheesh, 215
Nall, R. B., 37
Nash, John, 127, 138
National Eye Bank Center, 222
National Polio Foundation, 138
Navy Yard Hospital, 28
National Association of Cancer Centers, 221
National Cancer Institute, 174, 221, 225
Neal, Barry, 162
Neal, Michael, 173, 192, 219, 222
Neel, Mike, *197*
Neeley, E. A., 49
Neely, Charles, 162
Negro Baptist Association, 72
Neman, Felix R., 66
Nemir, Paul, Jr., 184
Nesbitt, E. E., 66
Netland, Peter, 177, 223
Neurosurgical Group of Memphis, 204
New Guide to Health; or Family Botanic Physician (Thomson), 6
Newborn Center, 161
Nezakatgoo, N., *215*
Nguyen, Trung T., 219
Nichopoulos, George, 176
Nienhuis, Arthur, 151, *203*
Nobles, Eugene R., 127, 210
Norris, Norwin B., 87
North Mississippi Health Services, 182
Novick, William, 184
nurse training, 52, *54*, 55

O'Brien, Thomas, 216
O'Donnell, Mary, 57
O'Kelly, Gertrude, *127*
Oakville Memorial Hospital, 72, 109, 142
Obstetrical and Gynecological Society, 96
Ockleberry, Billie, *140*
Odell, Dianne, 137
Odom, Guy, *105*
Ogle, Curtis, 122
Ogle, Evelyn, 122, *170*
Ogle, William 'Billy,' Jr., 122, 157, 170
Omer, George, Jr., 137
oncology, see cancer
OPMI VISU microscope, 140, 222
Orbis Flying Eye Hospital, 185
Orgill, Edward, *133*
OrthoMemphis, 192, 222
orthopedics, 109, *110*, 116, 123, 137, 142, 173-74, 180, 195, 205, 211, 224
Ostric, Elizabeth, 207
Otey, Paul, 37

Otis Warr Clinic, 172
Overall, G. W., 54
Overall-Knox Electrical Infirmary, 54
Overholt, Bergin F. "Gene," 189
Overman, Richard, 168
Overton Hospital, 28
Overton Park, 61
Owen, William F., Jr., 157
Owens, J. P., *79*

Page, Alfred, 134
Palermo, Joe, *69*
Pancreas Islet Transplant Laboratory. See University of Tennessee
Parsioon, Fereidoon, 217
Parsons, Charles Carroll, 41
Paster, Sam, 124
Pate, James W., 138, *167*, 169, 224
patent medicines, 15, 19-20, 38, 50, 103
Patterson, L. T., 50
Patterson, Richard, 201
Patterson, Russell, Jr., 119, 127
Patton, Georgia Lee, 57-58
Paul, Raphael, 130
Payne, Virgil, *69*
Payne, William King, 136, 162
Pearce, Iris A., 170, *170*
Pearce, Robert, 170
Pearson, Richard M., 193, 213
Peckham, C. T., 56
pediatrics, 123, 195, 200, 214, 216
Pegram, Robert H., 80
Pelz, Frederick, 219
penicillin, 112, 117, 129, 141, 175
Pennsylvania riverboat disaster, 13
Pepperman, David, 181, *182*
Perkins, Percy A., 80
Pettey and Wallace Sanitarium, 101
pharmacy, 19, 20, 62
Photopulos, Guy J., 174, 192
Physicians and Surgeons Building, 90
Pidgeon, J. Everett, *131*
Piggott, Jack, 134
Pillow, Gideon, 22
Pinkel, Donald, 116, *151*
Pistole, W. H., 64
Plesofsky, Jake, *69*
Plitman, Gerald, 162
Plough Foundation, 62, 195, 222, 223
Plough Laboratories, 62
Plough, Abe, 62-63
Plough's Antiseptic, *62*
polio, 93, 121, 129, 135-38, 141, 142, 148, 160
Poor Sisters of St. Francis, 55

Portera, Stephen G., 212
Porterfield, James, 180, *181*
Porter-Leath Home, *141*
Portis, Mary C., 221
positron emission tomography (PET), 226
Postlethwaite, Arnold, 163
Powell, Timothy J., 212
Powell, William Byrd, 16
Presley, Elvis, 128, 176-77
Presley, Elvis, Memorial Trauma Center. See Elvis Presley Memorial Trauma Center
Prewitt, Malcolm, *69*
Prezpriorka, Donna, 216
Price, Tommy, *166*
Pride, William T., 64, 96-97
primary care, 175, 181, 182, 188, 197, 198, 209, 217, 219
Prioleaux, T. E., 67
Pritchard, Elise, 130
psychiatry, 100-02, *110*, 116, 124, 145, 192
Public Health Cytology Laboratory, 146
Pulliam, H. N., 96
Pulsinelli, William, 144

Quinn, Arthur, 91
Quinn, Hugh, 16
Quintard, Charles Todd, 17, *19*

R. L. Sanders & Associates, 90
radiology, 91, 109, 153, 180-81, 205
Rago, Randall, 174
Ragsdale, John, 96
Raines, Newton Ford, 48
Raines, Richard, 186
Raines, Samuel L., *69*, 145, 193
Rains, B. Manrin, III, 219, *220*
Ramanathan, K. B., 180, 217
Ramey, Randy, 165
Ramsey, R. G., Sr., 73, 145
Ramsey, Lewis C., 114
Ranson, Richard, *105*
Ranson, Mrs. Richard, *105*
Rao, Bhaskar, *197*, 202
Raskind, Robert, *151*
Red Cross. See American Red Cross
Red Jack Stomach Bitters, *6*
Reed, Edward, 133, 156-57, 159, 162
Reed, Jarvis D., 174
Reed, Walter, 116, 228
Reeder, Robert, *151*
Regenisys, 194
Regional Biocontainment Laboratory, 229
Regional Medical Center. See The MED

Regional Newborn Center, *161*
Reinberger, James, 96-97
Reynolds, Stephen C., 213
Rhohr Pharmaceuticals, 194
Rice, Frank, 38
Rice, Stephen F., 57
Richardson, Arthur, 112
Richardson, Robert, 153, 186
Richardson, Robert L, III, 219
Riggs, Russell, *138*
Riggs, Webster, Jr., 153, 205
Riggs, Webster, Sr., 116, *138*, 122
Rightsel, Wilton, 180
Riverside Park, 61
Robards, Howell, 13, 16, 17
Robbins, Robert H., 114
Robbins, Samuel G., Jr., 212
Robert S. Pearce Chair in Community Medicine, 170
Robert, Martin, 65
Roberts, F. L., *136*
Robertson, Felix, 7
Robertson, James T., 173
Robertson, Jon H., 173, 216
Robinson, Bill, 165
Robinson, Kenneth, 181-82
Robinson, Walter W., 96, 97
Robinson, Wiley, 198
Robison, Leslie L., 224
Robison, Lowell B., 163
Rodd, Marco, 145
Roentgen, Wilhelm Conrad, 65
Rogers Surgical Infirmary, 52, 57
Rogers, Shepherd A., 52, 57
Rogers, William Bodie, 50, *51*, 52, 53, 56
Rogers, William E., 37, 49, 52
Rosafield, W. B., *86*
Rosamond, Eugene, 77, 90-91
Ross, L. R., 50
Rossett, N. Edward, 142
Rossville Health Center Clinic, 159
Roulhac, Christopher M., *158*
Roy, Shane, 168
Ruch, Robert, 122
Ruch, Walter, 122
Ruch, Walter, Jr., 122
Rudner, Henry G., Sr., *142*, 142
Rudner, Henry, G., Jr., 160
Rudolph, Wilma, 137
Runyan, John W. "Bill," Jr., 160, *160*
Russell, Melissa, *193*
Rychener, Ralph, 85, 122, *151*
Ryerson, Edwin, 94

Sacks, Harold, 218
Sage, Hubert, 88
Sam Walton Children's CT/MRI Center, 132
Samaha, Joseph K., 180, 220
Samant, Sandeep, 222
Sanders, R. L., 125, 126
Sanders, Sam, 69, 125
Sands, Chris, 198
Sanford, Conley H., 87, 97
Sanford, Jake C., 162
Sanford, James Blue, 80
Sappington, Mark Brown, 9, 10, 11
Sarner, Ferdinand, 42
Satterfield, William M. T., 114
Saunders, Dudley Dunn, 39
Savage, G. H., 77
Schanzer, Cathleen, 185
Schering Plough Corporation, 62
Schettler, Betty, 162
Schmeisser, Harry Christian, 97
Schmollinger, Jack, *131*
Schoettle, Glenn P., Jr., 213
Schreier, Phil, 96
Schroeder, Sister M. Rita, *166*, 186
Schwartzberg, Lee S., 174, 189, 221
Scott, Edward P., 180
Scott, Joseph, 191
Seager, Lloyd, 112
Searight, Wilson, 96
Seddon, Margaret Rhea, 194, *195*
Segerson, Edward C., *166*
Sellers, Ken, 224
Semmes-Murphey Clinic, 106, 107.122, 173, 204, 206, *222*, 225. (Later known as Semmes-Murphey Neurologic & Spine Institute)
Semmes, Raphael Eustace, 83, 86, 89, 103, 103-107, 173
Shala, M. Bashar, 217
Shanks, Lewis, 11, 17
Shea Clinic, 90, 143
Shea, Coyle, 182
Shea, John, III, 143
Shea, John J., Jr., 77, 80, 123, 128, *143*
Shea, John J., 77, 80
Shea, Monsignor J. Harold, *150*
Shea, Paul, 143
Shelby County Medical Examiner, 145
Shelby County Poor and Insane Asylum, 48
Shelby County Tuberculosis Society, 122
Shepard, Claudette, 171
Sherman, Roger, 224
Shields, R. T., 98

Shiloh, Battle of, 25-26
Shoat, Ann Hamilton, *127*
Shorb, Gary, 214
Shore, Dinah, 137
Shugart, Anita, 135
Siegel, Barry R., 219
Sievers, Rick, 177
Sills, Alan, 222
Sim, Francis, 49, 50, 56, 57
Simmons, Bryan P., 179
Simone, Joseph, 151
Simpson, W. L, 76-77
Sinclair, A. G., 49, 56, 57
Sister Rolandina, 166
Sisters of Charity, 30
Sisters of St. Francis, 166, 168
Sisters of St. Francis of Perpetual Adoration, 200
Sisters of the Holy Cross, 30
Sivley, Clarence L., 64
Sixty Years in Medicine (Boyd), 66
Slater, Edward C., 41
Sloas, David D., 221
Smeyne, Richard, *225*
Smiley, Linda M., 174, 192
Smith and Nephew, Richards, 194
Smith, Chapman, 165
Smith, Hugh M, 94-95
Smith, J. H., 96
Smith, Kirby L., 221
Smith, Vincent D., 219
Smith, William B., *136*
Smythe, Frank David, 80
Smythe, Frank Ward, 87
Society of Ophthalmology and Oto-laryngology, 77
Sorenson, Jeffrey M., 216
Southern Mothers Society, 24, 27
Southwestern Medical Reformer, 17
Spann, Jeanette, *109*
Speed, James Spencer, 92, *95*
Speight, William Oscar, Jr., 159, *159*
Speight, William Oscar, Sr., *72*, *133*, *159*
Speltz, Betty, 121
Spiers, Kathleen D., 174
Spiotta, Eugene J., Jr., 221
Spiotta, Eugene, Sr., 172, 219
Sprunt, Douglas H., 98, 147, *146*, 162
St. Andrew African Methodist Episcopal Church, 181
St. Clair, Frank, 220
St. Francis Hospital, 160, 168, 185, 186, *186*, 216-17, 227, 229
St. Francis-Bartlett Hospital, 216

St. Francis Healthcare, 216, 217
St. Joseph Aspirin for Children, 62
St. Joseph Hospital, *9*, 55-56, 93, 123, 155, 192, 200, *200*, 234
St. Joseph Hospital East, 166, *166*, 168, *186*, *224*
St. Joseph MedWise, 182
St. Jude Hospital, 116, 123, 132, 145
St. Jude Children's Research Hospital, 148, 150-51, 174, 176, *197*, 200-01, 202, 203, 205, 214, 221, 223, 225, 229; Comprehensive Care Center, 225
St. Jude Thaddeus, 148-50
St. Lazarus Episcopal Church, 41
Staffel, Gregory J., 143
Stanford, James Blue, 77, 80
Stanford, Robert A., 132
Steiner, Mitchell S., 205
stereotaxis, 213
Stern Clinic, 180, 220, 226
Stern, Neuton, S., 86, *86*, 87, 138, 139, 142, 219
Stern, Thomas N., 138, 139, *139*, 140, 153, 165
Stevenson, Edward, 127
Steward, John H., 67
Stewart, Marcus J., 95, 112, 116-17
Stewart, Walter, 69
Stollerman, Gene H., 162-63, 173, *175*
Storgion, Stephanie, 132
Strain, S. Frederic, 135, 142, 172
Straton, Leslie M., *135*
Stratton, Henry, Jr., 116
Stritch, Samuel, Cardinal, 150
Strobeck, John E., 180
Sturm, Dorothy, 98, *143*, 144
Sullivan, D. A., 69
Sullivan, Jay M., 180
Sullivan, Julian, 69
Sullivan, S. J., *69*
Sultana riverboat disaster, 32-33
Summers, C. K., *79*
Sutherland Heart Clinic, 220, 226
Sutherland Group, 138
Sutherland, Arthur, 138, 153, 165, 220
Sutton, Emily, 42
Swindell, R. R., 69
Swink, W. T., *79*
Symposium on War Medicine and Surgery, 112

Tabb, Thomas N., 212
Tabor, Owen, Sr., 173
Tacket, Hall S., 142, 172, *173*, 219

Tait, Lawson, 52
Tandy, George, 68
Tate, Richard H., 44
Tauer, Kurt, 174, 189, 221
Taylor, Arthur K., 17, 18
Taylor, Cindy, 182
Taylor, Ed, 172
Taylor, W. Zachary, 182, 221
Taylor, William W., 56
Tenet Healthcare Corporation, 186
Tenet Healthcare Systems, 185
TennCare, 181, 196, 198-200, 208-10, 227; and Medicare, 208-10
Tennessee Medical Society, 7, 39, 49
Tennessee Psychiatric Hospital and Institute (TPHI), 148
Tennessee Society for Internal Medicine, 142
Tennessee State Medical Association, 77, 84
Terhune, Ronald L., 219
Terrell, Cleveland A., 66, 67, 72
Thakus, Tapa, 198
Thomas F. Gailor Psychiatric Hospital and Diagnostic Clinic, *124*
Thomas, Danny, 145, 148-19, 150-51
Thomas, Rose Marie, 151
Thompson, A. I., 66
Thompson, Ed, *79*
Thompson, L. K., Jr., *166*
Thompson, Margaret E., 80
Thompson, Veerland, *197*
Thomson, Samuel, 6, *7*
Thornton, Gustavus B., 38, 45, 46, 59
Thummel, Robert, 38
TiGenix, 194
Tobey Children's Hospital, 138
Tobey, Frank, 151
Toombs, J. S., 64
Toombs, Percy Walthall, 91, 96, 97, *97*
Toombs, Reuben S., 97
Toombs, Robert E., 224
Toombs, Robert W., 95
Topp, Robertson, 11, 50
Towne, Carter, 186
Townes, Alex, 163
translational medicine, 195, 224-25
Translational Science Institute, 224
Transplant Clinical Pharmacotherapy Research Institute. See University of Tennessee
Transplantation, 167, 168, 169, 174, 192, 212, 214, 214
Transplants, bone marrow, 151, 179, 174;

corneal, 191; heart, 167, 212, 215; kidney, 168, 169, 214, 215; liver, 132, 214, 215; lung, 212; pancreas, 214, 215
Transylvania University, 6
Trezevant, Lewis, 11
Trimm, Robert, 132
Tri-States Baptist Hospital Association, 73
Tri-State Medical Association, 48, 49, 76
Trumbull, Merlin, 125
tuberculosis, 60, 93, 109, 118, 121, 122, 141, 142
Tuberville, Audrey, 177
Tuck, William J., 13
Tucker, William W., 10
Tullis, I. Frank, 142, 160
Tullis, Kenneth, 192
Turley, Hubert K., 96, 116, 124, 145
Turner, Benjamin F., 100-101
Turner, Carroll, 101
Turner, Ralph, 153
Turner, Randolph, 201
Turner-Gotten Sanatorium, 101
Twain, Mark, 14, 15
typhoid fever, 10, 20, 24, 31, 46, 49, 60
Tyrer, A. Roy, 122, 182

Underwood, N. T. "Ned," 69
Underwood, Robert B., 88
Union Hospital, 28
United States, Marine Hospital, 51; Naval Hospital, 116; Public Health Hospital, 107; Public Health Service, 51; Sanitary Commission, 30
University Medical Center Coordinating Council, 214
University of Nashville Medical Department, 16
University of Tennessee, 70, 92, 129, 147; admission of women to, 70, 169, 172; Alumni Service Professor Award, 169; Bowld Hospital, *129*, 148, 162, 163, 167, 168, 169, 174, 202, 214, 216; Cancer Center, 146, *174*; Cancer Institute, *174*, 221; Center of Health Sciences, *86*; College of Medicine, 16, 49, 64, 65, 67, 68-70, 75, 77, 84, 87, 97, 107, 108, 112, 114, 118, 119, 127, 129, 131, 133, 147, 157, 165, 169, 175, 180, 185, 188, 194, 201, 202, 205, 210, 211, 213, 215, 221, 224, 225, 226; Diggs-Kraus Sickle Cell Center, 144, 202; Department of Orthopaedics, 92; Doctors' football team, 68-69; Hamilton Eye Institute, 176, 223; Health Sciences Center, 93, 146-47;

Institute of Pathology, 147; Medical College, 16; Medical Department, 16; Medical Group, 138, 174, 213; Medical Units Student Awards, 181; National Alumni Association, 202, 214; Pancreas Islet Transplant Laboratory, 202; Translational Science Institute, 224; Transplant Clinical Pharmacotherapy Research Institute, 202
University of the South, 18
University of West Tennessee, 65-66, *158*
Upshaw, James J., 221
Upshaw, Jefferson D., 162
Upshaw, Jeremiah, 189
urology, 178, 192-3, 205
UT-Baptist Research Park, 195, *209*, 210, *211*
U.T. Docs football, 68-69
Utterback, Robert, 144

Vanderbilt Medical School, 67
Van Vleet Cancer Center, 133, 172
Varner, C. Ferrell, 125
Vaughan, James A, 80
Vera, Santiago, 169, 214
Veroza, Sam, 219
Veterans Administration Hospitals (VA), 99, 116, 117, 118, 124, 155, 158, 163, 167, 174, 175, 179, 214, 228
Vicksburg, Siege of, 31
Vieron, Leonidas N. 219
Vinicoff, Alan, *127*
Vinsant, Robert, 68
Virdell, Lewis, *79*
Vitreoretinal Foundation, 190-91
Voeller, Guy, 201, 202
Vogt, Val, 171
Voorhies, Alfred M., 57

Wake, Bob, 202
Walker, John C., 50 (late 1800s)
Walker, John C., 114 (World War II)
Walker, O. P., 96
Walker, William, 180
Wall, Herschel "Pat," *223*
Wallace, J. A., 148
Wallace, Robert, 224
Wallace, Walter, 101
Wallace, William L., 140
Waller, Robert, 214
Walls, Bruce, 148
Walsh, William K., 174
Walter Reed Commission, 60, 74

Walter Reed General Hospital, *88*, 118, 207
Wardlaw, Lee, 189,
Waring, George E., 46
Warr, Graham, 219
Warr, Otis S., Jr., 142, 219
Warr, Otis S., Sr., 90, 97, 108, 109, 138, *142*
Warren, Jeffrey S., 219
Washington Hospital, 28
Watkins, Edwin D., 64
Watkins, T. H., 159
Watridge, Clarence, 216
Watson, Norman M., 67, 72, *72*, *158*, 159
Wax, James A., *150*
Weaver, James, 16
Weaver, Jason, 216
Weaver, Virginia, 204, 217
Webb, Bill, 148
Weber, Bill, 173
Webster Hospital, 28
Webster, Robert, 203, 225
Weeks, A. Earle, 174, 189, 221
Weeks, George B., 31
Weiman, Darryl, 180
Weinel, Frank, *69*
Weir, Al, 182, 189
Weir, Alva B., III, 174, 216
Weiss, Joseph, 191
Weiss, Martin J., 219
Weiss, William T., 219
Wells, Van, 224
West Clinic, 174, 189, 221, 226
West Tennessee Cancer Clinic, 133
West Tennessee Regional Forensics Center, 145
West Tennessee Tuberculosis Hospital, 118
West Tennessee University, *158*
West, James, 186
West, Thomas H., 119, 127
West, William H., 162, *174*, 176, 189
Western Sanitary Commission, 30
Western State Hospital, 10
Wharton, A C, 199
Wheatley, Seth, 11, *11*
Wheeler, Benton M., 174
Wheelock, Bisson, *158*
Whitacre, Frank E., 97
White, A. A., 158
Whiteleather, Jack, 164
Whitlock, Lawrence, 219
Wilhelm, W. L., *69*
Wilkin, E. M., *72*
Willett, E. Miles, Jr., 56

Willett, E. Miles, Sr., 9, 56
Williams, James, 169
Williamson, S. B., 96
Williamson, Walker L., 96
Williford, Frank, 52
Wilson, Harwell, 112, 116, 126, 127, 1689
Wilson, Jim, 177
Wilson, John, 116
Wilson, President Woodrow, 88
Winchester, Marcus B., 11
WINGS, 174
Wise, Julius, 56
Witherington, J. B., 173, 219
Wolfe, Brad, 212
Wolfe, Rodney, 212
Wolford, David C., 180, 212
women in Memphis medicine, 30, 57-58, 109, 140, 162, 169-172
Women's Hospital Association, 56
Wood, Percy, 87, 96
Wood, Thomas O., 177, 222
Woodall, C. Jesse., 219
Woodside, Jeff, 202
Wooten, Hardy V., 17
Wooten, Mrs. Richard, 95
Wooten, Robert S., 221
World War I, 74, 78-89, 91, 98, 102, 103, 110, 112, 228
World War II, 93, 106, 109, 110-127, 128, *131*, 140, 155, 178, 196, 228
Works Projects Administration (WPA), 75, 107, 108
Wrenn, Earle, 131
Wright Medical Technology, 194
Wright, Daniel F., 17, 19
Wright, Frank O., 194, *195*
Wright, J. Leo, 139
Wright Sanatarium, 145
Wruble, Lawrence, 156-57, 189, 221

Yandell, Lundsford P., Jr., 20
Yandell, Lundsford P., Sr., 20, *21*
yellow fever, 34, 36-45, 51, 56, 57, 60, 74; epidemics, 34, 36-37, 39-45, 46, 49, 52, 56
York, Sara Conyers, 70, 169
Young, J. D., 142, 160
Young, Mark S., 219
Younger, Carl T., 219
Yunus, Furhan, 174, 216

Zafer, Ghany, 182
Zussman, Bernard M., 142

MEMPHIS MEDICINE

A HISTORY *of* SCIENCE AND SERVICE

INDEX TO MEMPHIS MEDICAL HISTORIES

Memphis Medical Society, The	232
Memphis and Shelby County Medical Alliance	234
Adams, R. Franklin, M.D.	235
Blotner, Adrian, M.D.	235
Baptist Memorial Health Care	236
Campbell Clinic Orthopaedics	238
Charles Retina Institute	240
Chase, Nancy A, M.D.	241
Clark, Winston Clark, M.D., Ph.D., F.A.C.S.	242
Coors, George A., M.D.	243
Crisler, Herman A., M.D.	244
DeFlumere, Charlotte S., M.D.	245
Fahhoum, Joseph, M.D., F.A.A.A.A.I., F.A.C.A.A.I.	245
Dirghangi, Banani, M.D.	246
Dirghangi, Jayanta, M.D.	247
Duckworth Pathology Group, Inc.	248
Eastmoreland Internal Medicine	251
Feild, James R., M.D.	252
Fleming, Julian G., M.D.	252
Graves, Lester R., Jr., M.D., Sam J. Cox, M.D., and Jack C. Sanford, M.D.	254
Hopkins, Jack T., M.D., FACC	253
Integrity Oncology	258
Jackson, Robert L., M.D.	259
Lakeside Behavioral Health Systems	259
Mays & Schnapp Pain Clinic and Rehabilitation Center	260
McCalla, Mary R., M.D.	262
Medical Anesthesia Group, P.A.	264
Memphis Dermatology Clinic, P.A.	266
Memphis Dermatology Society, The	263
Memphs Eye and Cataract Associates	268
Memphis Lung Physicians, P.C.	270
Memphis Neurology	271
Memphis Surgery Associates, PC	272
Merck/Schering-Plough	274
Methodist Le Bonheur Healthcare	278
Motley Internal Medicine Group	273
Mid-South OB/GYN PLLC	277
MidSouth Orthopedic Associates, P.C.	280
Palmieri, Genaro, M.D.	282
Parker, Autry J., M.D.	273
Parsioon, Fereidoon, M.D.	281
Pediatrics East, Inc.	284
Pediatric Surgical Group	286
Pender, John V., Jr., M.D., F.A.A.P.	285
Saeed, Salman, M.D., F.A.A.P.M.	291
Saint Francis Healthcare	288
Saint Francis Hospital-Bartlett	290
Saxton, Grady, Sr., M.D.	291
Semmes-Murphey Clinic	294
Shell, Dan H., III, M.D., F.A.C.S.	292
Shirazee, Syed H., M.D., F.C.C.P.	293
Southwind Medical Specialists	297
Tejwani, Indurani, M.D.	298
Tyrer, A. Roy, M.D.	293
University of Tennessee Health Science Center, The	300
UT Medical Group, Inc.	299
Wolf River Plastic Surgery	304
Woodbury, George R., Jr., M.D.	304
Wright Medical Technology, Inc.	302

In Memoriam

Knott, David H., M.D., Ph.D.	306
McGehee, John Lucius, M.D.	307
Moss, Thomas C., M.D.	308
Pearce, Iris A., M.D.	309
Wesberry, Jesse M., Sr., M.D.	310

MEMPHIS MEDICINE

A HISTORY *of* SCIENCE AND SERVICE

PICTURE CREDITS

THE PUBLISHER WOULD LIKE TO THANK THE following individuals for their assistance in procuring photographs: Lori Barksdale (International Children's Heart Foundation); Brent Baugus (Saint Francis); Greg Campbell (Baptist); Victor Carrozza (The Memphis Medical Society); Sheila Champlin, Rebecca Ennis and Dr. Richard Nollan (University of Tennessee); Dr. Cyril Chang; Dr. David Cunningham (Gamma Knife); Kay Daugherty (Campbell Foundation); Lori Estes (MECA); Ed Frank (University of Memphis); Dr. Jerre Freeman; Ruth Ann Hale and Joseph Martin (Methodist); Angie Herron and Frances Anderson (The MED); Diana Kelly and Darlene Brewer (MERI); John Lewis (Semmes-Murphey); John Mansfield (Assisi Foundation); Patricia and Jack McFarland; Sara Patterson and Katherine Whitfield (Le Bonheur); Dr. Edward Reed; staff of the Memphis Room, Memphis Public Library; Lee Thompson (Hamilton Eye Institute); Elizabeth Walker and Lin Ballew (St. Jude); Byron Wood (Charles Retina Institute).

The following abbreviations are used for sources from which a large number of images were obtained:

MRMPLIC—Memphis Room, Memphis Public Library and Information Center
SCUML—Special Collections, University of Memphis Libraries
UTHSL—University of Tennessee Health Sciences Library

FRONT MATTER: i (left) McCarver Postcard Collection, SCUML; (center) Dr. Thomas Gettelfinger; (right) McCarver Postcard Collection, SCUML. ii-iii photo © Jeffrey Jacobs Photography, courtesy Methodist Le Bonheur Healthcare. iv-v Methodist Le Bonheur Healthcare. vii Regional Medical Center at Memphis. ix St. Jude Biomedical Communications. CHAPTER 1: 2 Library of Congress. 3 Library of Congress. 4 (above) The Simon Rulin Bruesch Collection, Health Sciences Historical Collections, UTHSL; (below) SCUML. 5 SCUML. 6 (above) Library of Congress; (below) Arcadia Archives. 7 (above) National Library of Medicine; (below) *Dr. Goodenough's Home Cures and Herbal Remedies* (1904). 8 Library of Congress. 9 (above) MRMPLIC; (below) Library of Congress. 10 MRMPLIC. 11 SCUML. CHAPTER 2: 12 SCUML. 13 (both) MRMPLIC. 14 Library of Congress. 15 (both) Library of Congress. 16 (above) MRMPLIC; (below) Francis A. Countway Library of Medicine, Boston. 17 (both) Dr. Jere Freeman. 18 MRMPLIC. 19 Patricia McFarland. 20 Library of Congress. 21 (above) National Library of Medicine; (below) Library of Congress. CHAPTER 3: 22-23 MRMPLIC. 23 MRMPLIC. 24 Dr. B. J. D. Irwin Family Collection. 25 *Harper's Weekly.* 26 MRMPLIC. 27 MRMPLIC. 28 (above) Library of Congress; (below) Dr. B. J. D. Irwin Family Collection. 29 (both) Dr. B. J. D. Irwin Family Collection. 30 Library of Congress. 31 SCUML. 32 MRMPLIC. 33 (both) MRMPLIC. CHAPTER 4: 34 MRMPLIC. 35 MRMPLIC. 36 SCUML. 37 MRMPLIC. 38 (both) MRMPLIC. 39 MRMPLIC. 40 , MRMPLIC (photo by Jack McFarland). 41 (both) MRMPLIC. 42 (above) MRMPLIC; (below) William J. Armstrong Collection, Health Sciences Historical Collections, The University of Tennessee. 43 SCUML. 44 (above) Health Science Foundation Collection, MRMPLIC; (below) MRMPLIC. 45 MRMPLIC. CHAPTER 5: 47 MRMPLIC. 48 (both) MRMPLIC. 49 MRMPLIC. 50 *The Bright Side of Memphis*, G. P. Hamilton, 1908. 51 (above) *Press-Scimitar* staff photo, SCUML; (below) MRMPLIC. 52 O. F. Vedder, *History of the City of Memphis*, Vol. 2, MRMPLIC. 53 Health Science Foundation Collection, MRMPLIC. 54 (above) *Press-Scimitar* staff photo SCUML; (below) MRMPLIC. 55 (above) Health Science Foundation Collection, MRMPLIC; (below) MRMPLIC. 56 (above) Patricia McFarland Collection; (below) MRMPLIC. 57 MRMPLIC. 58 Health Science Foundation Collection, MRMPLIC. 58-59 Poland photo/*Press-Scimitar* Collection, SCUML. 59 O. F. Vedder, *History of the City of Memphis*, Vol. 2, MRMPLIC. CHAPTER 6: 60 *Commercial Appeal* staff photo, SCUML. 61 Levy Collection, MRMPLIC. 62 *Press-Scimitar* staff photo, SCUML. 63 Larry Coyne/*Press-Scimitar* staff photo, SCUML. 64 (above) Patricia McFarland Collection; (below) Health Science Foundation Collection, MRMPLIC. 65 (above) Mooney, *MidSouth and Its Builders*, MRMPLIC; (below) *The Bright Side of Memphis*, G.P. Hamilton, 1908. 66 (above) Health Science Foundation Collection, MRMPLIC; Meharry Medical Library Archives, MRMPLIC. 67 Health Science Foundation Collection, MRMPLIC. 68 (above) Health Science Foundation Collection, MRMPLIC. 69 *Press-Scimitar* staff photo, SCUML. 70 *Commercial Appeal* staff photo, SCUML. 71 MRMPLIC. 72 MRMPLIC. 73 (both) McCarver Postcard Collection, SCUML. 74 Colonel Joseph A. LePrince Collection, MRMPLIC. 75 Colonel Joseph A. LePrince Collection, MRMPLIC. 76 (above) *Memphis Evening Scimitar* Supplement, 1903, MRMPLIC; (below) MRMPLIC. 77 MRMPLIC. CHAPTER 7: 78 Library of Congress. 79 Levy Collection, MRMPLIC. 80 Signal Corps photo, Health Science Foundation Collection, MRMPLIC. 81 (above) Signal Corps photo, Health Science Foundation Collection, MRMPLIC; (below) Cassiday, The Gray Studio, *Press-Scimitar* Collection, SCUML. 82 Signal Corps photo, Health Science Foundation Collection, MRMPLIC. 83 Signal Corps photo, Health Science Foundation Collection, MRMPLIC. 84 The Simon Rulin Bruesch Collection,

Health Sciences Historical Collections, UTHSL. 85 (above) *Press-Scimitar* staff photo, SCUML; (below) Patricia McFarland Collection. 86 (both) *Commercial Appeal* staff photo, SCUML. 87 Patricia McFarland, *From Saddlebags to Science.* 88 (above) Jacobs photo, Seattle, Louis Levy Collection, MRMPLIC; (below) Library of Congress.. 89 Library of Congress. CHAPTER 8: 91 (above) *Commercial Appeal* staff photo, SCUML; (below) MRMPLIC. 92 Campbell Foundation. 93 Health Sciences Historical Collection, UTHSL. 94 Crippled Children's Hospital and School Collection, MRMPLIC. 95 (above) Campbell Foundation; (center and below) *Commercial Appeal* staff photo, SCUML. 96 (above) Dr. Webster Riggs, Jr.; (below) MRMPLIC. 97 (both) MRMPLIC. 98 *Press-Scimitar* staff photo, SCUML. 99 MRMPLIC. 100 Health Science Foundation Collection, MRMPLIC. 101 Dr. Thomas Gettelfinger. 102 (both) *Commercial Appeal* staff photo, SCUML. 103 Library of Congress. 104 Health Science Foundation Collection, MRMPLIC. 105 *Press-Scimitar* staff photo, SCUML. 106 *Commercial Appeal* staff photo, SCUML. 107 (above) MRMPLIC; (below) Health Science Foundation Collection, MRMPLIC. 108 (above) *Press-Scimitar* staff photo, SCUML; (below) MRMPLIC. 109 Health Science Foundation Collection, MRMPLIC. CHAPTER 9: 111 *Press-Scimitar* staff photo, SCUML. 112 (above) *Press-Scimitar* staff photo, SCUML; (below) The Simon Rulin Bruesch Collection, Health Sciences Historical Collections, UTHSL. 113 (both) *Press-Scimitar* Collection, SCUML. 114 (above) Signal Corps photo, Health Science Foundation Collection, MRMPLIC; (below) *Press-Scimitar* staff photo, SCUML. 115 (above) Larry J. Coyne/ *Press-Scimitar* staff photo, SCUML; (below) *Commercial Appeal* staff photo, SCUML. 116 (above) *Commercial Appeal* staff photo, SCUML; (below) *Press-Scimitar* staff photo, SCUML. 117 Campbell Foundation. 118 (both) MRMPLIC. 119 (above) *Commercial Appeal* staff photo, SCUML; (below) Ken Ross/ *Press-Scimitar* staff photo, SCUML. 120 *Press-Scimitar* staff photo, SCUML. 121 Tom Meanley/*Press-Scimitar* staff photo, SCUML. 122 Health Science Foundation Collection, MRMPLIC. 123 Campbell Foundation. 124 (above) *Commercial Appeal* staff photo, SCUML; (below) *Press-Scimitar* staff photo, SCUML. 125 (both) *Press-Scimitar* staff photo, SCUML. 126 (above) Official U.S. Navy photo, Dr. Robert Franklin Adams Collection; (below) Signal Corps photo, *Press-Scimitar* Collection, SCUML. 127 Dr. Robert Franklin Adams Collection. CHAPTER 10: 128-129 *Press-Scimitar* staff photo, SCUML. 130 (above) *Press-Scimitar* staff photo, SCUML; (below) *Commercial Appeal* staff photo, SCUML. 131 (above) *Press-Scimitar* staff photo, SCUML; (below) *Press-Scimitar* staff photo, SCUML. 132 Le Bonheur Children's Hospital. 133 *Press-Scimitar* staff photo, SCUML. 134 (above) *Press-Scimitar* staff photo, SCUML; (below) Sam Melhorn/*Press-Scimitar* staff photo, SCUML. 135 *Commercial Appeal* staff photo, SCUML. 136 (above) Campbell Foundation; (below) *Commercial Appeal* staff photo, SCUML. 137 (both) *Press-Scimitar* staff photo, SCUML. 138 Dr. Webster Riggs Collection. 139 Ken Ross/*Press-Scimitar* staff photo, SCUML. 140 (both) *Press-Scimitar* staff photo, SCUML. 141 Fred Payne/*Press-Scimitar* staff photo, SCUML. 142 (above) *Press-Scimitar* staff photo, SCUML; (below) Ken Ross/*Press-Scimitar* staff photo, SCUML. 143 *Press-Scimitar* staff photo, SCUML. 144 (above) Fred Payne/*Press-Scimitar* staff photo, SCUML; (below) Fred Payne/*Press-Scimitar* staff photo, SCUML. 143 *Press-Scimitar* staff photo, SCUML. 144 William Leaptrott/*Press-Scimitar* staff photo, SCUML; 145 (above) *Press-Scimitar* staff photo, SCUML; (below) The Dorothy Sturm Collection, Health Sciences Historical Collections, The University of Tennessee. 146 Gaokin/*Commercial Appeal* staff photo, SCUML. 147 *Commercial Appeal* staff photo, SCUML. CHAPTER 11: 149 St. Jude Biomedical Communications. 150 St. Jude Children's Research Hospital Collection, MRMPLIC. 151 (above) *Press-Scimitar* staff photo, SCUML; (below) St. Jude Biomedical Communications. 152 (both) *Press-Scimitar* staff photo, SCUML. 153 (above) *Press-Scimitar* staff photo, SCUML; (below) LBJ—White House Photo Office Collection, National Archives and Records Administration. 154 *Press-Scimitar* staff photo, SCUML. 155 *Press-Scimitar* staff photo, SCUML. 156 (above) James R. Reid/*Commercial Appeal* staff photo, SCUML; (below) The Memphis Medical Society. 157 (above) John P. George/ *Press-Scimitar* staff photo, SCUML; (below) Greg Campbell/Baptist Memorial Health Care. 158 Photo by B. W. Sims/Health Science Foundation Collection, MRMPLIC. 159 (above) *Press-Scimitar* staff photo, SCUML; (below) *Commercial Appeal* staff photo, SCUML.. 160 (above) *Press-Scimitar* staff photo, SCUML; (below) The Memphis Medical Society. 161 *Press-Scimitar* staff photo, SCUML. 162 Ken Ross/*Press-Scimitar* staff photo, SCUML. 163 *Press-Scimitar* staff photo, SCUML; James R. Reid/*Press-Scimitar* staff photo, SCUML. CHAPTER 12: 164 *Press-Scimitar* staff photo, SCUML. 165 Jack Gurner/*Press-Scimitar* staff photo, SCUML. 166 Saint Francis Hospital-Memphis. 167 *Press-Scimitar* staff photo, SCUML. 169 (above) Ken Ross/*Press-Scimitar* staff photo, SCUML; (below) William Leaptrott/*Press-Scimitar* staff photo, SCUML. 170 (above) The University of Tennessee Health Science Center; (below) William Leaptrott/*Press-Scimitar* staff photo, SCUML. 171 (above) The Memphis Medical Society; (below) *Commercial Appeal* staff photo, SCUML. 172 The Memphis Medical Society. 173 (above) Greg Campbell/Baptist Memorial Health Care; (below) Campbell Foundation. 174 (above) The Memphis Medical Society; (below) Greg Campbell/Baptist Memorial Health Care. 175 James R. Reid/*Press-Scimitar* staff photo, SCUML. 177 Les Cooper/Baptist Memorial Health Care. CHAPTER 13: 178 Louise M. Darling Biomedical Library, History and Special Collections, UCLA Digital Library. 179 Regional Medical Center at Memphis. 180 Greg Campbell/Baptist Memorial Health Care. 181 Ken Ross/*Press-Scimitar* staff photo, SCUML. 182 (above) Greg Campbell/Baptist Memorial Health Care; (below) *Commercial Appeal* staff photo, SCUML. 183 The Memphis Medical Society. 184 International Children's Heart Foundation. 185 *Press-Scimitar* staff photo, SCUML. 186 MRMPLIC. 187 (above) Dr. Thomas Gettelfinger; (below) Regional Medical Center at Memphis. 188 Dr. Thomas Gettelfinger. 189 Joseph Martin/Methodist Le Bonheur Healthcare. 190 (above) Charles Retina Institute; (below) Dr. David Meyer. 191 Memphis Eye and Cataract Associates. 192 William Leaptrott/*Press-Scimitar* staff photo, SCUML. 193 James R. Reid/*Press-Scimitar* staff photo, SCUML. 195 (above) *Commercial Appeal* staff photo, SCUML; (below)Saul Brown/*Press-Scimitar* staff photo, SCUML. CHAPTER 14: 197 St. Jude Biomedical Communications. 198 (above) The Memphis Medical Society; (below) University of Tennessee Health Science Center. 199 *Press-Scimitar* staff photo, SCUML. 200 McCarver Postcard Collection, SCUML. 201 The Memphis Medical Society. 202 The University of Tennessee Health Science Center. 203 St. Jude Biomedical Communications. 204 (above) Le Bonheur Children's Hospital; (below) The Memphis Medical Society. 205 Dr. Webster Riggs, Jr. 206 (both) Memphis Regional Gamma Knife Center. 207 Medical Education & Research Institute. CHAPTER 15: 209 The University of Tennessee Health Science Center. 210 Regional Medical Center at Memphis. 211 (both) Greg Campbell/Baptist Memorial Health Care. 213 Greg Campbell/Baptist Memorial Health Care. 214 Greg Campbell/Baptist Memorial Health Care. 215 (both) Joseph Martin/Methodist Le Bonheur Healthcare. 216 Dr. Thomas Gettelfinger. 217 The Memphis Medical Society. 218 (both) Le Bonheur Children's Hospital. 219 Greg Campbell/ Baptist Memorial Health Care. 220 Fred Payne/*Press-Scimitar* staff photo, SCUML; (below) The Memphis Medical Society. 221 (both) The Memphis Medical Society. 222 Greg Campbell/Baptist Memorial Health Care. 223 UT Hamilton Eye Institute. 224 The Simon Rulin Bruesch Collection, Health Sciences Historical Collections, UTHSL. 225 St. Jude Biomedical Communications. 226 Methodist Le Bonheur Healthcare. 227 (above) Dr. Thomas Gettelfinger; (below) Dr. Cyril F. Chang. 229 Dr. Tom Gettelfinger. 315 Greg Campbell. 324 Dr. Tom Gettelfinger. DUST JACKET: Front (above) SCUML; (below) St. Jude Biomedical Communications. Back (above) University of Tennessee Health Science Center, photo by Dr. Tom Gettelfinger; (below left) Baptist Memorial Health Care; (below right) photo by Greg Campbell, Greg Campbell Collection.

MEMPHIS MEDICINE
A HISTORY *of* SCIENCE AND SERVICE

PATRONS

BARBARA GEATER, M.D.

DR. AND MRS. GARY (JANE) KIMZEY

JAMES G. PORTERFIELD, M.D.

PRIMARY CARE SPECIALISTS, INC.

DR. CLAUDETTE J. SHEPHARD
